UNIVERSITY OF CALIFORNIA PUBLICATIONS

IN

BOTANY

Vol. 8, No. 1, pp. 1-138, plates 1-8   November 29, 1919

THE MARINE ALGAE OF THE PACIFIC COAST OF NORTH AMERICA

PART I

MYXOPHYCEAE

BY

WILLIAM ALBERT SETCHELL

AND

NATHANIEL LYON GARDNER

CONTENTS

Subclass I. MYXOPHYCEAE Stiz.

Thallophytes of simple construction possessing a blue coloring matter, known usually as phycocyanin, in addition to the chlorophyll; plants unicellular, or in colonies of regular or irregular form, or in longer or shorter, simple or branched filaments; cells discoid to globular, cuneiform or variously shaped, possessed of a distinct but very thin cell wall, a protoplast differentiated into a central chromatin containing body (nucleus, but without membrane) and outer color containing layer (cytoplasm and chromatophore?); granules (reserve materials?) of various kinds generally found in the cells, often arranged very definitely as regards certain cell walls; cells whether single, in colonies of regular or irregular form, or in filaments commonly surrounded by a more or less ample tegument of gelatinous to cartilaginous consistency, colorless or of various shades of yellow, brown, red or purple, structureless or stratified in various ways, commonly in the form of a "sheath" in the filamentous forms, the sheath enclosing one to several or many rows of cells (or trichomes). Multiplication: (1) by cell division, separating single cells (coccogonia) or short filaments (hormogonia); (2) by spores, either formed within a cell (gonidangium) in larger or smaller numbers (gonidia) or by differentiation as to size, shape and contents, and the formation of a thick outer coat (resting spores); sexual reproduction unknown.

Myxophyceae Stizenberger, Dr. Ludwig Rabenhorst's Algen Sachsens, 1860, p. 18.

Myxophykea Wallroth, Fl. Crypt. Germ., vol. 2, 1833, p. 4.

Gloesipheae Kuetz., Phycologia generalis, 1843, p. 179.

Phycochromophyceae Rabenh., Fl. Eur. Alg., vol. 1, 1864, p. 1.

Cryptophyceae Thuret, *in* Le Jolis, Liste des Alg. du Cherbourg, 1863, p .13.

Cyanophyceae Sachs, Lehrb. d. Bot., ed. 4, 1874, p. 248.

Schizophyceae Cohn, Ueber Thallophytsystem, 1879, p. 279.

The Myxophyceae, or Cyanophyceae, as they are perhaps better known, constitute a fairly compact and easily to be distinguished group of more simple and smaller plants. The cell structure is simple in that the protoplast lacks the distinct differentiation into cytoplasm, nucleus, and chromatophore found in other groups of the Thallophyta and in the Embryophyta, and there is certainly no distinct mitosis. The stature is slight as compared with that of plants of most other groups, the largest plants being only a few centimeters high. The morphological differentiation of the various genera and species is also simple. The reproductive methods are very simple indeed, being either purely vegetative or by non-sexual spores. Neither sexual reproduction nor zoospores are known in the Myxophyceae. Some filamentous species are motile, at least under certain conditions, and the hormogonia are said also, in the case of certain if not all species, to be motile, although largely of microscopic or very slight dimensions. The cells, colonies or filaments of most of the Myxophyceae occur in such numbers and so densely aggregated, or even agglutinated together as to form layers or masses of various shapes which are sufficiently conspicuous to be readily visible or even very noticeable. The Myxophyceae inhabit damp places or shallow waters even at fairly high temperature, and some non-marine members of the group even ascend, in thermal waters, to about 77° C or possibly higher. The Myxophyceae are often found in waters containing decomposing organic material. Some of those inhabiting calcareous waters are concerned in causing deposits of tufa or travertine, while some species bore their way into shells. Some species in silicious thermal waters, in turn, are closely associated with the deposit of sinter. Many species are epiphytic, some even are more or less deeply endophytic while some are associated with fungi in lichen formation.

The Myxophyceae do not seem to be closely related to the other groups of Phycophyta although there are certainly close resemblances to be seen between the vegetative conditions of certain unicellular forms of Myxophyceae and those of Chlorophyceae. A much closer resemblance is seen to be with various forms of the Schizomycetes or Bacteria, so close, in fact, that it is often difficult to decide whether a given cell or filament ought to be assigned rather to the one or to the other group. Yet there are differences, as West has clearly pointed out (1916, pp. 38, 39). It has been the custom, however, to group the Myxophyceae and Schizomycetes together in a group by themselves which Cohn has designated as the Schizophyta.

Key to the Orders

1. Vegetative multiplication by single cells (coccogonia): unicellular forms either single or in larger or smaller globular or irregular colonies, seldom filamentous.... .... Order 1. **Coccogonales** (p. 4)
1. Vegetative multiplication by groups of two to many cells (hormogonia); multicellular forms, distinctly filamentous ....Order 2. **Hormogonales** (p. 51)

Order 1.　COCCOGONALES atkinson

Thallus unicellular, associated into loose, more or less gelatinous families, or occasionally pluricellular, free floating or attached; multiplication by direct cell divisions or by division of specialized cells (gonidangia) into few to many small, non-motile cells (gonidia), or by both.

Atkinson, A college text book of Botany, 2d ed., 1905. p. 163.

Coccogoneae Thuret, Essai Class. Nost., 1875, p. 377.

Thuret (*loc. cit.*) was the first to emphasize the distinction between the two groups of Myxophyceae, or Nostochinées, as he designated them. The Coccogoneae were distinguished from the Hormogoneae (Thuret, *loc. cit.*) merely by name without further definite statement. In 1886 Bornet and Flahault (p. 323) stated that the separation between the two groups was made on account of their method of multiplication but without stating the exact nature of the method. The group was first definitely characterized by Kirchner (1898, p. 50) and it was defined again by West somewhat later (1916, p. 40). The proper distinction is that vegetative reproduction may take place by single cells (coccogonia) instead of by short rows of cells (hormogonia) and in general the members of this group may be distinguished as being unicellular (or in non-filamentous colonies) instead of being properly filamentous.

The majority of the species whether in filamentous form (as a few of them are) or unicells, or even the non-filamentous colonies, do definitely reproduce by vegetative cell division, separating cells which individually are capable of further growth. A few species, particularly those of the genus *Dermocarpa,* have no strictly vegetative method of multiplication, reproducing only by gonidia, yet in every other way are closely allied to the rest of the genera included under Coccogonales and must, therefore, be placed in this order.

The order is usually divided into two families, viz., Chroococcaceae and Chamaesiphonaceae.

KEY TO THE FAMILIES.

1. Multiplication by vegetative cell division (coccogonia) or by encysted cells (resting spores); strictly non-filamentous......Family 1. **Chroococcaceae** (p 5)
1. Multiplication by vegetative cell division (coccogonia) and by endogenous gonidia; unicellular, in globular or irregular colonies and, at times, distinctly filamentous..Family 2. **Chamaesiphonaceae** (p 20)

FAMILY 1. CHROOCOCCACEAE NAEG.

Cell solitary or associated into families of indefinite shapes and sizes, free floating or attached; cell wall usually thin, surrounded by a more or less copious gelatinous or mucilaginous, at times highly colored, tegument; protoplast homogeneous or granular; multiplication wholly by vegetative cell divisions in one, two, or three planes, the cells separating immediately, or remaining in contact for a longer or shorter period of time after which the teguments dissolve and liberate them.

Naegeli, Gatt. einz. Algen, 1849, p. 44.

Naegeli, in founding the family Chroococcaceae, included some eight genera, all of which, by common agreement, remain within the limits of the family. Later writers added genera some of which have been separated to form the family Chamaesiphonaceae of Borzi. Various distinctions have been attempted between these two families but it seems to us that the only satisfactory basis, at present at least, is that of the lack of gonidia-formation in the Chroococcaceae and its presence in the Chamaesiphonaceae. This was, at least, one of the important distinctions in the mind of Borzi (1882, p. 312). Later writers have found difficulties in this (cf. West, 1916, p. 41) because of possibly greater prevalence of gonidia-formation among members of the Myxophyceae than had been previously described. Some of the described cases, however, must, as it seems to us, be more definitely substantiated and more carefully studied, to constitute a really decisive objection. In this account the presence or absence of gonidia, together with the presence or absence of certain other characteristics to be emphasized under Chamaesiphonaceae, will be used in determining the membership of each of the families of the Coccogonales.

KEY TO THE GENERA.

1. Cell division in 1 plane; plants 1-2 celled 2
1. Cell division in 2 planes forming a plate-like colony ....3. **Merismopedia** (p 8)
1. Cell division in 3 planes forming solid colonies.. 3
    2. Cells globular without distinct tegument......................1. **Synechocystis** (p 6)
    2. Cells longer than broad or thick, tegument present neither thick nor stratified...2. **Synechococcus** (p 7)
1. Cells within the colony with distinct, often stratified teguments.......................... 4
1. Cells within the colony without distinct or stratified teguments.......................... 5
    4. Colonies irregular in shape not forming a distinct thallus................................
    ....4. **Chroococcus** (p 9)
    4. Colonies forming a distinct, lobulated, hollow thallus............5. **Placoma** (p 11)
1. Colonies free, or only slightly attached, cells without definite arrangement........
    ....6. **Anacystis** (p 12)
1. Colonies epiphytic or partially endophytic, cells arranged in more or less distinct vertical rows...7. **Chlorogloea** (p. 15)

1. **Synechocystis** Sauvageau

Cells spherical, free, separating soon after division, with very thin walls, and blue-green protoplasts; cell division in but one plane; teguments wanting.

Sauvageau, Sur les algues d'eau douce, 1892, p. cxv, pl. 6, fig. 2.

The genus *Synechocystis* is to be distinguished by its spherical cells, elongating slightly just before division, dividing in only one plane, and absolutely devoid of tegument. Its nearest relatives are the species of *Synechococcus*.

Synechocystis aquatilis Sauvageau
Plate 1, fig. 7

Cells globose, 5–6μ diam., single or geminate, cell wall hyaline and very thin; protoplast verdigris green.

Forming an irregular layer in pools of salt marsh. Alameda, California.

Sauvageau, *loc. cit.*; Gardner, Cyt. studies in Cyan., 1906, pp. 239, 268, 280, pl. 26, figs. 40–44; Collins, Holden and Setchell, Phyc. Bor.-Amer. (Exsicc.), no. 1206.

The description given above is translated from Forti (1907, p. 26) and is Sauvageau's description of the plant found at Hammam-Salahin, Algeria. It may be seriously questioned whether the plant distributed by Gardner from the dripping sides of a water tank at Berkeley, California, which was submitted (in duplicate) to Sauvageau by F. S. Collins and received his sanction, is the same as no. 1552 Gardner which was found forming an irregular layer in pools of a salt marsh at Alameda, California. The cells of the latter specimen are slightly smaller than those of the Berkeley specimen or as given for the type. The cells of the Alameda plant are 4–4.5μ in diameter in the strictly spherical stage and show no tegument even on being treated with methylene blue.

2. **Synechococcus** Naeg.

Cells elliptical or cylindrical, single or adhering in pairs for some time after cell division, free from enveloping jelly; color blue-green or yellowish; cell wall thin; division in but one plane; tegument lacking, or at least extremely thin.

Naegeli, Gatt. einz. Alg., 1849, p. 56.

The genus *Synechococcus* is closely related to the genus *Synechocystis* and the two seem to intergrade. They are both unicellular, in the strict sense, with divisions in a single plane. They are both lacking in tegument, but some species of *Synechococcus* seem to have a light enveloping jelly and approach the genus *Gloeothece*. As in other cases among the Chroococcaceae, it seems best to retain the distinction as generic even though it be slight.

The type species of *Synechococcus* is *S. elongatus* Naeg. from the Katzensee near Zurich in Switzerland.

Synechococcus curtus Setchell

Plate 1, fig. 6

Cells single or united by strands of transparent jelly, slightly elongated, with obscure cell wall, and pale bluish-green protoplast; cells 3μ diam., 6μ long just before dividing, the pair of daughter cells up to 8μ long before separating.

Growing in warm salt water. Oakland, California.

Setchell, *in* Collins, Holden and Setchell, Phyc. Bor.-Amer. (Exsicc.), no. 1351, March, 1907; Gardner, Cyt. studies Cyan., November 10, 1906, p. 239 (MS. name; no descr.).

The species of *Synechococcus* are to be distinguished from one another by their diameters and by the relation of proportion between these and their lengths. The present species is of intermediate diameter as these range among the species of the genus, and very short in proportion. Some of the cells approach spherical and consequently resemble *Synechocystis,* but the majority are at least somewhat longer than broad even in stages just succeeding division.

3. **Merismopedia** Meyen

Cells spherical, or narrow ellipsoidal, with thin cell walls, and firm, hyaline, structureless teguments becoming diffluent and binding the cells together; division regularly in two planes, building up colonies, at first square, or regularly rectangular, later often irregular in outline through injury or failure of certain cells to continue division.

Meyen, Wiegman's Arch. f. Naturgesch., Jahrg. 5, vol. 2, 1839, p. 67 (*nomen nudum*); Kuetzing, Phyc. Gen., 1843, p. 163 (descr.). *Gonidium* Ehrenb., Infusionsth., 1838, p. 59 (*nomen nudum*); Meneghini, Syn. Desm. huc Cog., 1840, p. 213 (descr.). *Agmenellum* de Brébisson, De quelq. nouv. genres d'Alg., 1839 (May), p. 2, Dict. univ. d'hist. nat., vol. 1, 1841, p. 187; Trevisan, Prospetto della Flor. Eug., 1842.

The plants which have passed undisputedly under the name of *Merismopedia* for so many decades form a well defined and recognizable generic unit. When the history of the name is investigated, it is found, unfortunately, to possess no valid status. *Merismopedia* was used by Meyen simply as part of a binomial and with no definite diagnosis in 1839, and was probably published comparatively late in the year. It seems, therefore, to be a *nomen nudum.* It was first accompanied by a proper diagnosis in Kuetzing's *Phycologia generalis* in 1843. In the meantime two other names were suggested, *Gonidium* by Ehrenberg and *Agmenellum* by de Brébisson. The name *Gonidium* is also to be held as a *nomen nudum* since it was used by Ehrenberg as a subgeneric designation under *Gonium.* He remarks (1838, p. 59), however, that he would, if he understood the plants better, place both *Gonium tranquillum* and *G. glaucum* in a new genus *Gonidium,* but he prefers to leave them under *Gonium.* In 1840 Meneghini took up and described *Gonidium* as a genus, on the ground, as he distinctly states, that it antedated the genus *Agmenellum* of de Brébisson.

Agmenellum was published as a genus by de Brébisson in May, 1839, with complete diagnosis and naming of the type species, *A. quadruplicatum* Bréb. It seems, therefore, that *Agmenellum* is the first valid publication and ought to be adopted, but following the principle enunciated by the various International Congresses that names of undisputed use for a long series of years may be conserved, we have retained *Merismopedia* until proper action may be taken.

Merismopedia Gardneri (Collins) Setchell

Plate 8, fig. 7

Plants up to 3 cm. or more long, delicate, membranaceous, very irregular in form, more or less lacerated and folded, 7–8μ thick; cells closely placed, cylindrical, 6–6.5μ long, 3.5–4μ in diam.; protoplast bluish green, homogeneous.

Floating in pools of salt water, salt marshes of Alameda, California.

Setchell, *in* Gardner, Cyt. studies Cyan., 1906, p. 239 (*nomen nudum*). *Prasiola Gardneri* Collins, *in* Collins, Holden and Setchell, Phyc. Bor.-Amer. (Exsicc.), no. 1185 (1904), Green Alg. N.A., 1909, p. 221 (in note).

Merismopedia Gardneri has been collected only once in pools of salt water on a marsh in Alameda, California, but it was plentiful at that time. Since then the type locality has been filled in with earth. It showed a remarkable resemblance to species of *Prasiola,* but was decidedly of a more bluish tint, which is even more pronounced in dried specimens. The cytological structure agrees well with that of other Myxophyceae and lacks the characteristics of the cells of *Prasiola.* It is gigantic for a *Merismopedia,* surpassing even the *Merismopedia convoluta* Bréb., hitherto known as the largest member of the genus.

4. Chroococcus Naeg.

Cells single or united into small colonies of 2, 4, or 8 cells, spherical, or more or less flattened when in colonies; contents blue-green, violet or yellowish; tegument comparatively thin although distinct, firm, uniform or nearly so, not gelatinous, that of the original cell enclosing the entire small colony; division in three planes.

Naegeli, Gatt. einz. Alg., 1849, p. 45.

The genus *Chroococcus,* as established by Naegeli in 1849, was distinctly set off and used in practically the exact sense of the present time. The type of the genus, as designated by Naegeli, is the *Pleurococcus rufescens* Bréb. published as a manuscript name by Kuetzing (1846, p. 9) under his *Protococcus rufescens.* Some difference of opinion seems to exist as to the present status of this type. Rabenhorst (1868, p. 28) uses the name of de Brébisson in his account of the ''Chlorophyllophyceae'' and West (1916, p. 93) restores the genus *Protococcus* and reckons *P. rufescens* (Bréb.) Kuetz. among the Chlorophyceae. An examination of the type specimen in Herb. Kuetzing, for which privilege we are indebted to Dr. Weber-van Bosse, demonstrates clearly that the plant of de Brébisson is a *Chroococcus* in the sense of Naegeli. Naegeli, however, evidently derived his conception of the genus from *Chroococcus rufescens* var. *turicensis* Naeg., as described and figured by him (1849, p. 46, pl. 1, fig. 1). All of the stages figured by Naegeli are to be found in the type material, so that it seems likely that no considerable differences are to be found between the species and its variety. From Naegeli's figures and from the species other than the type referred to *Chroococcus* by Naegeli, there can be no doubt as to his conception of the genus.

The genus *Chroococcus* is compared by Naegeli (1849, p. 53) with *Gloeocapsa* Kuetz. and *Aphanocapsa* Naeg., who states that while they are closely related to one another, they are distinctly set off from the other genera whose descriptions follow in Naegeli's monograph. The whole difference, as Naegeli says, lies in the fact that in *Chroococcus* the teguments are thin, in *Gloeocapsa* thick and firm, while in *Aphanocapsa* they are thick but so soft that they coalesce into a structureless jelly. All other characteristics are secondary, and there occur forms which might with equal reason be referred to one or another of the three genera. It may seem best, at some time, to unite *Chroococcus* with *Gloeocapsa,* but, for the present, we shall follow the custom and retain both as distinct genera.

Chroococcus turgidus (Kuetz.) Naeg.

Cells spherical or ellipsoidal when single, or angular from mutual compression when in colonies, 1–4, rarely 8, in colonies, 13–25μ, rarely 40μ diam., tegument comparatively thick, often somewhat lamellate, hyaline, cell wall thin, not distinct; protoplast vivid blue-green, homogeneous, or becoming faded more or less, and granular.

Growing in fresh and in brackish water along high-tide line, Alaska to central California.

Naegeli, Gatt. einz. Alg., 1849, p. 46; Setchell and Gardner, Alg. N.W. Amer., 1903, p. 179. *Protococcus turgidus* Kuetzing, Tab. Phyc., vol. 1, 1846, p. 5, pl. 6, fig. 1.

The type of the species is to be found in Herb. Kuetzing and shows the plant to be mingled with fragments of a moss and consequently as occurring in fresh water. The type locality is not given with definiteness but is stated in the *Species algarum* (1849a, p. 198) to be Germany. Single cells of the type are about 8μ in diameter, but with the teguments are $12–16\mu$ in diameter. The teguments show distinct stratification and colonies of eight cells, of total diameter of 40μ, are frequent. Altogether the type specimens correspond exactly to Kuetzing's figures and description.

Chroococcus turgidus is not properly a marine species but an inhabitant of fresh water. The forms occurring in the brackish water of pools of the salt marshes and just above high water mark are larger, with variable color of the protoplast. Most, if not all of them, are more properly to be referred to the following variety.

Chroococcus turgidus f. submarinus Hansg.

Plate 1, fig. 14

Single cell spherical, protoplast $22–26\mu$ diam., tegument $5–6\mu$ thick; two-celled colony $40–45\mu$ diam., $50–60\mu$ long, cells angular; extremely variable in color in the same collection.

Growing on *Chaetomorpha californica* Collins, in rock pools along high-tide line, at present credited to a single locality, Laguna Beach, Orange County, California, but probably extending along the entire coast in suitable habitat.

Hansgirg, Beiträge Kennt. Meeresalgen, 1889, p. 6; Collins, Holden and Setchell, Phyc. Bor.-Amer. (Exsicc.), no. 1551.

Although a single locality is quoted, it seems probable that the forma *submarinus* really includes all the specimens found in strictly brackish water.

5. **Placoma** Schousb.

Colonies spherical or lobed, solid or hollow, with cells arranged more or less radially, at least toward the surface, and quaternate; tegument of colony structureless, special teguments of cells more or less stratified as in *Gloeocapsa;* division in three planes.

Schousboe, *in* Bornet and Thuret, Notes Algol., fasc. 1, 1876, p. 4.

Placoma is a genus of three known species founded on *P. vesiculosa* Schousb. from "Tingin" on the coast of Morocco. There are now three known species, one of which (*P. africana* Wille) is an inhabitant of fresh water while *P. vesiculosa* Schousb. and *P. violacea* S. and G. are marine. M. A. Howe (1914, p. 11) has indicated the presence of a species on the coast of Peru, but has not assigned to it any definite specific name.

Placoma is closely related to *Entophysalis, Oncobyrsa* and *Chondrocystis*. From *Entophysalis*, it differs in its more hemispherical type of thallus, from *Oncobyrsa* in its less strictly and regularly radiating rows of cells, while from *Chrondrocystis*, it differs in the possession of Gloeocapsoid teguments enveloping the cells.

Placoma violacea S. and G.

Plate 4, fig. 1

Thallus microscopic, 175–250μ diam., irregular to somewhat spherical, verrucose, tegument pale violet; cells 3.5–4μ diam., in groups of 2–8, usually of 4, mostly without order in the interior, but arranged radially toward the periphery, angular when young, becoming spherical later, special teguments very distinct and violet toward the surface, hyaline below; protoplast homogeneous, pale blue-green.

Forming a sort of pulverulent or verrucose black layer, intermixed with other Myxophyceae, on logs along high water mark. Cape Flattery, Washington.

Setchell and Gardner, *in* Gardner, New Pac. Coast Alg. III, 1918*a*, p. 456.

Placoma violacea resembles *P. africana* Wille (1903, p. 90), a fresh water species from South Africa. Besides the difference in habitat, *P. violacea* has larger cells, aggregated into larger colonies, and possesses a violet colored tegument.

6. **Anacystis** Menegh.

Cells spherical to oval, with very thin cell walls and blue-green or violet protoplast, associated into small spherical or more or less lobed and irregular colonies, and embedded within a copious gelatinous tegument, within which they are either uniformly distributed or segregated in small groups; cell divisions in all planes.

Meneghini, Conspect. Alg. Eugan., 1837, p. 6 (in part), cenni sulla organogr. e fisiol. delle Alg., 1838, p. 25 (sp. excl.), Monogr. Nost. Ital., 1842, p. 92 (emend.); Reichenbach, Nom. gen. pl. syst., 1841, p. 18; de Brébisson, Dict. univ. d'hist. nat., vol. 1, 1841, p. 417; Endlicher, Gen. Pl., suppl. III, 1843, p. 11; Kuetzing, Tab. Phyc., vol. 1, 1846, p. 7 (as subgenus of *Microcystis*), Sp. Alg., 1849a, p. 209. *Microcystis* Kuetzing, Tab. Phyc., vol. 1, 1846, p. 7 (in part). *Polycystis* Kuetzing, Tab. Phyc., vol. 1, 1846, p. 7 (as subgenus of *Microcystis*), Sp. Alg., 1849a, p. 210 (in part).

In selecting the proper generic name for the group of species which has been for some years included under the name of *Polycystis* Kuetzing and more recently still under the name of *Microcystis* Kuetzing, it has seemed desirable to scrutinize closely the literature bearing on the subject. *Microcystis* was proposed by Kuetzing in 1833 to include ten species (1833a), now variously referred, but no one of which remains in the more recent revisions of the genus. Furthermore the variety of algae, fungi, and gemmae included under the original *Microcystis* precludes any preponderance of species of one genus or general idea. The first species in the list, *Microcystis Noltii* Kuetz., would now be referred to *Euglena* as the type specimen clearly proves. In 1843 (p. 170), and again in 1845 (p. 148), Kuetzing enumerated four species under his *Microcystis, M. Noltii* still being included and placed first. No one of the four is unmistakably a *Microcystis* in the sense of later authors, *M. Noltii* and *M. olivacea* being species of *Euglena, M. parasitica* being doubtful but probably a *Microcystis* of the later usage, while *M. icthyoblabe,* as used here, is largely, at least, what is now called *Clathrocystis aeruginosa* (cf. also Nordstedt, 1911, p. 264). In 1846 in the first volume of the *Tabulae Phycologicae* (p. 7) Kuetzing separated his genus *Microcystis* into three, retaining in *Microcystis* only *M. olivacea, M. austriaca* and *M. Noltii.* These are all, seemingly at least, species of *Euglena.* In 1849, in his *Species Algarum* (pp. 208, 209) Kuetzing retained the same species, adding to them *M. minor,* a very similar species. It seems, therefore, that the final conception of Kuetzing points toward *Euglena* rather than toward the species included by later writers. Most writers, however, have employed Kuetzing's genus *Polycystis* founded in 1846. This is the second of the two segregates from *Microcystis* and is fairly definite in its content. Kuetzing assigns it three species, viz., *P. elabens, P. aeruginosa* and *P. icthyoblabe. P. aeruginosa* is now separated as *Clathrocystis aeruginosa,* but the other two species certainly seem cogeneric.

Polycystis might properly be chosen for the generic name were it not for the earlier *Anacystis* of Meneghini. *Anacystis* was first published in 1837 (p. 6) to include *A. marginata* and *A. botryoides*. It was again published by Meneghini in 1838 (p. 25) but without mentioning *A. marginata* or any of the species related to it. Finally in 1842 (p. 92) Meneghini emended the genus on the type of *A. marginata*.

Anacystis was recognized by Reichenbach in 1841 (p. 18) and by de Brébisson (1841, p. 417) in the same year. It was also included by Endlicher among the genera of algae in 1843. In 1846 (p. 7), Kuetzing designated it as the first of his segregates from *Microcystis* to include his *M. marginata* and *M. parasitica* and in 1849 (cf. 1849a, p. 209) he used it with nearly the same content, adding only a doubtful species.

Our conclusions are that *Microcystis* as finally revised by Kuetzing is practically synonymous with *Euglena* and that *Anacystis* and *Polycystis* are to be applied to the same group of species. Of these *Anacystis* clearly has the priority of publication.

Anacystis elabens (Bréb.) S. and G.

Plate 4, fig. 6

Cells oblong-ellipsoidal, 3.5–4μ by 5–6μ, closely packed together in more or less globose or globose-elliptical colonies 60–80μ broad, aggregated into botryoidal masses of fair size (up to 500μ), pale bluish or verdigris green.

Among filamentous algae. Pacific Beach, near San Diego, California. Mrs. M. S. Snyder.

Setchell and Gardner, *in* Gardner, New Pac. Coast Mar. Alg. III, 1918a, p. 455, pl. 38, figs. 6, 7; Collins, Holden and Setchell, Phyc. Bor.-Amer. (Exsicc.), no. 2251. *Micraloa elabens* de Brébisson, *in* Meneghini, Monogr. Nostoch. Ital., 1842, p. 104. *Microcystis elabens* Kuetzing, Tab. Phyc., vol. 1, 1846, p. 6, pl. 8. *Polycystis elabens* Kuetzing, Sp. Alg., 1849a, p. 210; Farlow, N. E. Algae, 1881, p. 28.

This is a true *Anacystis* of the *Polycystis* or botryoidal type. It has been compared with the type specimen of Kuetzing's *Polycystis elabens* and found to agree very closely although the cells average very slightly larger (0.5–1.0μ in each diameter). They are decidedly smaller than those of the type specimen of *Polycystis pallida* (Kuetz.) Farlow. There are differences in cell dimensions as well as those of the colony to separate *A. elabens* from both *Microcystis litorea* Hansg. and *Anacystis Reinboldii* Richter.

7. **Chlorogloea** Wille

Cells spherical or oval, with firm, structureless, gelatinous tegument; by division in three planes building up irregular colonies of indefinite size, with cells arranged in more or less distinct vertical or radial rows; reproduction by cell division; new colonies produced by disintegration and separation of the cells of the old colonies.

Wille, Algol. Not. I–VI, 1900, p. 5.

The genus *Chlorogloea* was founded by Wille to include two plants which he considered as most probably identical. The one is *Palmella(?) tuberculosa* Hansgirg (1892, p. 240), an epiphyte on various filamentous marine algae in the neighborhood of Ragusa in Dalmatia in the upper portion of the Adriatic Sea. The other, and the plant studied by Wille, was found at Mandel and at Dröbak on the south coast of Norway. These specimens were found on *Laminaria digitata, Rhodochorton Rothii* and even on the exoskeletons of Bryozoa. The Norway specimens were considered by Wille as specifically and generically identical with those of Hansgirg. Wille's description was drawn chiefly from the plants on *Rhodochorton* found at Dröbak. The position of the genus seems clearly to be with the Chroococcaceae, since no gonidangia with gonidia have been detected.

The genus *Chlorogloea* consists, at present, of four described species, *C. tuberculosa* (Hansg.) Wille (type), from the northern Adriatic Sea, south Norway, and Peru, *C. endophytica* M. A. Howe from Peru, and the two species described recently from the Pacific Coast of North America. *Chlorogloea* resembles *Oncobyrsa*, but is characterized by very much smaller cells without distinct special teguments and forming less distinctly spherical or lobed colonies. It also approaches some species seemingly more properly referred to *Hyella,* from which it differs vegetatively in not having its cells arranged in definite filaments of two sorts as is the case in *Hyella*. *Chlorogloea* also differs, so far as known, in not producing gonidangia and gonidia as is the case in *Hyella*. In the absence of gonidia it is difficult in some cases, at least, to determine with satisfaction to which genus a certain plant may best be referred. The experience of Howe and of Gardner points towards the existence of a number of these obscure epiphytes and endophytes on red algae, bryozoa, etc., which need careful search and study before they can be added to the flora and the complexity of their relationships unravelled.

Wille (*loc. cit.*) has stated that cell division takes place only in one plane. This does not seem to be the case in the species studied by us, nor does it seem possible that the colonies described by Wille could have been built up by such restricted direction of cell division. Howe (1914, p. 14), who has made the most careful study of the species of this genus, says of his *C. epiphytica,* "cell divisions apparently occur in various planes."

1. Chlorogloea tuberculosa (Hansg.) Wille?

Cells in young colonies arranged in rows from the substratum outward, 2μ long, 1–1.5μ diam., yellowish green.

Epiphytic on *Cladophora.* Bairds Point, Straits of Juan de Fuca, British Columbia.

Wille, Algolog. Not. I–VI, 1900, pp. 2–5, pl. 1, figs. 1–6; Setchell and Gardner, Alg. N.W. Amer., 1903, p. 182. *Pringsheimia scutata* f. *cladophorae* Tilden, Amer. Alg. (Exsicc.), no. 382 (in part?). *Palmella ? tuberculosa* Hansg., Neue Beitr. zur Kenntniss der Meeres·algen, 1892, p. 240, pl. 6, fig. 9.

It is extremely doubtful whether *Chlorogloea tuberculosa* should be retained as a member of the algal flora of the Pacific coast of North America. In an earlier paper (cf. Setchell and Gardner, 1903, p. 182) we referred here with some doubt a plant distributed under the name of *"Pringsheimia scutata* f. *cladophorae"* by Tilden (Amer. Alg., no. 382). It was our impression at that time that there was to be found in the specimen examined a small celled plant occurring in tubercular masses, in which the cells were arranged in more or less distinct vertical rows. We also noticed certain large rounded cells which we suggested might be young gonidangia. More recently M. A. Howe (1914, p. 12) has studied another of the Tilden specimens and has failed to find any trace of a *Chlorogloea* present but a unistratose layer of smaller cells intermingled with larger cells. Howe suggests that the Tilden name "was intended to apply to some species of *Dermocarpa."* We, also, have made further studies of the Tilden plant and have found several seemingly distinct things, not all of them identifiable. The principal plant is what we have named and described as *Xenococcus Cladophorae* (cf. Gardner, 1918a, p. 461, pl. 38, fig. 8, and also below), and we believe this to have been the plant of Tilden. Further search has, however, failed to reveal a true *Chlorogloea*, although the impression still prevails with us that one may be present.

2. Chlorogloea conferta (Kuetz.) S. and G.

Plate 2, fig. 6

Colonies forming tubercular masses of indefinite shape and size; cells angular, 0.8–1.2μ diam., slightly longer than the diameter, embedded in a dense, copious, gelatinous matrix of light yellow color, arranged in no definite manner; contents very pale blue-green with hyaline center; reproduction vegetative; cell divisions in all directions.

Growing on *Rhodochorton Rothii* in company with *Dermocarpa hemispherica* and *Dermocarpa suffulta* along high-tide level in shaded places. Moss Beach, San Mateo County, California.

Setchell and Gardner, *in* Gardner, New Pac. Coast Alg. II, 1918, p. 432, pl. 36, fig. 6. *Palmella conferta* Kuetzing, Phyc. Gen., 1845, p. 149, Tab. Phyc., 1846–49, p. 12, pl. 16, fig. 4. *Pleurocapsa conferta* (Kuetz.) Setchell, Alg. novae I, 1912, p. 229.

Chlorogloea conferta has been the subject of discussion among algologists for some time and differences of opinion that have been expressed are probably due to different interpretations as to what plant Kuetzing had in mind when he described *Palmella conferta,* the descriptions being brief and the type material being a mixture of small plants. We have been enabled to examine a small portion of his type material growing on *"Callithamnion Rothii"* through the courtesy of Dr. Weber-van Bosse. We find two species of Myxophyceae very intimately associated on the host plant. One consists of masses of very small cells embedded in a firm gelatinous matrix, varying much in shape and size, of a pale yellowish green color, wholly or only in part surrounding the filaments of the host plant. The other consists of cells, distributed either in small groups or singly, from 5–15μ in diameter, of bright blue-green color, and very frequently surrounded entirely by the preceding. Doubtless it was one of these which Kuetzing took as the type of his *P. conferta,* and it remains to decide which one. Careful measurements of the specimens at hand compared with the measurements given by Kuetzing, "1/700'" gross," leads to the conclusion that his measurements refer to the gelatinous, smaller-celled form, whose cells we find to measure about 1μ in diameter, instead of the larger, more conspicuous form. This form seems to be very close to, if not identical with, the plant commonly associated with the same host plant on both the Atlantic and the Pacific coasts of the United States. It is also usually accompanied by other species of Myxophyceae.

The genus *Palmella,* as now restricted, includes only forms of Chlorophyceae, hence it was necessary to reject that generic name and adopt another for this widely distributed plant which seems to be on the border between the Bacteria and the Myxophyceae. Setchell (1912, p. 229), with special reference to the larger cells, has referred the plant of Kuetzing to *Pleurocapsa,* but the smaller-celled plant is not properly to be referred to that genus. Wille has created the genus *Chlorogloea* to receive a plant very similar to ours, and known as *Palmella ? tuberculosa* Hansgirg (1892, p. 240) and this generic name, *Chlorogloea,* is consequently adopted.

Chlorogloea conferta differs from *C. tuberculosa* in the size of the cells, in their arrangement, and in the number of planes of cell divisions, *C. tuberculosa* dividing in but one plane according to Wille, although it is difficult to understand just how the colony arises if this be true. The other species of Myxophyceae found on the material of Kuetzing's type seems to be a species of *Pleurocapsa,* though being immature it is not safe to attempt to place it. Possibly this is the plant which has been considered to be a form of *P. amethystea* Rosenvinge (1893, p. 967).

2. Chlorogloea lutea S. and G.

Plate 2, fig. 1

Colonies extremely variable in shape and size, spreading over the surface of the host by cell divisions in two planes, divisions in the third horizontal plane forming a cushion of cells up to 100μ thick, the cells at first being arranged more or less in vertical rows, but the radial arrangement being soon destroyed by divisions in other planes, the outer portion of the colony having cells arranged in no definite order; by horizontal divisions certain cells from the lower side of the colony penetrate into the host, forming crooked, branched filaments increasing in length by apical growth; cells of these filaments soon begin to divide in other planes than horizontal, producing masses of cells, encroaching on one another, and finally coalescing into a solid mass in the central portion of the colony; cells 0.9–1.5μ diam., angular, nearly quadrate; terminal cells penetrating the host, up to 4μ long;

cell walls hyaline; cell contents pale yellowish green, homogeneous.

Growing on the stipitate portion of *Iridaea minor* J. Ag. in the lower littoral belt. Carmel Bay, Monterey County, California. May, 1916.

Setchell and Gardner, *in* Gardner, New Pac. Coast Alg. II, 1918, p. 434, pl. 36, fig. 1.

Just where to align a plant like *Chlorogloea lutea* is a problem more or less perplexing. It starts to develop on the surface of the host plant, and if the cuticle of the host is sound in that particular place, it developes a dense mass of cells of considerable dimensions before penetrating into the interior. The plant at first spreads out over the host by cell divisions in two or more planes. It is impossible to state the size of a single plant or colony, for in some places on the host it is continuous for several millimeters. This may, however, be due to the coalescence of several colonies. Later cell divisions in the horizontal plane increase the mass in thickness, up to 100μ or more. At first the cells are arranged in vertical rows, but soon this arrangement is destroyed by false branching or by divisions in other planes. The cells are mostly quadrate in this mass, being somewhat spherical in its outer portion. Sooner or later at various points from the underside of the horizontal layer certain cells are able to penetrate through the cuticle and make their way among the cortical filaments of the host. These penetrating filaments branch. The terminal cells become two or more times as long as the other cells. Growth in length of these filaments is apical. Enlargement of the cells back of the growing points, and divisions in other planes soon produce groups of cells which, encroaching on one another, form a solid mass, the cortical cells of the host at times completely disappearing from the area which they occupy. Plants penetrate to a depth of 200μ.

Chlorogloea lutea very closely resembles in some ways, particularly in its method of development within the host, *Hyella socialis* found growing with it, and also described below. The method of behavior of both within the host is the same. We have not, however, seen *H. socialis* extending beyond the surface of the host as *C. lutea* does. *H. socialis* thus appears to be wholly endophytic, although, on account of the absence of gonidia, we consider the plants in this collection to be immature, and we are unable to say positively what might have developed later.

The habitat and method of development of *Hyella socialis* are so similar to those of *Hyella endophytica* Börgesen (1902, p. 525), which forms gonidangia near the surface of the host, that, nothwith-standing the absence of gonidangia, there can be but little doubt that it belongs to the genus *Hyella*. On the other hand, the genus *Chlorogloea* never has gonidangia, and the species *lutea* is so assigned because of the absence of gonidangia and because of its resemblance to *C. tuberculosa* Wille in the size, color, shape, and early arrangement of the cells in a colony. But that species is wholly epiphytic, and does not have radiating, branching filaments. Furthermore, Wille states that cell division is in one plane, "nach einer Richtung des Raumes" (1900, p. 4, pl. 1, figs. 4–6). This method in the increase of the number of cells of the plant, or colony as it may be called, certainly is different from that of *Chlorogloea lutea*, which has cell division in at least three planes.

Howe (1914, p. 13) has described a plant which is very similar to *Chlorogloea lutea*. He placed it in the genus *Chlorogloea*, and since it was wholly within the frond of the host, he called it *C. endophytica*, stating that it was very close to *Chlorogloea tuberculosa* (Hansg.) Wille, but differs "in its endophytic habit, and in its less distinctly seriate arrangement of the cells, the softer gelatinous walls soon allowing the cells to become inordinate." He stated further that "cell divisions apparently occur in various planes." *Chlorogloea lutea* differs from *Chlorogloea endophytica* in being only partially endophytic, in having smaller cells, in building up larger colonies, and distinctly having cell divisions in all directions.

FAMILY 2. CHAMAESIPHONACEAE BORZI

Thallus unicellular, distinct or associated into non-filamentous families, or pluricellular individuals which are either filamentous or which are differentiated into a more or less chroococcoid basal prostrate portion and an upright filamentous portion; multiplication by transformation of the whole or of a part of the protoplast of a cell, gonidangium, into gonidia either by simultaneous or by successive divisions in some species, or by both vegetative cell divisions and formation of gonidia.

Borzi, Note alla morfol. e. biolog. delle Alghe ficocr., vol. 14, 1882, p. 312; West, Algae, vol. 1, 1916, p. 41.

Borzi founded the family of the Chamaesiphonaceae with the special view of including the fresh water species referred by him to *Chamaesiphon, Clastidium* and *Cyanocystis*. He mentions *Dermo-*

carpa as a marine genus also to be included. The presence of gonidangia, "coccogonia," whose contents break up into from four to many
gonidia, "conidia," is especially mentioned as characteristc. If, however, it is to be taken as established that gonidia also occur in a large
number of genera both of the Coccogonales and Hormogonales, this distinction will prove unsatisfactory. As remarked under Chroococcaceae,
some, if not all, of these cases need substantiation and more careful
study, especially those credited to various Hormogonales. It seems
most likely that they may occur in *Gomphosphaeria* and this genus has
been transferred on this account to the Chamaesiphonaceae, since it
also resembles, in the shape of its cells, *Dermocarpa*. Gonidia have
occurred also in plants resembling closely *Gloeocapsa crepidinum*
Thuret, but which have been placed below, equally satisfactorily,
under *Pleurocapsa*. Most of the Chamaesiphonaceae are epiphytic
and many are attached by a more or less differentiated basal portion,
but these characters, either separately or taken together, seem sufficient to distinguish this family from the Chroococcaceae.

KEY TO THE GENERA.

1. Cells not in filaments.. 2
1. Cells in filaments... 5
　　2. Cells solitary or in layers... 3
　　2. Cells in floating globular masses....................13. **Gomphosphaeria** (p. 49)
3. Layers more or less uniform... 4
3. Layers irregular or complex...10. **Pleurocapsa** (p. 36)
　　4. Cells destitute of vegetative division.....................8. **Dermocarpa** (p. 21)
　　4. Cells dividing in 2 planes.............................9. **Xenococcus** (p. 30)
5. Gonidangia basal at surface of substratum..............11. **Hyella** (p. 40)
5. Gonidangia terminal above surface of substratum.............12. **Radaisia** (p. 45)

8. **Dermocarpa** Crouan

Cells most frequently epiphytic, spherical, ovoid, pyriform to
narrowly cuneate, occasionally existing singly but mostly aggregated
into dense clusters, often so crowded as to distort their form; protoplast homogeneous or finely granular, with blue-green, brownish or
violet color; cell wall comparatively thick, mostly homogeneous,
hyaline; no multiplication by vegetative cell division; reproduction
wholly by the formation of gonidia.

Crouan, Note sur quelques algues marines nouvelles, 1858, p. 70.

Dermocarpa was founded by the brothers Crouan (1858, *loc. cit.*)
on *D. violacea,* a species growing on fragments of crockery at Brest.
Nothing further seems to have been done in the study of the genus

until Reinsch (1874, p. 15) founded his genus *Sphaenosiphon* as a member of the "Melanophyceae" and described a number of species, some showing gonidial stages (*loc. cit.* pl. 25, fig. 2c, 2d, 3c). The genus was properly placed and illustrated by Bornet and Thuret in the *Notes algologiques* (1880, pp. 73–77, pl. 26, figs. 3–9). Since then the number of species has been increased to over twenty and there are doubtless many more awaiting discovery by careful search and study.

The cells of *Dermocarpa* vary from spherical, or almost hemispherical, to obovate and pyriform, or even elongated linear, but the genus possesses one seemingly certain distinction in that its cells undergo no vegetative division, reproducing only by gonidia. This it shares only with *Cyanocystis* Borzi. The difference between the two genera is in the manner of the dehiscence of the gonidangia. In *Cyanocystis* the gonidangia are circumscissile, the top separating as a lid by a smooth transverse rupture. In some species of *Dermocarpa*, the entire wall or, at least, the greater portion of it, dissolves, releasing the gonidia, but in other species, the wall ruptures by a slit at the apex. Whether these distinctions are sufficient for separating the two genera will be left for future study and decision.

KEY TO THE SPECIES.

1. Cells solitary (or at most slightly gregarious).. 2
1. Cells aggregated into dense layers... 4
   2. Adult cells (especially gonidangia) hemispherical....1. **D. hemisphaerica** (p 22)
   2. Adult cells (especially gonidangia) nearly spherical.. 3
   2. Adult cells (especially gonidangia) ovoid to pyriform....4. **D. suffulta** (p 26)
3. Adult cells about 25μ in diam..................................3. **D. sphaeroidea** (p 26)
3. Adult cells 8–16μ in diam...2. **D. sphaerica** (p 24)
   4. Adult cells (especially gonidangia) narrowly to broadly pyriform..............
   ....5. **D. fucicola** (p 27)
   4. Adult cells (especially gonidangia) spherical to narrowly cuneate.................
   ....6. **D. pacifica** (p 27)
   4. Adult cells (especially gonidangia) very variable in shape in same layer........
   ....7. **D. protea** (p 28)

1. **Dermocarpa hemisphaerica** S. and G.
Plate 3, fig. 21

Cells epiphytic, solitary, hemispherical, attached by the flat plane surface, 18–21μ diam. at the base, 10–13μ high; contents of cell bright blue-green, homogeneous; cell wall hyaline, moderately thick, homogeneous, reproduction by successive divisions of the whole protoplast, forming spherical gonidia 0.8–1.2μ diam.

Growing on *Rhodochorton Rothii* in moist, shaded places along high-tide level, or even above. Moss Beach, San Mateo County, California. The above locality is the one from which the type material has been obtained, but the plant has been observed growing on the same host at a number of different localities along the California coast.

Setchell and Gardner, *in* Gardner, New Pac. Coast Alg. II, 1918, p. 438, pl. 37, fig. 21; Collins, Holden and Setchell, Phyc. Bor.-Amer. (Exsicc.), no. 2253.

Dermocarpa hemisphaerica is a somewhat aberrant plant. The individuals are small, mostly solitary, and multiplication is wholly by means of gonidia. In size and distribution on the host it resembles *Xenococcus* in the early stages of development of that genus, but differs from it in not having increase by means of vegetative cell divisions. In its solitary habit and method of reproduction it resembled *Cyanocystis,* but differs from that genus in its method of escape of gonidia. Those of *D. hemisphaerica* escape through a small opening at the apex of the gonidangia, and those of *Cyanocystis* escape by means of a circumscissile splitting of the gonidangia. It seems more nearly to fulfil the requirements of the genus *Dermocarpa,* as now generally understood than of any other genus of Chamaesiphonaceae. It differs, however, from previous conceptions of species of that genus in the method of the formation of gonidia. They are formed in this species by successive divisions of the contents of the gonidangia, those in earlier described species being formed by simultaneous division. In this respect it is like some species of *Pleurocapsa* as well as more recently described species of *Dermocarpa.*

Dermocarpa hemisphaerica is commonly associated with *Chlorogloea conferta* (Kuetz.) S. and G. and *Dermocarpa suffulta* S. and G. The early stages of the development of *D. hemisphaerica* and *D. suffulta* are very similar to each other, but they soon differentiate into their characteristic shapes, and at maturity are very readily distinguishable. *Chlorogloea conferta* is very frequently also present and in such abundance as to completely cover up the other two species, hence this condition along with the presence of diatoms and other foreign material, has made the separation of these forms somewhat difficult. The type material is comparatively free from foreign substances and the plants of *D. hemisphaerica* and *D. suffulta* are both fruiting, are abundant, and comparatively free from *Chlorogloea.* This condition has made it possible to trace the life history of each of these three species, establishing beyond a doubt that they are not to be considered stages in the life history of a single form.

Dermocarpa hemisphaerica seems closely related to *Pleurocapsa amethystea* Rosenvinge (1893, p. 967). This is especially true of the early vegetative stage and of the beginning of spore formation (*loc. cit.*, p. 968, figs. C, D). Later developments of *P. amethystea* depart from that of *D. hemisphaerica* (cf. Rosenvinge, *loc. cit.*, figs. E, F, G).

2. **Dermocarpa sphaerica** S. and G.

Plate 5, fig. 14

Cells solitary or contiguous, spherical, 8–16μ diam., pale blue-green; cell wall thin, hyaline; protoplast finely granular; gonidangia spherical, 8–16μ diam.; gonidia angular at first, becoming spherical at maturity, 2.5–3μ diam., formed by simultaneous division of the protoplast, escaping by dissolution of the entire gonidangial wall.

Growing on various species of algae in the littoral belt, frequently in salt marshes. Ranging from Whidbey Island, Washington, to central California. The type material was found growing on a species of *Lyngbya*, Lands End, San Francisco, California.

Setchell and Gardner, *in* Gardner, New Pac. Coast Alg. III, 1918*a*, p. 457, pl. 39, fig. 14. *Xenococcus Schousboei* Setchell and Gardner, Alg. N.W. Amer., 1903, p. 180; Collins, Holden and Setchell, Phyc. Bor.-Amer. (Exsicc.), no. 554 (not Thuret, *in* Bornet and Thuret, 1880, pp. 73–77, pl. 26, figs. 1, 2).

Dermocarpa sphaerica is the plant which has commonly passed for *Xenococcus Schousboei* on the Pacific Coast of North America. Examination of a bit of the type material of Schousboe's *Coleonema arenifera,* upon which Thuret founded the genus, shows that the Pacific Coast plant belongs to a different genus. *Xenococcus Schousboei,* as described and figured by Thuret, has increase in the number of individuals by means of vegetative cell divisions. This is the condition in which Schousboe found the type material, the plants being young and the gonidial stage having not yet appeared. Our plant which has passed under the name of *Xenococcus Schousboei,* does not divide vegetatively, and hence belongs to the group of which *Dermocarpa* may be taken as a type, in which there is increase only by the formation of gonidia, instead of the group of which *Pleurocapsa* may be taken as a type, in which there is increase by both vegetative cell divisions and by gonidia.

Kirchner (1898, p. 58) and Forti (1907, pp. 119, 120) recognize five genera of Chamaesiphonaceae which have no vegetative cell divisions and which reproduce exclusively by gonidial formation. These are *Cyanocystis* Borzi (1882, p. 314), *Dermocarpa* Crouan (1858, p. 70), *Clastidium* Kirchner (1880, p. 195), *Chamaesiphon* A. Br. and Grun. (cf. Rabenhorst, 1865, p. 148), and *Godlewskia* Jancz. (1883, p. 227). Kirchner separates the first three genera from the last two because of the simultaneous division of the whole protoplast of the gonidangium into gonidia, whereas in the last two genera the gonidia are abstricted successively from the apex of the gonidangium, the basal portion always remaining sterile. Forti adopts this arrangement except the statement as to the simultaneous division of the protoplast.

It is necessary to modify Kirchner's statement in regard to the simultaneous division of the protoplast in *Dermocarpa* since we have found in several species that the gonidia result from successive divi· sions in different planes. This condition is notably true of *D. protea* S. and G. A further exception to the statement of Kirchner must be made in *D. Leibleiniae* (Reinsch) B. and Th. and in *D. suffulta* S. and G., in that only a portion of the protoplast of the gonidangium is converted into gonidia, the basal portion in each species uniformly remaining sterile. There is but little likelihood of confusing *Dermocarpa* with either *Chamaesiphon* or *Godlewskia*, on account of their cylindrical shape, and method of formation and liberation of gonidia, or even with *Clastidium*, which is narrow and cylindrical and possesses a seta at the outer end, but in attempting to place a species like *D. sphaerica*, or *D. sphaeroidea* S. and G., the question arises as to which genus, *Dermocarpa* or *Cyanocystis,* if both of these are to be considered valid genera, should receive them. The method of escape of gonidia in *D. sphaeroidea* has not been determined, but in *D. sphaerica* the whole wall of the gonidangium dissolves and the group of gonidia is left free in position. The only distinction between *Cyanocystis* and *Dermocarpa*, as brought out by both Kirchner and Forti, is that of the method of escape of gonidia. In the former they escape by a circumscissile rupture and in the latter by a dissolution of the apex of the gonidangium. If the method of escape of the gonidia is to be taken as sufficient for generic distinction, it will be necessary to create another genus for *D. sphaerica* which does not conform to either of the above methods. It seems preferable in this case to refrain from extending the number of genera, but to reduce it rather, and since *Dermocarpa* is the older genus to adopt that to receive our species, and reduce *Cyanocystis* in case further study makes it necessary or desirable.

3. **Dermocarpa sphaeroidea** S. and G.

Plate 2, fig. 7

Cells spheroidal or slightly obovate, somewhat angular when compressed, solitary, or mostly grouped together in small clusters, up to 25μ diam.; cell wall hyaline, homogeneous, thin, 1.5μ thick; cell contents pale blue-green at maturity, finely granular; gonidangia 18–25μ diam., the whole cell contents dividing into small spherical gonidia.

Growing on *Porphyra perforata* forma *lanceolata* along high-tide level. Lands End, San Francisco, California. April, 1917.

Setchell and Gardner, *in* Gardner, New Pac. Coast Alg. II, 1918, p. 440, pl. 36, fig. 7.

This species of *Dermocarpa* probably grows on other hosts than that mentioned above in the same locality. Groups of cells resembling those of this species have been observed intermixed with *Radaisia* and *Pleurocapsa* growing on small species of *Enteromorpha* and *Ulva,* but it is difficult to determine the separate members in such a mixture. The material found on *Porphyra* was free from other forms. It seems to occupy an intermediate position, as regards its gregarious habit, between *D. fucicola* Saunders and *D. suffulta* S. and G., the former growing in compact colonies and the latter almost singly.

4. **Dermocarpa suffulta** S. and G.

Plate 2, fig. 9

Cells solitary or loosely associated into small groups, ovoid, pear-shaped or sometimes stipitate, 17–20μ long, 10–14μ diam. at the larger end; contents bright blue-green; cell wall hyaline, homogeneous; gonidangia formed from the upper part of the cell, leaving a cone-shaped, sterile, basal portion; gonidia 8–12 in a gonidangium, 4–6μ diam.

Growing on *Rhodochorton Rothii* near high-tide limit in shaded places on rock. Moss Beach, San Mateo County, California.

Setchell and Gardner, *in* Gardner, New Pac. Coast Alg. II, 1918, p. 440, pl. 36, fig. 9.

These plants were found growing in moderate abundance in company with *Dermocarpa hemisphaerica* S. and G. and with *Chlorogloea conferta* (Kuetz.) S. and G. Of all the known species it seems most closely related to *Dermocarpa Leibleiniae* (Reinsch) B. and Th. It differs from that species in being narrower in general, in having fewer and larger gonidia, and in having a larger part of the cell changed into a gonidangium. Nearly half of *D. Leibleiniae* remains sterile.

It is also closely related to *D. solitaria* very recently described by Collins and Hervey (1917, no. 2155 MS and 1917*a*, p. 17) from Bermuda. Both species have the same habit of growth on the host, being mostly solitary instead of being aggregated into dense clusters as are most species of *Dermocarpa*. *D. solitaria*, however, is much longer than *D. suffulta*, being up to 75μ long.

5. **Dermocarpa fucicola** Saunders

Plate 8, figs. 5, 6

Cells densely crowded into groups of irregular shape and indefinite size, up to 12 mm. across, of very dark olive green to purplish violet color, 22–30μ diam., 40–60μ high at maturity, broadly pyriform to balloon-shaped, narrowing into a stipe-like portion below; gonidia formed by the simultaneous division of the whole protoplast, 4–5μ diam.

Growing on many different species of algae. Common along the whole coast from Puget Sound, Washington, to southern California.

Saunders MS., *in* Collins, Holden and Setchell, Phyc. Bor.-Amer. (Exsicc.), no. 801, Alg. Harriman Exped., 1901, p. 397, pl. 46, figs. 4–5; Setchell and Gardner, Alg. N.W. Amer., 1903, p. 181; Collins, Holden and Setchell, Phyc. Bor.-Amer. (Exsicc.), no. 1251.

The above description is based upon authentic material taken from our fascicles of *Phycotheca Boreali-Americana*, specimens distributed by Saunders under no. 801. It is presumably from material of this distribution that Saunders drew his original description of the species. These specimens are just coming into maturity, as shown from the scarcity of the gonidangia. The measurements are based upon the mature cells, that is the gonidangia, very few of which measure 40μ long, the majority being between 50μ and 60μ long. We express some doubt as to whether the gonidia are formed by simultaneous or by successive divisions of the protoplast. No indications of successive divisions were seen, all of the gonidia noticed being completely formed.

6. **Dermocarpa pacifica** S. and G.

Plate 3, figs. 22–24

Cells aggregated into colonies up to 200μ diam., varying in shape from nearly spherical, broadly ovate or pear-shaped, to narrowly wedge-shaped, 30–45μ long, 20–35μ diam.; cell walls thick, hyaline, homogeneous; cell contents bright blue-green or olive green when young, changing to brownish when old; gonidia numerous, 2μ diam.

Growing on *Chaetomorpha aerea* in a tide pool near high-tide limit. Cypress Point, Monterey County, California. January, 1917.

Setchell and Gardner, *in* Gardner, New Pac. Coast Alg. II, 1918, p. 439, pl. 37, figs. 22–24.

The plants of this species were found associated with *Xenococcus Chaetomorphae* and the two species were so abundant as to give the host plant a very dark color. The shapes of the cells of *D. pacifica* are determined, to a certain extent, by their position on the host and by their age. The younger cells of *C. aerea* are cylindrical, but they become quite torulose at maturity. This change in the shape of the host cells modifies the form of certain cells of the epiphyte. If the gonidia of the *Dermocarpa* happen to locate at the cross walls when the host cells are young, increase in the size of the cells of both the host and the epiphyte causes the cells of the epiphyte to become much crowded and thus assume a narrow wedge shape. As the cells of the host plant mature and begin to disintegrate the younger *Dermocarpa* cells that have had more room in which to expand are liberated and become broadly oval or even spherical. The cell contents become much darker and brownish at maturity, and when mounted in glycerine and acetic acid change to purple. The contents of the whole cell change into gonidia by simultaneous division.

7. **Dermocarpa protea** S. and G.

Plate 4, figs. 4, 5

Cells extremely variable in shape and size, broadly pyriform to narrowly cuneate, 40–120μ long, 6–40μ diam. at the apex, 3–7μ at the base; cell wall hyaline, 2–3μ thick; protoplast homogeneous, light blue-green; gonidia 3–3.5μ diam. formed by successive divisions of the protoplast.

Growing on *Spongomorpha* sp. West coast of Whidbey Island, Washington.

Setchell and Gardner, *in* Gardner, New Pac. Coast Alg. III, 1918a, p. 456, pl. 38, figs. 4, 5.

A single specimen of this species of *Spongomorpha* with the above epiphyte growing upon it has thus far been collected. The terminal portions of the host were so thickly clothed with the epiphyte as to give them a very decidedly dark appearance. Microscopic examination showed that the colonies were unusually variable in shape and in size, some capping the filaments, others completely surrounding and obscuring them for some distance, and still others small and widely separated. Material scraped from the filaments and magnified revealed a surprising variation in the shape and size of the vegetative cells, as well as of the gonidangia. Among the collection one could select cells in both vegetative and reproductive conditions which would conform to the description, in shape and in size, of each of several well known species; but since there is such perfect gradation in lengths and in widths, both in purely vegetative cells and in the gonidangia, ranging in size of mature cells from 6μ to 40μ wide in the upper parts, and from 40μ to 120μ long, it is impossible to segregate them into species. Considered as a species *D. protea* represents the largest known species of *Dermocarpa,* and has the greatest range of variation in size and in shape of the cells. Plate 4, figs. 4, 5 represents some of these found among the gonidangia.

Dermocarpa protea is an excellent example of the formation of gonidia by successive and progressive divisions of the gonidangia to form gonidia. This method is represented in the gonidangia irrespective of shapes and sizes, and is another evidence in support of their all belonging to a single species.

Kirchner (1898, p. 58) in his key to the genera places *Dermocarpa* in the group which produces the gonidia by simultaneous division of the protoplast. In Cytological Studies in Cyanophyceae (1906, p. 281) Gardner expressed the opinion that *Dermocarpa fucicola* Saunders produces gonidia in this manner. We have since examined authentic material of that species and, although the gonidangia are very sparse, it appears to form them by simultaneous division; but having examined other collections from our coast which seem to be of the same species, judging from the shape and size of the cells, which clearly show that the formation of gonidia is by successive divisions of the protoplast, some doubt may still be entertained as to which method *D. fucicola* follows. This subject must have more careful study and may prove to be a more stable character than shapes and sizes of cells upon which to establish species.

Dermocarpa prasina (Reinsch) B. and Th. has been credited to our coast by Saunders (1898, p. 397) as occurring on "*Sphacelaria racemosa arctica*" and "*Sphacelaria cirrhosa*" and extending from Puget Sound to the Shumagin Islands in Alaska. It seems to us unlikely that this species should be represented in these waters, but we have

no specimens of *Sphacelaria* from them, nor have we seen any specimens of Saunders. The *Dermocarpa prasina* of Setchell (1899, p. 54)
is indefinite and undoubtedly mostly what was later named *D. fucicola*
by Saunders.

9. **Xenococcus** Thur.

Cells spherical or more or less angular due to mutual pressure
when closely aggregated, scattered or collected into a continuous
stratum, usually epiphytic; cell contents pale blue-green or dark
violet, homogeneous; reproduction by cell division in two planes, and
by formation of gonidia.

Thuret, Essai Class. Nost., 1875, p. 373 (nomen nudum), *in* Born.
and Thur., Notes Algol., vol. 2, 1880, pp. 73–75 (description of type);
Hansgirg, Physiol. u. Algol. Studien, 1887, p. 111 (lim. mut.), Prodr.
Algenfl. Böhm. II, 1892a, p. 128.

The genus *Xenococcus* was founded by Thuret in 1880 (Notes
Algol., vol. 2, pp. 74, 75) upon *X. Schousboei* Thuret as the type
species, but no distinct and definite generic diagnosis was given.
Thuret had already mentioned the genus in 1875 (Essai Class. Nost.,
p. 373), but neither described it nor mentioned a type species. Hansgirg (1887, p. 111) discussed *Xenococcus* and its limits, but set the
limits beyond those instituted by Thuret and repeated these later in
what is probably the first formal diagnosis of the genus (1892a, p. 128).
Bornet (in 1889 and 1892) reduced *Xenococcus* under *Dermocarpa*
Crouan, but Kirchner (1898, p. 58) restored it to generic rank. The
species of *Xenococcus,* as was understood by Thuret, differed from
those of *Dermocarpa* particularly in their lack of gonidangia. When
Batters discovered gonidangia and gonidia in *Xenococcus Schousboei,*
it seemed necessary to reduce the species under *Dermocarpa,* and this
was suggested by Bornet. Kirchner, however, brings forward the fact
that *Xenococcus Schousboei* differs from the species of *Dermocarpa*
in that the cells divide vegetatively, and consequently restores *Xenococcus* to independent rank. Achille Forti (1907, p. 133) states that
the vegetative division is in three directions, but Thuret (*in* Bornet
and Thuret, 1880, p. 75) emphatically says that they divide perpen-
dicularly to the surface of the substratum and in this direction only.
An examination of a portion of the type material indicates cell division in two directions, perpendicular to one another and to the surface
of the substratum.

1. Xenococcus acervatus S. and G.

Plate 5, fig. 13

Cells wholly epiphytic, dividing in two planes perpendicular to the host, building colonies at first one cell deep, later confusedly heaped up, of indefinite extent; cells angular at first, soon becoming spherical or pear-shaped, 3–6μ diam., cell wall thin, hyaline; protoplast homogeneous, pale blue-green; gonidangia unknown.

Growing in great profusion on *Enteromorpha* sp. in salt marsh pools. San Francisco Bay, California.

Setchell and Gardner, *in* Gardner, New Pac. Coast Alg. III, 1918*a*, p. 459, pl. 39, fig. 13. *Pleurocapsa amethystea* var. *Schmidtii* Collins, *in* Collins, Holden and Setchell, Phyc. Bor.-Amer. (Exsicc.), no. 1704.

The first publication of the name *P. amethystea* var. was by Börgesen (Mar. Alg. Fäer., 1902, p. 524), Johannus Schmidt having identified one of Börgesen's species from the Fäeröes as belonging there. Börgesen stated that Schmidt would comment on the species later in Helgi Jónsson's paper. Jónsson's paper appeared in 1903 but Schmidt mentions only a plant from Iceland under the above name ("*P. amethystea* var."). Collins (*loc. cit.*) considers our plant to be of the same variety as the Iceland plant and gives it a varietal name without further comment.

In the absence of gonidangia it is not at present possible to give a complete comparison of our plant with the description of the Greenland plant, *P. amethystea,* of Rosenvinge (1893, p. 968), nor with the variety of that species growing in Iceland, determined by Schmidt (*in* Jónsson, 1903, p. 378). There are no specimens of either of these available at present for comparison. Rosenvinge's figures and description of *P. amethystea* show plainly that the cells divide in three planes, which places it with the genus *Pleurocapsa*. Schmidt states that the

Iceland plant differs from the Greenland plant only in color, but forbears naming it on that character alone. The Iceland plant seems, therefore, certainly to be a *Pleurocapsa*. Our plant divides vegetatively in but two planes, and, accepting Thuret's understanding of his genus *Xenococcus*, it belongs to that genus rather than to *Pleurocapsa*. A brief discussion of the structure and relation of these two genera is given in Gardner's New Pacific Coast Marine Algae II, 1918, p. 436.

Xenococcus acervatus differs from *P. amethystea* in the number of planes of vegetative cell divisions, in the shape and size of the cells and in their color, the color of the latter being "sordide violacea," and the former pale blue-green. None of the seven illustrations of Rosenvinge (*loc. cit.*) resembles very clearly any phases of the development of our plant except A, the surface view of a group of vegetative cells.

At times the cells as viewed in the median plane of the host plant are piled up several cells deep, as though they had arisen by horizontal divisions. If this were the case, our plant would properly belong to the genus *Pleurocapsa*. This does not seem to be the case, however, as the cells above the surface layer are very generally spherical, apparently independent, and very variable in size. They appear rather to be gonidia in various stages of development, that have come to rest on the surface layer, or in some cases, seem to have grown in position where they were formed. The cell walls are decidedly gelatinous, which is conducive to holding the colonies together.

2. **Xenococcus Gilkeyae** S. and G.

Plate 5, fig. 11

Cells solitary or aggregated into small colonies, spherical when solitary, angular and more or less elongated in colonies, 4–7μ, rarely 9μ diam.; cell wall inconspicuous, hyaline; protoplast light blue-green; gonidangia of the same shape and size as the cells; gonidia 0.8–1μ diam., formed by successive divisions of the protoplast.

Growing on the filaments of *Elachistea* sp. which is epiphytic on *Fucus* sp. Lower littoral belt. Sitka, Alaska.

Setchell and Gardner, *in* Gardner, New Pac. Coast Alg. III, 1918*a*, p. 462, pl. 39, fig. 11.

Having vegetative cell divisions in but two planes perpendicular to the substratum, *Xenococcus Gilkeyae* is a typical member of the genus. It is an exceedingly delicate species but the type material being in excellent vegetative and reproductive conditions is clearly definable. The gonidia appear to be the results of two lines of development of the vegetative cells. Some gonidia seem not to divide vegetatively after coming to rest, but continue to increase in size until maturity is reached, then by a few successive internal divisions the whole protoplast is progressively converted into gonidia. In some cases the first division takes place horizontally cutting off a small portion of the base of the protoplast, which in some instances seems to remain sterile, at least the whole upper part is converted into gonidia before the basal portion is (pl. 5, fig. 11*a*). In other cases the first division is through the center of the cell. Other gonidia develop for a time after coming to rest, then divide vegetatively several times, generating small colonies before advancing to the gonidial stage (pl. 5, fig. 11*b*).

3. **Xenococcus Cladophorae** (Tilden) S. and G.

Plate 4, fig. 8

Plants forming a more or less continuous layer one cell deep, or occasionally associated in small groups; cells variously shaped, angular, prismatic, spheroidal or pyriform, $8–15\mu$, occasionally 22μ diam.; cell walls prominent, hyaline, homogeneous, often diffluent; protoplast pale blue-green, homogeneous; gonidangia the same shape and size as the vegetative cells; gonidia formed by successive divisions of the protoplast, $1.5–2\mu$ diam.

Growing on *Cladophora* sp. in a tide pool, Baird Point, Strait of Juan de Fuca, British Columbia.

Setchell and Gardner, *in* Gardner, New Pac. Coast Alg. III, 1918*a*, p. 461, pl. 38, fig. 8. *Pringsheimia scutata* f. *Cladophorae* Tilden, Amer. Alg. (Exsicc.), no. 382, type. *Chlorogloea tuberculosa* Setchell and Gardner, Alg. N.W. Amer., 1903, p. 182 (in part); Tilden, Minn. Alg., vol. 1, 1910, p. 46 (in part).

So far as we are able to discover on the available specimens of *Cladophora* distributed by Miss Josephine Tilden under no. 382 of her *American Algae,* there are no epiphytes present belonging to the Chlorophyceae. On the contrary, there is a mixture of species belonging to the Myxophyceae.

One of these, a very small-celled form, very sparse in the distribution at our disposal, as suggested by us (1903, p. 182), seems to be closely related to *Chlorogloea tuberculosa* (Hansg.) Wille. The material is too sparse and scattered to admit of a very positive determination as to what species it really may be. There are also groups of well defined vegetative cells of a species of *Pleurocapsa*. Judging from its present vegetative stage it seems to be undescribed. There are, however, no gonidangia that we have been able to discover, and as the material is very scanty we forbear naming it at present. Howe (1914, p. 12) states that he is unable to find in Miss Tilden's distribution, mentioned above, any member of the Chlorophyceae that he could interpret as being her *Pringsheimia scutata* forma *Cladophorae*, under which name no. 382 was distributed, but suggested that possibly the species referred to was that of a *Dermocarpa*, thus indicating that genus as being present in the material he examined. By far the most abundant species present in our material is no one of these but is the plant described above. Some specimens of the host are much contorted by its presence. This species seems most likely to be the one upon which was based *Pringsheimia scutata* forma *Cladophorae*, and is the plant, in part at least, later described and figured as *Chlorogloea tuberculosa* by Miss Tilden (Minn. Alg., 1910, p. 46, pl. 2, fig. 42).

4. Xenococcus pyriformis S. and G.

Plate 5, fig. 12

Colonies small, single or occasionally confluent, young cells somewhat angular, pyriform to subspherical at maturity, 10–15μ diam., 12–20μ long; protoplast bright blue-green; cell wall conspicuous, dense, hyaline; gonidangia the same shape and size as the cells; gonidia 2.8–3.5μ diam., formed by successive divisions of the protoplast.

Growing on *Rhodochorton Rothii* on rock ledge along high-tide level and above. Cape Arago, at the entrance to Coos Bay, Oregon.

Setchell and Gardner, *in* Gardner, New Pac. Coast Alg. III, 1918a, p. 463, pl. 39, fig. 12.

This species having cell divisions in only two planes perpendicular to the substratum, thus forming colonies only one cell deep, is to be placed under *Xenococcus* rather than under *Pleurocapsa*, and, while its pyriform cells suggest species of *Dermocarpa*, it differs from members of that genus in having vegetative cell division. In size and shape of cells it differs from any other species of *Xenococcus*, as yet described.

5. **Xenococcus Chaetomorphae** S. and G.

Plate 2, figs. 2–4

Vegetative cells extremely variable in shape and size, some spherical, some very angular and of nearly equal dimensions, and some long and narrow, tapering to sharp points at both ends as seen from above, up to 25μ in vertical diameter, the narrower cells up to 45μ long; gonidangia the same shape as the vegetative cells; gonidia formed by successive divisions of the whole protoplast; color bright blue-green.

Growing on *Chaetomorpha aerea* in a pool near high-tide limit. Cypress Point, Monterey County, California.

Setchell and Gardner, *in* Gardner, New Pac. Coast Alg. II, 1918, p. 436, pl. 36, figs. 2–4.

This species was found associated with *Dermocarpa pacifica,* and it is somewhat difficult to distinguish the two species. Both species produce gonidia in great abundance, and plants of both species may be found in all stages of growth at the same time. They are very abundant, intermingled, and as seen from above are very angular on account of lateral pressure resulting from growth. There appears certainly to be vegetative growth in the *Xenococcus,* as the continuous areas covered by it, comprising many hundreds of cells, seem far too great to have arisen from a single group of gonidia; also should the gonidia escape singly they could never by chance become so uniformly and closely associated as they often are to be found. The cells of *Xenococcus* in all stages of growth are extremely angular and very variable in shape and size; sometimes being much crowded at the cross walls of the host plant, the tendency is to elongate vertically. Frequently groups of cells seem to have started to grow on the cell walls of young cells of the host between the cross-walls, and as the host cell elongates the cells of the epiphyte seem to elongate abnormally in the direction of the long diameter of the host (pl. 2, fig. 2). There are no particular shapes and sizes of gonidangia, since any of the long narrow cells, small angular cells, or large spherical cells may be transformed into gonidangia. The gonidia are formed by successive divisions of the whole of the protoplast (pl. 2, fig. 4). On the whole, it seems almost certain that we have here two plants of different genera closely and intimately associated, and it certainly is not possible to distinguish the individuals of the two in all stages of their development.

10. **Pleurocapsa** Thur.

Cells spherical or angular, and many sided by pressure, united into more or less gelatinous colonies of various shapes and sizes arising by cell divisions in three planes; protoplast blue-green, olive green, yellowish or violet; reproduction by vegetative cell divisions and by gonidia formation.

Thuret, *in* Hauck, Meeresalg. Deutschl., 1885, p. 515.

The genus *Pleurocapsa* while named by Thuret was published by Hauck (*loc. cit.*). The type specimen of the species *Pleurocapsa fuliginosa* Hauck, the type of the genus, was collected at Trieste as is evidenced by the label of the specimen in Herb. Hauck at present in the possession of Dr. Anna Weber-van Bosse, through whose kindness it has been possible to examine a fragment. The species properly to be referred to *Pleurocapsa* are to be distinguished from those of *Xenococcus* by forming a less definite layer and by division in three directions, while from those of *Gloeocapsa,* they differ chiefly by forming gonidia.

The genus *Pleurocapsa,* as given by Forti (1907, pp. 120–123), is credited with ten species, of which one occurs on limestone walls, four are marine, while five are of fresh water. The fresh water species referred to *Pleurocapsa,* especially those described by Hansgirg in his Physiologische und algoloische Mittheilungen (1890), are not readily to be reconciled with the description and distinctions as indicated above. They need further study. From the figures it seems that their affinities may possibly be rather with *Radaisia,* or possibly even with *Chamaesiphon.* Certainly no such filamentous arrangement of the cells is to be found in *Pleurocapsa fuliginosa* Hauck or the other marine species now referred to the genus.

KEY TO THE SPECIES.

1. Teguments (enclosing jelly) hyaline... 2
1. Teguments (enclosing jelly) yellowish brown........3. **P. entophysaloides** (p 38)
    2. Cell wall hyaline..1. **P. fuliginosa** (p 36)
    2. Cell wall brownish..2. **P. gloeocapsoides** (p 37)

1. **Pleurocapsa fuliginosa** Hauck

Colonies forming a thin, dark crust on wood and rocks, each colony being 50–100µ diam., cells 5–20µ diam., with colorless membrane and homogeneous, golden, reddish brown or sordid violet contents.

Growing on piles. Seattle, Washington.

Hauck, Meeresalg. Deutschl., 1885, p. 515, fig. 231; Setchell and Gardner, Alg. N.W. Amer., 1903, p. 181.

The type of *Pleurocapsa fuliginosa* Hauck was collected at Trieste and has been examined by us as mentioned above. Further remarks on this type will be found under *P. entophysaloides* S. and G. The result of our study has been to retain the Seattle plant under this name, for the present at least, since it resembles the type material very closely. We did not, however, find any of the larger colonies described and figured by Hauck and the cells, with thick teguments, measure up to 30μ in diameter, which is half again greater than the maximum measurements given by Hauck. For reasons given later we feel uncertain as to the exact form and development of the gonidangia and gonidia in this species.

2. Pleurocapsa gloeocapsoides S. and G.

Plate 5, figs. 15, 16

Colonies associated into soft, glistening, gelatinous masses, 1–2 mm. thick; cell divisions regularly in three planes; cells globose when single, angular from mutual pressure in colonies; protoplast 4–8μ diam., homogeneous, pale blue-green; cell wall conspicuous, brownish; colonies of 2–8 cells enclosed in an ample, homogeneous, hyaline, soft, gelatinous tegument; gonidia 2–4 formed in unchanged vegetative cells, 2.5–3.5μ diam.

Growing on a water-soaked log, at the margin of a salt marsh. Alameda, California.

Setchell and Gardner, *in* Gardner, New Pac. Coast Alg. III, 1918*a*, p. 465, pl. 39, figs. 15–17. *Gloeocapsa crepidinum* Collins, Holden and Setchell, Phyc. Bor.-Amer. (Exsicc.), no. 1151 (not of Thuret).

The material of the above mentioned distribution was collected in 1913 in the same locality in which the type material of *P. gloeocapsoides* was collected in October, 1917. The gonidia, if present in the original collection, were overlooked, and the close resemblance of the species in the vegetative stage to Thuret's *Gloeocapsa crepidinum,* so well and amply figured in Bornet and Thuret (Notes Algol. I, 1876, pl. 1), led the authors of the above mentioned exsiccatae to place it in that species. Now that well formed, typical gonidia are found in great abundance it has seemed best to remove it from the Chroococcaceae and place it in the Chamaesiphonaceae, if we are to adhere to the well recognized distinction between these two families. The formation of gonidia is not unique in this species. It is similar to that reported as taking place in *Pleurocapsa entophysaloides,* and in other undescribed species which we have observed. In all of these species the contents of all of the cells of a colony seem to divide simultaneously. Any and all of the vegetative cells may produce gonidia without change in size. This is also true of all species of *Dermocarpa, Gomphosphaeria,* etc. The cell being small the number of gonidia from a cell is small, viz., two to four, and it has not been determined positively whether the two divisions necessary to form four gonidia take place successively or simultaneously. In most of the species in other genera which we have studied the divisions take place successively, and this method may possibly be the one followed by all species. Wille (1906, p. 21) has described and figured what appears to be the same method of gonidial formation in the European *Gloeocapsa crepidinum* Thur., which he finds near the biological station at Drontheim. He does not look upon this stage in the life history as being the gonidial stage comparable to that of the Chamaesiphonaceae, but speaks of it as the "*Aphanocapsa*-Stadium." He concludes that these small cells float away, finally lodge in tide pools, and become the so-called marine species of *Aphanocapsa,* e.g., *A. marina* Hansg. The facts as we find them here seem hardly sufficient to warrant such conclusions concerning our species.

There are certain difficulties in the way of a satisfactory placing of this form that have led to a disposition of it as a species of *Pleurocapsa.* Were it not for the presence of gonidia, it might be placed fairly satisfactorily under *Gloeocapsa crepidinum* Thuret. The discovery of similar gonidia in what Wille refers to *Gloeocapsa crepidinum* in Norway, strengthens this position. There are, however, possibilities that Wille's plant may not be the same as that of Thuret and that ours may be different from both. It may be that all *Gloeocapsa* species may be found ultimately to form gonidia, but it does not seem very probable. In view of the various uncertainties, it seemed best for the time, at least, to refer the California plant to *Pleurocapsa.*

3. **Pleurocapsa entophysaloides** S. and G.

Plate 4, figs. 9, 10; plate 7, fig. 30

Plants forming a dark colored, pulverulent, somewhat mucilaginous stratum, 1–4 mm. thick; single cells spherical, 8–10μ diam., in colonies very angular, 4–8μ diam., forming spherical or variously lobed, frequently entophysaloid colonies, 40–60μ, up to 200μ diam., by cell divisions without rupturing the original tegument; teguments firm, slightly mucilagenous on the surface, yellowish brown, 2μ thick; protoplast homogeneous, light blue-green; gonidia formed in unchanged vegetative cells of either the small or the large colonies, 3–3.5μ diam.

Growing on rocks in the upper littoral belt. Carmel Bay, Monterey County, California.

Setchell and Gardner, *in* Gardner, New Pac. Coast Alg. III, 1918*a*, p. 463, pl. 41, fig. 30. *Pleurocapsa fuliginosa* Collins, Holden and Setchell, Phyc. Bor.-Amer. (Exsicc.), no. 704 (not of Hauck).

Pleurocapsa entophysaloides is closely related to *P. fuliginosa* Hauck (1885, p. 515, fig. 231) but differs from it in having slightly larger colonies which are often entophysaloid (pl. 7, fig. 30), in having a blue-green protoplast, a yellowish brown tegument, in having slightly larger vegetative cells and in having gonidia develop in the small vegetative cells. In this last statement concerning the size of the cells, we are assuming that Hauck intended to include the large spherical cells which he figures, and which would ordinarily be considered gonidangia, in his measurement "Zellen 5–20μ dick." We have examined a bit of the type material of his *P. fuliginosa* and find that the vegetative cells of our species are slightly larger than the cells which we consider to be the vegetative cells of the type. Hauck does not mention gonidangia but he figures, in addition to groups of two, four, to many cells, four large spherical cells, one of which is filled with undoubted gonidia. The presence of these two forms and sizes of cells brings up the question as to whether or not we are dealing with a single species or with two species, and if the latter, the two species probably belong to different genera. We have been puzzled with several similar mixtures collected on our coast. We are of the opinion that we have to reckon with two species in these cases and in the case of *P. fuliginosa*. Now that we have discovered gonidia in these colonies of small vegetative cells in at least two species of *Pleurocapsa*, viz., *P. entophysaloides* and *P. gloeocapsoides*, the evidence in favor of considering that such mixtures as mentioned above belong to two species is stronger. Either we may take this view of the matter or we may note that we have two lines of development in a single species. One starts with a single cell, e.g., a gonidium, and, after enlarging to mature size, divides in three planes successively, more or less at right angles to each other, the process continuing until a smaller or larger colony is built up, according to the species, the cells all remaining within a common tegument; then at maturity each small cell of the colony divides into a few gonidia, the whole mass of teguments and cell walls dissolves, thus setting free the gonidia. The other line of development starts likewise with a single cell. This continues to increase in size without division until maturity is reached when it is many times larger than the vegetative cells in the other method of development, after which the whole protoplast divides into gonidia, either by successive or by simultaneous divisions. These questions need further observation and study before any decision of value can be made concerning them.

11. Hyella B. and F.

Plants forming tangled masses of indefinite expansion, boring into shells of mollusks or into other algae; primary or basal filaments mainly extending horizontally, one or more rows of cells enclosed within a sheath, often very much crumpled and contorted, with frequent true branching, composed of a series of cells practically independent of one another; short secondary filaments composed principally of longer and narrower cells enclosed in a separate sheath arising from the basal filaments; cell divisions in all directions; reproduction is accomplished by the escape of vegetative cells from the sheaths, and by gonidia formed by the successive divisions of the contents of gonidangia developed on short branches of the basal filaments or by the modification of cells of the basal filaments.

Bornet and Flahault, Note sur deux nouveaux genres d'algues perforantes, 1888, p. 163 (p. 3, Repr.).

The type species is *Hyella caespitosa* B. and F., but the type specimen is not definitely designated. The two localities explicitly indicated are those of Croisic in Brittany, on the Bay of Biscay and Cette in Languedoc, on the Gulf of Lyons. Since the locality mentioned especially as the source of the material is Croisic (Bornet and Flahault, Sur quelques plantes vivant, 1889, p. 3), we have assumed this as the type locality.

The genus *Hyella* is dimorphous and presents an interesting and, to some extent, puzzling morphology. It is distinctly filamentous and in general appearance suggests the Stigonemataceae. At first a layer is formed at the surface of plant or shell, which appears chroococcoid but which seems really to be filamentous of the complex type of *Stigonema*. From the inner face of this layer branches are given off which penetrate into the substratum and which are more simple in their structure. These in turn give off true branches. The structure thus far is stigonemataceous. There are no hormogonia, however, but seemingly coccogonia (cf. Bornet and Flahault, *loc. cit.*, p. 22). There are also distinct gonidangia whose contents divide into gonidia, such as are found in the Chamaesiphonaceae.

While the general appearance of *Hyella caespitosa* B. and F. differs very much from that of *Radaisia Gomontiana* Sauv., yet it is by no means a simple matter to place certain Californian plants recently described. Gardner (1918, p. 432) has stated these difficulties and has outlined his reasons for placing some under *Hyella,* as follows: "These forms with erect filaments, more or less branched and distorted, arising from basal filaments on the surface of the substratum, and growing into it, and having gonidangia at their bases near the surface of the host, have been assigned to *Hyella.*"

The genus *Hyella* is credited by Forti (1907, pp. 124–127) with five species, the type (*H. caespitosa*) and three others being shell borers, while the fifth burrows(?) into certain red algae. Of the four species thus far detected on the Pacific Coast of North America, two are shell borers and two burrow into membranous red algae.

KEY TO THE SPECIES.

1. Boring into shells... 2
1. Partially endophytic on larger algae... 3
   2. Erect filaments usually less than 10μ in diam...........1. **H. caespitosa** (p 41)
   2. Erect filaments 10–14μ in diam..................................2. **H. Littorinae** (p 42)
3. Filaments branching dendroidally...4. **H. socialis** (p 44)
3. Filaments seldom branching...3. **H. linearis** (p 43)

1. Hyella caespitosa B. and F.

Colonies appearing at first as minute discolorations on shells, later becoming much expanded and confluent, frequently covering the entire outer part of the shell, causing it to be very rough; erect filaments usually parallel, 5–6μ, even up to 10μ diam., 100–200μ long, lower cells shorter, and upper ones several times longer than the diameter and sometimes branching; sheaths hyaline; horizontal filaments much branched, composed of spherical or angular cells, often several rows in a sheath; cell contents yellowish olive green or blue-green; gonidangia formed by the modification of cells towards the base of the filaments into large oval to pyriform gonidangia, gonidia numerous, spherical, about 2μ in diameter.

Boring into the shells of oysters and other mollusks. Central to southern California.

Bornet and Flahault, Note sur deux nouveaux genres d'algues perforantes, 1888 (p. 2, Repr.), Sur quelques plantes vivant, 1889, vol. 36, pl. 10, figs. 7–9, and pl. 11; Collins, Holden and Setchell, Phyc. Bor.-Amer. (Exsicc.), no. LI.

In placing this species among our Pacific Coast algae, we do so with some hesitation. The specimens of oyster shells distributed (Phyc. Bor.-Amer., Exsicc., no. LI) as well as others from the same locality at other dates, show definitely a *Hyella* and some even with young gonidangia, but no such typically well-fruited specimens as is desirable for certainty. Specimens possibly the same have also occurred at San Pedro, California, but unfortunately without gonidangia. This species is, therefore, listed in the hope that further information may be obtained.

2. Hyella Littorinae S. and G.
Plate 3, figs. 19, 20

Erect filaments numerous, straight, parallel, simple or sparsely branched, 75–85μ long, 10–14μ diam.; growth in length by division of apical cell; basal cells of filaments angular, 4–6μ diam., divisions in all planes; terminal cells of erect filaments cylindrical, 20–30μ long; cell contents blue-green, homogeneous; cell wall or sheath hyaline, homogeneous; 2.5–3.5μ thick; gonidangia unknown.

Growing on *Littorina planaxis* Nutt. along high-tide level on rocks. Common along the coast of California, but probably grows wherever this species of *Littorina* occurs. The type material is from Carmel, Monterey County, California.

Setchell and Gardner, *in* Gardner, New Pac. Coast Alg. II, 1918, p. 441, pl. 37, figs. 19, 20; Collins, Holden and Setchell, Phyc. Bor.-Amer. (Exsicc.), no. 2255.

Reproduction by gonidia has not been observed in this species, but the vegetative characters are so similar to those of *Hyella caespitosa* growing on the shells of oysters and other mollusks as to leave room for little doubt as to its close affinity with that species. The prostrate filaments, characteristic of *Hyella caespitosa,* have not been observed. Gonidangia may not develop and reproduction may be wholly vegetative in the same manner as vegetative reproduction is reported to occur in *Hyella caespitosa.* We base our opinion on the fact that we have examined shells at all seasons of the year and from different localities and have not found gonidia. The basal cells divide in several planes into numerous small, angular cells approximating gonidia. The filaments resulting from these divisions resemble those figured by Bornet and Flahault (1889, pl. 11, fig. 2). The small, angular cells probably escape and develop into new plants. They can hardly be considered as having formed in gonidangia, because a few cells just above the basal cells divide likewise though progressively to a less degree, forming clavate filaments at maturity (pl. 3, fig. 19).

3. **Hyella linearis** S. and G.

Plate 2, fig. 8

Thalli dark blue-green, penetrating into the host, 350–450μ long; cells of the inner ends of the filaments smallest, 4–6μ diam., 3.5–4.5μ long, gradually increasing in size to 12μ in diam. toward the periphery of the host; cell walls thin, hyaline; cell divisions in one plane at first, building up more or less tortuous, rarely branched filaments; later cell divisions in three planes, often decidedly oblique, most abundant toward the periphery of the host, building up clavate filaments; gonidangia at the surface of the host, 14–20μ diam.; gonidia numerous, 1μ diam.

Growing on *Prionitis* sp. Sunset Beach, near the mouth of Coos Bay, Oregon. This is the type locality and only one plant of *Prionitis* has been found infested by this species of Myxophyceae.

Setchell and Gardner, *in* Gardner, New Pac. Coast Alg. II, 1918, p. 442, pl. 36, fig. 8.

This species seems closely related in form to *Hyella socialis* S. and G., but differs principally in the size of the cells, and in having the filaments nearly straight and rarely branched. The early stages in the development have not been observed. There is probably a prostrate layer of cells developed first, characteristic of the genus *Hyella*, and the erect penetrating filaments arise from that. The discovery of *Hyella socialis* was made by observing the presence of small warts or excrescences growing on *Prionitis*. *Hyella* was found to be uniformly associated only with the warts. We presume that the presence of a foreign plant may have stimulated the cells of *Prionitis* thus producing the abnormal growths at the points of infection. The rapid growth of the cells of *Prionitis* would have the effect of disturbing the horizontal layer of *Hyella* and of dispersing its erect filaments.

4. Hyella socialis S. and G.

Plate 2, fig. 5

Filaments penetrating into the host, 200–300μ long, tortuous, branching dendroidally, at first cell divisions only in one plane, later dividing in all directions, producing groups of cells within the original sheath; each group resulting from the divisions of a single cell in the filament, becoming practically independent, similar to groups of *Gloeocapsa* cells, the largest groups being nearest the surface of the host; cells very angular, irregular in shape and size, 4–6μ diam., terminal penetrating cell 7–9μ long, decidedly conical; cell walls hyaline, soft; cell contents bright blue-green; gonidangia unknown.

Growing in the stipitate portion of *Iridaea minor* J. Ag. in the lower littoral belt. Carmel Bay, Monterey County, California, May, 1916.

Setchell and Gardner, *in* Gardner, New Pac. Coast Alg. II, 1918, p. 443, pl. 36, fig. 5.

The basal filaments of the plants, if present at all, cannot be thoroughly worked out with the material at hand. Further study will be required to elucidate this point. The early stages of the development of this species are unknown. No gonidangia were present and the host plant was nearing maturity. It is possible that no gonidia are developed, and that the dissolution of the sheath of the groups of vegetative cells free them at the time that the host is beginning to disintegrate and these vegetative cells locate on younger hosts.

We have placed this species in the genus *Hyella* on account of its close similarity in some of its vegetative characters and in its penetrating habit to those of *H. caespitosa* B. and F. (1888), the original species of the genus. They represent cell divisions as taking place in all directions at the base of the erect filaments. In surface view the cells arising from divisions of the cells in the basal filaments appear in more or less isolated groups in their illustrations. The few cells near the base of the erect filaments are less divided, and the terminal cells of the erect filaments remain undivided. The erect filaments thus become somewhat isolated, and clavate in form. The filaments of *H. socialis* appear to be formed in the same manner although they are crooked and branched. In the absence of knowledge concerning the prostrate portion of the thallus, characteristic of the typical *Hyella*, and in the absence of gonidangia, this species must remain somewhat in doubt as to its generic positon.

12. **Radaisia** Sauv.

Plants forming small, compact masses of cells on the surface of
the host, consisting of a more or less parenchymatous basal layer
formed by repeated divisions in vertical planes of a single cell, and
of simple or complex, erect, more or less branched filaments arising
from the basal layer by horizontal divisions and usually very compact and more or less coalescent; gonidangia arising by modification
of terminal cells of the erect filaments.

Sauvageau, Sur le *Radaisia,* 1895, p. 373 (p. 2, Repr.), pl. 7, fig. 1.

The genus *Radaisia* was founded by Sauvageau (*loc. cit.*) to
receive two species, *R. Gomontiana* Sauv. and *R. Cornuana* Sauv., of
which the first, both because of its position in the article and because
it is described and figured with gonidangia, may be taken as the type.
It is a marine species, epiphytic upon species of *Fucus,* and was found
in the southwest of France, at Biarritz and at Guethary.

The type shows a minute cushion, composed of vertical and slightly
radiating filaments which are simple or occasionally fastigiately dichotomously branched, whose terminal cells are transformed into large,
spherical, ovoid or pyriform gonidangia. Of the development, and
manner of the cell divisions Sauvageau says nothing. The second
species is an inhabitant of fresh water, growing on submerged stones
in rapidly flowing springs, but was not provided with gonidangia.

From the figures and description of the adult stages of *Radaisia
Gomontiana* and from the study of our *R. Laminariae,* which seems
certainly to be referred to the same genus, it appears that the divisions
of the cells take place variously and successively in one, two, and three
directions. Apparently a cushion may and does arise, as a rule, from
a single cell (gonidium). By divisions taking place in two directions,
perpendicular to the substratum, a layer of cells is formed. The cells
of this layer divide, after a time, in the third direction, i.e., parallel
to the substratum, producing the filaments which form a cushion out
of the layer. The filaments by occasional division in other directions
than parallel to the substratum initiate branches and consequently
the radiate structure of the cushion arises. The transformation of the
terminal cells into large and conspicuous gonidangia completes the
structure of the typical species.

We have followed the distinction set forth by Gardner (1918,
p. 432) in distinguishing *Radaisia* from *Hyella. Radaisia,* then, will
include those forms with erect filaments, simple or complex, closely

compact, more or less parallel, not at all or but slightly branched, arising from basal filaments on the surface of the host and extending away from it, and with the gonidangia on the outer free ends.

1. Radaisia Laminariae S. and G.
Plate 3, figs. 14–16

Prostrate portion of the plant consisting of compact, radiating filaments, dichotomously or subdichotomously branched, forming a closely compact thallus circular in outline, up to 300μ diam.; cells of prostrate filaments quadrate, 4–4.5μ diam., giving rise by horizontal divisions to closely compact, erect filaments; the whole thallus 30–40μ thick; gonidangia terminal on the erect filaments, spherical, or slightly oval, 8–9μ diam.; gonidia 0.8μ diam. formed by simultaneous division; color bright blue-green.

Growing on the terminal portion of the blades of *Laminaria Sinclairii*. Fort Point, San Francisco, California. This locality, so far as we know, is the only one in which this species has been observed. It probably has a much wider distribution. The host plant extends from Vancouver Island, British Columbia, to the vicinity of Point Conception, California.

Setchell and Gardner, *in* Gardner, New Pac. Coast Alg. II, 1918, p. 444, pl. 37, figs. 14–16; Collins, Holden and Setchell, Phyc. Bor.-Amer. (Exsicc.), no. 2254.

Radaisia Laminariae we have taken to represent a typical form of the genus, having a single layer of cells for the base which gives rise to perpendicular, parallel, unbranched filaments extending away from the host, and each bearing a single terminal gonidangium. The whole cushion or colony is in reality a single plant resulting from the growth of a single gonidium. The plants of this species start from a single cell, and by a few divisions a small group of cells is formed; around the margin the cells begin to arrange themselves radially in rows, and by divisions in two planes, radial and tangential, a circular disk, one layer of cells deep, is formed. The marginal cells enlarge tan-

gentially and are cut into two cells by a radial wall; the resulting
"filaments" are of equal growth as a rule, and the forking thus
becomes dichotomous (pl. 3, fig. 16). Plants often become so closely
associated as to form a continuous layer over the surface of the host.
Even under this condition they have no effect upon the host, so far
as death and disintegration of the cells are concerned, and hence they
are strictly epiphytic. The erect filaments result from the horizontal
division of the prostrate or basal cells. They form very dense, compact masses, their cell walls adhering firmly (pl. 3, fig. 15). The
gonidangia are numerous, spherical or slightly oval, being transformed
terminal cells of the erect filaments (pl. 3, fig. 14).

2. Radaisia subimmersa S. and G.

Plate 3, figs. 12, 13

Thalli small, inconspicuous, irregular in outline on the surface of
the host, growing on the cuticle, or in small surface cavities; prostrate
or basal layer composed of angular cells 3–5µ diam., arranged irregularly, giving rise to erect filaments, parallel at first, later spreading
somewhat at the free distal ends, 35–45µ long; cell division on the
free surface portion in one plane, within the host in three planes,
building up oval masses; cells in erect filaments 3–5µ long, 1.5–2.5µ
wide, older cells frequently becoming pear-shaped; cell contents homogeneous, blue-green; gonidangia terminal, spherical, 4–6µ diam., producing 6–8 gonidia by simultaneous division.

Growing on *Rhodymenia* sp. Carmel Bay, Monterey County,
California. The host plant was collected by Gardner (no. 3350). It is
a *Rhodymenia* and possibly *R. Palmetta*, at least it belongs to the
Palmetta group. It was cast ashore and slightly faded, thus making
the *Radaisia* groups appear distinct, which led to their discovery.
The host is not uncommon along the California coast, and the epiphyte
is to be expected in other localities.

Setchell and Gardner, *in* Gardner, New Pac. Coast Alg. II, 1918,
p. 446, pl. 37, figs. 12, 13.

This species of *Radaisia*, like others, seems to be epiphytic at first
until the basal layer of cells is produced. Plants start from a single
cell on the cuticle of the host, at least some of them have been so
observed. By repeated divisions, sometimes perpendicular to the long
diameter of the cell, but usually quite oblique, a single layer of cells
of varying shapes and sizes is built up (pl. 3, fig. 13). The cell walls
of this basal layer are transparent, seem to be gelatinous, so that the

cells soon become more or less freed and independent of each other. Horizontal divisions begin and short filaments, three to five cells long, are formed above the surface of the host. Meanwhile it seems possible that the prostrate layer has begun to dissolve the cuticle and the underlying cortical cells, and division of the basal cells may have begun forming filaments which push into the host, completely dissolving the host cells as they penetrate (pl. 3, fig. 12).

The plants upon which this species is founded are scarcely mature, the gonidangia being very rare. The depth to which they may penetrate the host is thus far rather uncertain. It is not at all unlikely that they may penetrate to a greater depth than that reported in the diagnosis above. It certainly seems that the whole of the cells of the host is actually absorbed, or at least destroyed, as far as the parasite travels.

Radaisia subimmersa may be looked upon as being on the border line between *Radaisia* and *Hyella* so far as its relation to the host is concerned, being partially internal and partially external, and possibly growing in both directions from the original basal layer. The filaments, however, being more or less parallel, and having the gonidangia on the outer free ends, are characters which have led us to place it in the genus *Radaisia*.

3. Radaisia clavata S. and G.

Plate 3, figs. 17, 18

Plants forming microscopic, deep blue-green cushions on the surface of the host, up to 100μ diam., more or less fan-shaped in median section; filaments very closely compact, 70–100μ long, sparsely branched near the outer ends; cell walls 4–5μ diam. at the base, 7–8μ above, 3–4μ long, cell divisions often irregularly oblique, cell walls thin, hyaline; protoplast homogeneous; gonidangia 8–9μ diam., terminal, hemispherical; gonidia angular, 1–1.5μ diam., formed by simultaneous division.

Growing on *Gymnogongrus linearis*, in the lower littoral belt. Lands End, San Francisco, California. This is the only known locality in which this plant grows. The host plant is common along the California coast and extends as far north as the Straits of Juan de Fuca. It is not at all unlikely that it may be found on the same host in other localities.

Setchell and Gardner, *in* Gardner, New Pac. Coast Alg. II, 1918, p. 445, pl. 37, figs. 17, 18.

Radaisia clavata departs slightly from *R. Laminariae* in having the filaments occasionally branched, and the cross cell divisions often decidedly irregularly oblique.

4. Radaisia epiphytica S. and G.

Plate 3, figs. 10, 11

Thalli forming cushions on the surface of the host, nearly circular in outline as seen from above, 250–350μ across, 50–60μ thick; prostrate or basal layer formed by dichotomous or subdichotomous branching around the margin; erect filaments loosely adherent, composed of groups of angular cells, 3–5μ diam., produced by cell divisions in three planes; protoplast homogeneous, blue-green; terminal and subterminal cells of the vertical groups transformed into gonidangia with slight increase in size; gonidia angular, 1.8–2.4μ diam., formed by simultaneous division of the protoplast.

Growing on *Iridaea minor* J. Ag. in the lower littoral belt. Carmel, Monterey County, California.

Setchell and Gardner, *in* Gardner, New Pac. Coast Alg. II, 1918, p. 447, pl. 37, figs. 10, 11.

Only a single antheridial plant of *Iridaea* has been observed with this epiphyte growing upon it, but doubtless it is common and probably grows on other forms of the same species. *Iridaea minor* is usually rather dark colored, which makes the presence of the *Radaisia* somewhat obscure, and probably accounts for its not having been observed previously.

The gonidangia of *R. epiphytica* depart somewhat from the typical form of the genus. Since the erect filaments are complex in each sheath, resulting from vertical as well as horizontal divisions, often irregular and oblique, the gonidangia are also complex, consisting of a group of transformed terminal cells instead of a single terminal cell as is usually the case. The original cell wall or sheath persists as the erect filaments elongate and become complex. Likewise when the terminal cells metamorphose into gonidangia the original sheath of a group of cells persists, and a compound gonidangium is the result.

13. Gomphosphaeria Kuetz.

Cells associated into definite, small, irregularly more or less spherical, solid, floating masses; cells usually obovoid, tapering within, rounded or, at the beginning of division, obcordate without, supported on radiating, dichotomous, gelatinous stalks; protoplast blue-green, yellowish or reddish, homogeneous; gonidia numerous, arising in the otherwise unaltered cells (gonidangia) through successive divisions.

Kuetzing, Alg. Decade XVI, 1836, no. 151.

The type of the genus *Gomphosphaeria* is *G. aponina* Kuetzing collected at Abano near Padua, in waters of 36° R. (45°C). The structure of the colonies has not been generally understood, but has been clearly described and illustrated by Schmidle (1901, pp. 16–20, pl. 1, figs. 1–5). We have been able to confirm the account of this structure and the development of the colony. The structure and development seem to relate this genus rather to *Xenococcus* and other Chamaesiphonaceae than to any of the *Chroococcaceae* and the discovery of gonidia (cf. Schmidle, *loc. cit.*, and Zukal, 1894, p. 259, pl. 19, figs. 9, 10), which we, however, have not as yet observed, gives additional reason for placing it in the Chamaesiphonaceae as we have done.

There are, at present, three species referred to *Gomphosphaeria*, of which the type, *G. aponina* Kuetzing, is regarded as being the most common and widely distributed.

Gomphosphaeria aponina Kuetz.

Plate 1, figs. 2, 3

Cells associated into microscopical blue-green colonies up to 90μ diam., tegument hyaline, moderately thick, slightly lamellate, cells clavate or pear-shaped to obcordate, radially arranged in the colonies, 10μ long, 4–5μ diam., pedicellate with radiating, short, thick stalks, dividing in two directions; gonidia numerous, globular, about 2μ in diameter, successively formed.

Floating among other algae in a pool in a salt marsh. Whidbey Island, Washington.

Kuetzing, Alg. Decade, XVI, 1836, no. 151, Tab. Phyc., vol. I, 1846–1849, p. 22, pl. 31, fig. 3; Setchell and Gardner, Alg. N.W. Amer., 1903, p. 180.

Gomphosphaeria aponina is found in warmer and cooler fresh waters as well as in brackish water of pools in salt marshes and may be rather widely distributed along our coast although we are able to cite definitely only one locality. The specimens agree perfectly with those distributed by Kuetzing in his *Decades* (no. 151), as cited, and also with the type material in Kuetzing's herbarium.

ORDER 2. HORMOGONALES ATKINSON

Thallus pluricellular, filamentous, with or without a sheath, cylindrical, or tapering to a hair, unbranched, or with false or true branching; filaments single, or several within a common sheath, or united into a more or less gelatinous mass; all of the cells of the filaments similar, or with occasional specialized cells called heterocysts; multiplication by the filament breaking into segments a few cells long, called hormogonia, through the death of certain cells, and by the formation of resting spores.

Atkinson, A college text book of botany, ed. 2, 1905, p. 163.

Hormogoneae Thuret, Essai Class. Nost., 1875, p. 377.

As previously stated under Coccogonales, Thuret was the first to divide the Myxophyceae into two groups and it seems best to consider these as orders. The Hormogonales differ from the Coccogonales in being uniformly filamentous, either with or without branches, and reproducing vegetatively by short pieces of two, to three, to many cells instead of separating single cells for this purpose. The hormogonia are formed by the definite splitting off of these small lengths of the filaments, usually in a series, the hormogonia being separated from one another by the death of intermediate cells. Spore formation, when present, is by means of cells changing their shape and size to a greater or less extent and forming a thick, smooth, or at times variously roughened, outside coat. Gonidia formation is described for certain of the Hormogonales, but needs further observation and study before being accepted as equivalent to that of the Chamaesiphonaceae.

After careful consideration of the various views, it has seemed best, for reasons given below, to divide the Hormogonales into two suborders, Homocystineae and Heterocystineae.

KEY TO SUBORDERS AND FAMILIES.

1. Cells alike, neither differentiated into hair cells nor heterocysts......................
.....................(Suborder 1. **Homocystineae**) Family 3. **Oscillatoriaceae** (p 53)
1. Cells of two sorts, vegetative and either hair cells or heterocysts, or both...............
....(Suborder 2. **Heterocystineae**) 2
   2. Trichomes alike at both ends, branching wanting.......................................
....Family 4. **Nostocaceae** (p 89)
   2. Trichomes unlike at opposite ends... 3
3. Branching false, hairs usually present, abundant..Family 5. **Rivulariaceae** (p 93)
3. Branching true, hairs (in our genera) present, occasional or abundant..................
....Family 6. **Stigonemataceae** (p 109)

SUBORDER 1. HOMOCYSTINEAE nobis

Filaments either floating free or in layers, usually not attached to the substratum at either end; trichomes of cells very little different from one another, not provided with heterocysts or tapering into a hair, simple or branched, with or without a sheath enclosing one or more trichomes; propagation by hormogonia; spores unknown; gonidia ?.

Homocysteae Bornet and Flahault, Tab. syn. d. Nost. fil. Het., 1885, p. 197, Rev. I, 1886, p. 325; Gomont, Monogr. des Oscill., 1892, p. 289 (1893, p. 27, Repr.).

It has seeemd best to continue the practice, introduced by the French phycologists, of separating the Hormogonales into two suborders, the Homocystineae and the Heterocystineae. The presence or absence of heterocysts is so nearly an exact dividing character and the exceptional cases so few and so readily dealt with, that it seems unnecessary, as well as decidedly undesirable, to change the designations. Gomont (1899, pp. 30–33) has discussed the question, and further details will be given in the discussion of the Heterocystineae. The Homocystineae, as here defined, include all the Hormogonales having neither heterocysts nor terminal hairs. It forms a compact, well defined, seemingly natural group.

FAMILY 3. OSCILLATORIACEAE HARVEY

Trichomes pluricellular, straight, arcuate, more or less uncinate at the apices, or spirally twisted, simple, or rarely with false branching, cylindrical, or slightly tapering at the apices, all of the cells similar in shape and in function, with or without a sheath, single or plural within a sheath; multiplication by means of motile(?) hormogonia.

Harvey, Ner. Bor.-Amer. III, 1858, p. 96 (in part); Kirchner, Schizophyceae, *in* Engler and Prantl, Natürl. Pflanzenfam. I, 1*a*, 1898, p. 61. *Oscillarideae* Gray, Nat. Arr. Br. Pl., vol. 1, 1821, p. 80.

It seems best to recognize a single family under the Homocystineae and this was done in 1898 by Kirchner (*loc. cit.*) who is quoted above, therefore, as final authority, although the name in its present form was used as early as 1858 by Harvey (*loc. cit.*), but in much more extended sense. In the sense of Kirchner, Oscillatoriaceae include both the Vaginarieae and the Lyngbyeae of Gomont (1893, p. 28.

Repr.). The family of the Oscillatoriaceae may be subdivided into several subfamilies, such as the Schizotricheae, Lyngbyeae, Oscillatorieae, and Spirulineae.

SUBFAMILY 1. SPIRULINEAE FORTI

Filaments without a sheath, more or less regularly or irregularly coiled in a laxer or closer spiral.

Forti, *in* De-Toni, Syll. Alg., vol. 5, 1907, p. 145. *Spirulinoideae* Gomont, Monogr. des Oscill., 1892, p. 248 (1893, p. 116, Repr.) (lim. mut.).

We have followed Forti (*loc. cit.*) both as to the form of the name of this subfamily and as to including *Arthrospira* as well as *Spirulina* although Gomont (*loc. cit.*) includes only *Spirulina*. From our point of view, both these genera closely approach species of *Oscillatoria* as will be shown in detail below.

14. **Arthrospira** Stizenb.

Trichomes multicellular, cylindrical, evaginate, loosely and regularly coiled, usually of relatively large diameter and large spirals, and comparatively short and with few coils; dissepiments distinct, apices slightly or not at all tapering, terminal cell rounded, calyptra absent.

Stizenberger, *Spirulina* und *Arthospira, in* Hedwigia, vol. 1, 1852, p. 32.

The type species is *Arthrospira Jenneri* (Hass.) Stizenb., a fresh water plant from Tunbridge. *Arthrospira* is intermediate between *Oscillatoria* and *Spirulina*. Some species of *Oscillatoria*, such as *O. Bonnemaisonii* Crouan are loosely, but regularly, spiral. The coarser species of *Arthrospira* approach these. On the other hand, the looser and laxer species of *Arthrospira*, e.g., *Arthrospira miniata* (Hauck) Gomont which is slender and with the dissepiments obscure, approach *Spirulina*. In certain slender but straight species of *Oscillatoria*, e.g., *O. amphibia* Ag., the dissepiments are as obscure as in species of *Spirulina* until they are revealed by treatment with strong solutions of chromic acid (cf. Gomont, 1893, pl. 7, figs. 4, 5). In the same fashion, treatment with chromic acid reveals the dissepiments in species of *Spirulina* (cf. also Gardner, 1917, p. 377).

Key to the Species.

1. Cells 7–9μ in diam., 5–7μ long..1. **A. maxima** (p 54)
1. Cells 14.5–16μ in diam., 2.5–3.5μ long................2. **A. breviarticulata** (p 55)

1. **Arthrospira maxima** S. and G.

Plate 8, fig. 3

Trichomes 7–9μ diam., forming an open regular spiral of 3–8 turns, 40–60μ diam., 70–80μ between the turns, slightly tapering at the ends; cells 5–7μ long, not constricted, with numerous, coarse, angular, refringent granules frequently crowded at the partitions; end walls of terminal cells rounded, slightly thickened; color verdigris green.

Found floating among other species of Myxophyceae in a warm salt-water pond. Known only from a single locality, viz., in a hot, salt-water pond, Key Route pool, in Oakland, California, on the shore of San Francisco Bay.

Setchell and Gardner, *in* Gardner, New Pac. Coast Alg. I, 1917, p. 377, pl. 33, fig. 3; Collins, Holden and Setchell, Phyc. Bor.-Amer. (Exsicc.), no. 2259.

The salt water of this pond was pumped from the bay and used for condensing the steam of a power house, the same water being used repeatedly, thus it became alternately heated and cooled. This treatment subjected the floating algae to the unusual condition of extremes of heat and cold under which it thrived abundantly.

2. **Arthrospira breviarticulata** S. and G.

Plate 7, fig. 26; plate 5, fig. 18

Trichomes very loosely twisted into a more or less regular spiral, and much knotted and contorted, 14.5–16μ diam., comparatively short, slightly torulose, not tapering at the apices; cells 2.5–3.5μ long; protoplast pale steel-blue, with few scattered, somewhat angular granules; cell walls thin, distinct, terminal cell very convex, outer wall not thickened.

Growing on rocks and among *Cladophora trichotoma* in tide pools along high-tide level. Lands End, San Francisco, California.

Setchell and Gardner, *in* Gardner, New Pac. Coast Alg. III, 1918*a*, p. 466, pl. 39, fig. 18, and pl. 41, fig. 26; Collins, Holden and Setchell, Phyc. Bor.-Amer. (Exsicc.), no. 2258.

Arthrospira breviarticulata is distinguished from all other species of the genus by its much greater length, and greater diameter, by its being loosely coiled, and by its relatively much shorter cells. It approximates the spirally coiled species of *Oscillatoria*, yet seems to be referable rather to *Arthrospira*.

15. **Spirulina** Turp., emend. Gard.

Trichomes multicellular, cylindrical, evaginate, loosely or tightly coiled into a more or less regular spiral; apex of trichome usually not tapering; dissepiments obscured; terminal cell rounded, without calyptra; protoplast homogeneous or granular.

Turpin, *in* Dict. des sciences nat. de Levrault, vol. 50, 1827, p. 309; Gardner, New Pac. Coast Alg. I, 1917, p. 379.

It has been shown (Gardner, 1917, pp. 377–379) that there is no fundamental difference between the genera *Spirulina* and *Arthrospira*, both having multicellular, spirally twisted filaments. We have retained the two genera as a matter of convenience, since the species are few in each genus and well known, and since they fall rather naturally into two groups as regards size. We have retained *Spirulina* to include the small, more or less tightly coiled forms with inconspicuous cross walls, usually described as "unicellular," and *Arthrospira* to include the large, comparatively short, loosely coiled forms with more or less conspicuous cross walls.

The genus *Spirulina* was founded by Turpin in 1827 (*loc. cit.*) with the type species named *Spirulina oscillarioides*, a name placed among the "Species inquirendae" by Gomont (1893, p. 275, Repr.). Judging from his figures (*loc. cit.* Planches, 2ᵉ. partie, regne organise, "oscillariees," figs. 3, 3*a*, 3*b*, 3*c*) Turpin's plant was a mixture of fresh water species of *Spirulina*, although some of his figures resemble species of *Spirogyra*, but without cross walls. This appearance is doubtless due to the idea of Turpin that the spiral trichome is enclosed in a tube or thread of mucilage.

Gomont (1893, p. 269, Repr.) has adopted the name *Spirulina* of Turpin (1827) instead of the earlier *Spirogyra* of Link (1809, p. 20) which seems to have been founded on a species of *Spirulina* Turpin, giving his reasons in full. Since Gomont's Monographie des Oscillariées has been made the starting point for the nomenclature of "Nostocaceae homocysteae" by the Brussels Congress (cf. Actes de IIIᵐᵉ· Congrès Intern. de Botanique, Bruxelles, 1910, vol. 1, p. 103), and since also the *Spirogyra* of Link 1820 has been placed by action of the same body (*loc. cit.*, p. 109) among the "nomina conservanda" as against the "*Conjugata*" of Vaucher 1803, it seems necessary as well as desirable to follow Gomont.

Key to the Species.

1. Spiral regular, loosely coiled..1. **S. major** (p 56)
1. Spiral regular, tightly coiled..2. **S. subsalsa** (p 57)

1. **Spirulina major** Kuetz.

Plate 1, fig. 5

Plants in mass bright blue-green, trichomes pale blue-green, more or less flexuous, 1.2–1.7μ diam., twisted into a fairly loose, regular spiral, 2.5–4μ diam., with a distance of 2.7–5μ between the turns.

Growing in brackish water. Puget Sound, Washington, to central California.

Kuetzing, Phyc. Gen., 1843, p. 183; Gomont, Monogr. des Oscill., 1892, p. 271, pl. 7, fig. 29 (1893, p. 271, Repr.) ; Setchell and Gardner, Alg. N.W. Amer., 1903, p. 182.

The species of *Spirulina* are to be distinguished from one another by four different criteria: (1) by the varying diameter of the trichome; (2) by the horizontal amplitude of the spiral; (3) by the distance of the turns of the spiral from one another, i.e., by the looseness or tightness of the spiral; and (4) by the regularity or irregularity of the spiral. Usually the trichomes show decided uniformity within the species, but occasionally certain trichomes are found showing variation of a profound and perplexing character. *Spirulina major*, for example, is usually to be distinguished from *Sp. Meneghiniana* Zanard., which approaches it closely in dimensions, by its regular spiral, which is, however, lax in both species. From *Sp. subsalsa* Œrst., *Sp. major* is to be distinguished by its lax spiral, which is tight in *Sp. subsalsa*. A specimen from the warm salt waters of the Key Route pool in Oakland (Gardner, no. 3237) shows tightly coiled and regular loosely coiled turns in the same trichome, seemingly being partly *Sp. major* and partly *Sp. subsalsa*. Such specimens throw doubt on the usual criteria of species.

2. **Spirulina subsalsa** f. **oceanica** (Crouan) Gomont

Trichomes forming a thin blue-green or yellowish green stratum on mud, or floating among other algae, 1μ diam., twisted into a nearly straight, regular, tightly coiled spiral.

Growing in pools of brackish water. Whidbey Island, Washington, and San Francisco Bay, California.

Gomont, Monogr. des Oscill., 1892, p. 254, pl. 7, fig. 32 (1893, p. 274, Repr.); Setchell and Gardner, Alg. N.W. Amer., 1903, p. 182; Collins, Holden, and Setchell, Phyc. Bor.-Amer. (Exsicc.), no. 954. *Oscillaria oceanica* Crouan, Algues marines du Finistère, no. 324 (in part).

The smaller form of the species is the only one which has occurred to us in our territory, but seems typical and is probably widespread in shallow pools of brackish water.

SUBFAMILY 2. OSCILLATORIEAE FORTI

Trichomes always simple, straight or at times in a loose more or less regular but not especially distinct spiral, destitute of a distinct sheath.

Forti, *in* De-Toni, Syll. Alg., vol. 5, 1907, p. 147.

We have followed Forti in the name and limits of this subfamily. It is decidedly more restricted than the subtribe Oscillarioideae of Gomont (1892, p. 95, 1893; p. 115, Repr.).

16. **Oscillatoria** Gomont

Trichomes free, often forming into dense tangled masses, cylindrical, without a sheath, or at times with a delicate, more or less mucous sheath, smooth or constricted at the partitions, not moniliform, straight or arcuate, not spirally twisted; apices straight or uncinate, more or less tapering, terminal cell of some species with thickened membrane.

Gomont, Faut il dire *Oscillatoria* ou *Oscillaria*, 1891, p. 273. *Oscillatoria* Vaucher, Hist. des Conferv., 1803, p. 165 (lim. mut.).

Gomont (1891, p. 273) has apparently discussed the generic name and its relation to *Oscillaria* so thoroughly as to demand no especial consideration here. As used at present and generally accepted its limits are considerably changed from those proposed by Vaucher. The type of the genus seems clearly to be *Oscillatoria princeps* Vaucher (*loc. cit.*, p. 190) and the type locality (of this species) is Crevin, near Geneva, Switzerland. Of the other eleven species assigned by Vaucher to this genus Gomont (1893, pp. 256–266, Repr.) lists nine as doubtful, and two, one each, to be referred to the genera *Beggiatoa* and *Microcoleus* respectively.

Oscillatoria is a genus containing somewhat over one hundred recognized species, the great majority of which are inhabitants of fresh water or growing in moist situations. The species usually listed in a marine flora are largely, at least, to be found in brackish water and consequently in the pools or other bodies of water in salt marshes. In such situations the water is warmed by the air and by the sun and is of decidedly higher temperature than that of the adjacent ocean or larger body of strictly salt water.

The species of *Oscillatoria* are to be distinguished from those of *Phormidium* and of *Lyngbya* by being destitute of a sheath and, in general, this method of distinguishing them is both certain and satisfactory. Certain species seemingly of *Oscillatoria*, however, show (or develop?) a sheath under certain circumstances (especially *O. sancta* and *O. limosa*) while the trichomes of *Lyngbya*, and even more so those of *Phormidium*, escaping from the sheath, take on for a time exactly the appearance of being trichomes of *Oscillatoria*. The naked trichomes of certain species of *Phormidium* as well as those of *Microcoleus,* not only imitate the appearance of trichomes of *Oscillatoria* but also possess the power of movement characteristic of certain species of *Oscillatoria*.

Section 1. Principes Gomont

Trichomes straight, curved or spiral at or toward the upper end, slightly, if at all, attenuated, apices obtuse; cells very short.

Gomont, Monogr. des Oscill., 1892, p. 201 (1893, p. 221 Repr.)

1. *Oscillatoria limosa* Ag.

1. **Oscillatoria limosa** Ag.

Trichomes aeruginous or more or less olive green, in a blackish-aeruginous layer, on drying often a steel-blue black, straight, on drying rigid and fragile, not at all torulose, 11–20μ, commonly 13–16μ thick, apices straight, slightly if at all attenuate, not capitate; cells 3–6 times shorter than the diameter, 2–5μ long; dissepiments frequently granulate; apical cell convex above showing a slightly thickened membrane.

Chiefly in fresh water, but occasionally found floating in pools or ditches in salt marshes where the water is slightly brackish. Whidbey Island, Washington.

Agardh, Disp. Alg. Suec., 1812, p. 35; Gomont, Monogr. des Oscill., 1892, p. 210, pl. 6, fig. 13 (1893, p. 230, Repr.); Setchell and Gardner, Alg. N.W. Amer., 1903, p. 183.

This species has only a slight claim to being considered marine, at least so far as our territory is concerned. It has occurred to us in the locality noted above, but it may possibly be met with among *O. nigro-viridis, Lyngbya semiplena,* species of *Microcoleus,* etc., as it was in the Whidbey Island locality. It may be that young and naked trichomes of *Lyngbya aestuarii* may be mistaken for this species.

Oscillatoria ornata (Kuetz.) Gomont has been credited to the marine flora of our coast on the strength of a specimen distributed in the Phycotheca Boreali-Americana (no. 1604). There has been some confusion of specimens under this number, the locality being undoubtedly correct, but the statement on the label reads: "Among *Cladophora* sp.," referring to another specimen collected at the same general locality. Neither specimen, however, shows, upon careful reëxamination any filaments of *O. ornata,* and we seriously doubt the existence of *O. ornata* at the locality mentioned. Our specimens under no. 1604 Phycotheca Boreali-Americana show two hormogonial species, viz., *Hydrocoleum lyngbyaceum* Kuetz. and *Microcoleus confluens* S. and G.

Section 2.   Margaritiferae Gomont

Trichomes constantly torulose, apex obtuse, scarcely attenuated, very long arcuate, rarely either straight or totally spiral; maritime species, diameter moderate to thick.

Gomont, Monogr. des Oscill., 1892, p. 202 (1893, p. 222, Repr.).

    2. *O. Bonnemaisonii* Crouan
    3. *O. margaritifera* (Kuetz.) Gomont
    4. *O. nigro-viridis* Thwaites
    5. *O. Corallinae* (Kuetz.) Gomont

2. **Oscillatoria Bonnemaisonii** Crouan

Trichomes usually intermixed with various other aquatic Myxophyceae, somewhat regularly or loosely twisted, very flexible, slightly torulose, 18–36μ diam., dilute olive green; apices not at all or only slightly tapering, with end wall of terminal cell convex and not thickened; cells 3–6μ long; protoplasm with fine granules uniformly distributed.

Growing on mud in salt marshes and floating in pools of salt water. Puget Sound, Washington, to central California.

Crouan, *in* Desmazières, Pl. crypt. de France, 2 ser., no. 537, 1858 (*fide* Gomont); Gomont, Monogr. des Oscill., 1892, p. 215, pl. 6, figs. 17, 18 (1893, p. 235, Repr.); Setchell and Gardner, Alg. N.W. Amer., 1903, p. 183 (small form); Collins, Holden, and Setchell, Phyc. Bor.-Amer. (Exsicc.), no. 1707.

According to Gomont (*loc. cit.*) this species has a considerable variation in diameter, viz., from 18μ to 36μ, although not so great as that given for *O. princeps*, which is 16μ to 60μ (Gomont, 1892, p. 207). In general, no single collection shows any such variation in the diameter of the filaments and it seems possible that there may be necessity for more careful study of these seemingly polymorphous species. What seems to be good *O. Bonnemaisonii* has occurred at Whidbey Island, Washington (cf. Setchell and Gardner, 1903, p. 183) and about San Francisco Bay, California (cf. Phyc. Bor.-Amer., nos. 1707*a, b,* and *d*).

3. **Oscillatoria margaritifera** (Kuetz.) Gomont

Trichomes bright olive green, forming dark agglomerated masses, or intermixed with other algae, very flexible, torulose, 17–29μ diam., straight, arcuate at the ends and slightly attenuate, obtuse; length of cells 3–6μ; apical cell capitate, furnished with a slightly convex calyptra; protoplasm with numerous large granules distributed along the cross walls.

Growing on mud and in pools in salt marshes. San Francisco Bay, California.

Gomont, Monogr. des Oscill., 1892, p. 161, pl. 6, fig. 19 (1893, p. 236, Repr.); Collins, Holden and Setchell, Phyc. Bor.-Amer. (Exsicc.), no. 1708. *Oscillaria margaritifera* Kuetzing, Tab. Phyc., vol. 1, 1846, p. 31, pl. 43, fig. 10.

What seems to be this species has been encountered several times in the salt marshes about San Francisco Bay, California. The calyptrate tip is well shown in our specimens.

Under no. 1604 in Phycotheca Boreali-Americana this species is said to occur in the specimen distributed. We fail to find it in the specimens in our copies, there having been some confusion and mistakes of identification as already indicated above under *O. ornata.*

4. **Oscillatoria nigro-viridis** Thwaites

Plants forming extensive dark olive green layers; trichomes olive green, of moderate length, nearly straight, fragile, torulose, 7–11μ diam., slightly curved at the extremities; apex slightly tapering; end cell wall very convex, slightly thickened; cells 3–5μ long, with two rows of granules at the cross walls.

Growing in pools in salt marsh. Whidbey Island, Washington, and on the shores of San Francisco Bay, California.

Thwaites, *in* Harvey, Phyc. Brit., Synopsis, vol. 3, 1849, p. xxxix, no. 375, pl. 251, fig. A; Gomont, Monogr. des Oscill., 1892, p. 217, pl. 6, fig. 20 (1893, p. 237, Repr.); Setchell and Gardner, Alg. N.W. Amer., 1903, p. 183.

This is another more or less puzzling species of the "Margaritiferae" section, but certain specimens both from Puget Sound and from the San Francisco Bay region seem referable to it. There is some departure from the descriptions in some specimens. In filaments whose cells are actively dividing, the length of the cells may be as short as one-sixth as long as broad, but are mostly one-third the diameter. The filaments are mostly short arcuate in the California specimens, and while some filaments show distinctly the transverse rows of granules at the dissepiments, others seem nearly, if not quite, to lack them.

5. **Oscillatoria Corallinae** (Kuetz.) Gomont

Trichomes aggregated into a thin, dark, olive green stratum or intermixed with other small algae, aeruginous, densely intertwined, flexuous, torulose, 6–10μ diam., apices long and moderately arcuate, very slightly attenuate; cells 2–3 times shorter than the diameter, protoplast sometimes with sparsely scattered granules, not crowded at the cross walls; apical cell subcapitate, with slightly thickened convex terminal wall.

Forming a thin stratum on rocks and on or among other small algae. Lands End, San Francisco, California, Gardner, no. 1634; Moss Beach, San Mateo County, California, Gardner, no. 2978; Pacific Grove, Monterey County, California, Setchell, no. 5490*b*.

Gomont, Essai Class Nost., 1890, p. 356 (in part), Monogr. des Oscill., 1892, p. 218, pl. 6, fig. 21(1893, p. 238, Repr.). *Leibleinia Corallinae* Kuetz., Sp. Alg., 1849*a*, p. 276.

Oscillatoria Corallinae resembles the other members of the same section of *Oscillatoria*, but it is smaller than any except the last, from which it differs by not having granular dissepiments. Its general appearance is that of a very small *O. Bonnemaisonii*, but it is much smaller than even the most slender forms referred to that species. It is usually, if not always, epiphytic.

Section 3. Aeqüales Gomont

Trichomes not attenuate above, straight or arcuate; cells at least one-third as long as broad up to equally long; slender species of not over 11μ in diameter; largely fresh water.

Gomont, Monogr. des Oscill., 1892, p. 202 (1893, p. 222, Repr.).

6. *O. amphibia* Ag.
7. *O. geminata* Menegh.

6. **Oscillatoria amphibia** Ag.

Trichomes forming a thin layer at the bottom of quiet water or floating on the surface, color pale copper acetate, straight or somewhat contorted in age, fragile, not constricted at the partitions, 2–3μ diam.; terminal cells not attenuate nor capitate, with outer cell wall rounded; cells 2–3 times as long as the diameter, up to 8μ long; protoplast homogeneous except with usually 2 granules at the dissepiments which are obscure.

Growing in pools in salt marshes, usually in company with other species of *Oscillatoria*. Puget Sound, Washington and San Francisco Bay, California.

Agardh, Aufzählung, *in* Flora, vol. 10, 1827, p. 632; Gomont, Monogr. des Oscill., 1892, p. 221, pl. 7, figs. 4, 5 (1893, p. 241, Repr.).

Judging from the figures of Gomont (*loc. cit.*), *Oscillatoria amphibia* does not show dissepiments until the protoplasts are dissolved out with strong acids. What seems to be this species, although in some cases varying slightly from it, has occurred to us in salt marshes. The species usually occurs in few or isolated filaments intermixed with other homocysted forms.

7. **Oscillatoria geminata** Menegh.

Trichomes pale aeruginous color, matted together into a thin, sordid, yellowish green stratum, occasionally circinate, slightly fragile, constricted rather deeply at the dissepiments, 2.3–4μ diam., 2.3–16μ

long; protoplast with a few large, scattered, refringent granules; dissepiments conspicuous, non-granular, apical cell convex, calyptra none.

Growing among other species of *Oscillatoria* in a salt water pond, B street, Oakland, California.

Meneghini, Consp. Algol. Eugan., 1837, p. 9; Gomont, Monogr. des Oscill., 1892, p. 222 (1893, p. 242, Repr.).

The type specimens of *O. geminata* were from the thermal waters of the Euganean springs, and what seems to be the same species or, at least, very close to it, occurs also in thermal waters in the Yellowstone National Park. The brackish water specimens referred here have the cells short, yet within the proportions given. Otherwise they seem typical. They were mixed with *Oscillatoria brevis* var. *neapolitana* (Kuetz.) Gomont.

Section 4.   Attenuatae Gomont

Trichomes decidedly attenuate above, more or less acutely uncinate or flexuous, not definitely spiral (*O. chalybea* at times excepted); cells longer or shorter than the diameter, never very short; plants commonly slender, not over 13μ thick, inhabitants of fresh and thermal, more rarely of saline waters.

Gomont, Monogr. des Oscillar., 1892, p. 203 (1893, p. 223, Repr.).

8. *O. laetevirens* (Crouan) Gomont
9. *O. acuminata* Gomont
10. *O. brevis* Kuetz.
11. *O. Okeni* (Ag.) Gomont
12. *O. chalybea* (Mert.) Gomont

8. **Oscillatoria laetevirens** (Crouan) Gomont

Trichomes associated into a delicate membrane of deep blue-green color, the single filaments being of a yellow-green color, straight, delicate, slightly constricted at the dissepiments, 3–5μ diam.; apices straight or uncinate, with the terminal cells obtuse or subacute; cells 2.5–5μ long; protoplasm with fine, evenly distributed granules.

Growing in salt marsh pools. San Francisco Bay, California.

Gomont, Monogr. des Oscill., 1892, p. 226, pl. 7, fig. 11 (1893, p. 246, Repr.). *Oscillatoria laetevirens* Crouan, Liste des algues marines Finistère, 1860, p. 371 (nomen nudum), Fl. du Finist., 1867, p. 112 (descr.).

This species approaches *O. brevis* Kuetz. but is somewhat smaller, with proportionally longer cells and a different arrangement of granules. It never has the occasional swollen cells so fairly characteristic of *O. brevis*. It is present in seemingly typical form in salt marsh pools about San Francisco Bay. The plant from Whidbey Island, referred here with some doubt (cf. Setchell and Gardner, 1903, p. 184), on further examination seems to be referable rather to *O. brevis* var. *neapolitana*.

9. **Oscillatoria acuminata** Gomont

Plate 1, fig. 11

Trichomes entwined in a dense thin layer of a verdigris color, delicate, straight, slightly constricted at the partitions, 3–6μ diam.; a few apical cells attenuated to a sharp point, not uncinate; length of cells 5–8μ; protoplast finely granular, with the granules frequently crowded at the dissepiments.

Floating in warm salt water. Oakland, California.

Gomont, Monogr. des Oscill., 1892, p. 227, pl. 7, fig. 12 (1893, p. 247, Repr.); Collins, Holden and Setchell, Phyc. Bor.-Amer. (Exsicc.), no. 1303.

The type specimen of *Oscillatoria acuminata* was collected at the Euganean springs, but has since been reported from other waters in several widely separated localities. Our specimens from warm or almost hot pools of brackish waters agree too nearly with the description and figures of Gomont to allow of separation. The plants distributed under no. 1303 of the Phycotheca Boreali-Americana, which are from the vicinity of hot salt waters, have shorter cells than the type.

10. **Oscillatoria brevis** Kuetz.

Filaments forming dense olive green layers or intermixed with other species of algae, of blue-green color, comparatively short and straight, rather fragile, 4–6.5μ diam., not constricted at the dissepiments, briefly and subacutely attenuate, uncinate or slightly twisted; cells very short, 1.5–3μ long, with an occasional refringent torulose cell interspersed; protoplasm with small granules evenly dispersed throughout the cell.

Growing among other algae in salt marsh pools. San Francisco Bay, California.

Kuetzing, Phyc. Gen., 1843, p. 186; Gomont, Monogr. des Oscill., 1892, p. 215, pl. 6, figs. 17, 18 (1893, p. 235, Repr.); Collins, Holden and Setchell, Phyc. Bor.-Amer. (Exsicc.), no. 1158.

This species seems to be represented with us, in brackish water, by the typical form, although more frequently by the var. *neapolitana* (Kuetz.) Gomont. Gomont (*loc. cit.*) describes *O. brevis* as having "dissepimenta non granulata," but his figures show them granulate.

10a. Oscillatoria brevis var. neapolitana (Kuetz.) Gomont

Trichomes larger than the type, 5–6.5μ diam.; with uncinate or twisted apices.

Growing in salt marshes. San Francisco Bay, California.

Gomont, Monogr. des Oscill., 1892, p. 229, pl. 7, figs. 14, 15 (1893, p. 249, Repr.); Collins, Holden and Setchell, Phyc. Bor.-Amer. (Exsicc.), no. 1304. *Oscillaria neapolitana* Kuetzing, Phyc. Gen., 1843, p. 185. *O. laetevirens* Setchell and Gardner, Alg. N.W. Amer., 1903, p. 184 (not of Gomont).

This variety is the more common form of *O. brevis* with us and is found in characteristic form.

11. Oscillatoria Okeni (Ag.) Gomont

Trichomes forming a dark blue-green mass, straight, manifestly constricted at the joints, .5.5–9μ diam.; apices long and gradually tapering, uncinate-arcuate with blunt apical cells; length of cells 2.7–4.5μ; end cell up to 8μ, neither capitate nor calyptrate; protoplasm filled with fine granules.

Growing in hot salt water. Oakland, California.

Gomont, Monogr. des Oscill., 1892, p. 232, pl. 7, fig. 18 (1893, p. 252, Repr.); Setchell and Gardner, Alg. N.W. Amer., 1903, p. 184; Collins, Holden and Setchell, Phyc. Bor.-Amer. (Exsicc.), no. 1605. *Oscillaria Okeni* Agardh, Aufzählung, *in* Flora, vol. 10, 1827, p. 633.

The type specimen of *Oscillatoria Okeni* came from thermal waters at Carlsbad. The specimen from the hot salt waters of the Key Route pool seems typical. A reëxamination of the specimens from Whidbey Island referred to this species (Setchell and Gardner, 1903, p. 184), shows them to be *O. chalybea* (Mert.) Gomont.

12. Oscillatoria chalybea (Mert.) Gomont

Trichomes pale steel-blue, dark blue-green in mass, fragile, usually straight, but at times slightly spirally twisted; moderately constricted at the joints, 8–13μ diam.; briefly, or long and gradually attenuate, one to a few apical cells uncinate; cells quadrate to 2–3 times shorter than the diameter, protoplast finely granular, with few, scattered, angular, refringent granules; dissepiments non-granular; apical cells obtuse, non-capitate, non-calyptrate.

In ponds of brackish water. Whidbey Island, Washington and West Berkeley, California.

Gomont, Monogr. des Oscill., 1892, p. 232 (1893, p. 252, Repr.). *Oscillaria chalybea* Mertens, *in* Jürgens, Algae aquaticae, Decas XIII, no. 4, 1822 (determined by Gomont). *O. Okeni* Setchell and Gardner, Alg. N.W. Amer., 1903, p. 184 (not of Agardh).

This species is widely distributed, occurring in fresh waters, thermal waters, and brackish waters. It is found in salt marshes both in the Puget Sound region and in the region of San Francisco Bay. The form found in both these regions is the typical form (var. *genuina* Gomont). The tip, while characteristic, varies in its more sudden or more gradual attenuation, and the cells are frequently equally long and thick. Certain filaments seem straight at the tip and bear some likeness to *O. simplicissima* Gomont.

SUBFAMILY 3. LYNGBYEAE KUETZ.

Trichomes simple or branched, each enclosed within a more or less firm or diffluent gelatinous sheath.

Kuetzing, Phyc. Gen., 1843, p. 179 (char. emend.); Forti, *in* De-Toni, Syll. Alg., vol. 5, 1907, p. 217 (lim. mut.). *Lyngbyoideae* Gomont, Monogr. des Oscill., 1892, p. 95 (1893, p. 115, Repr.).

We have quoted Forti under the subfamily Lyngbyeae because his idea of the limitations of the subfamily is nearest to, although not exactly, ours. We include the genus *Plectonema* in addition to those named by Forti. Our subfamily differs from the subtribe Lyngbyoideae Gomont by the inclusion of *Phormidium*.

17. *Phormidium*
18. *Lyngbya*
19. *Plectonema*
20. *Symploca*

17. **Phormidium** Kuetz.

Trichomes many celled, single within a sheath, simple, frequently attenuate and uncinate at the apices; sheath distinct, hyaline, sometimes mucous, more or less diffluent, in a few species obscure.

Kuetzing, Phyc. Gen., 1843, p. 190.

Kuetzing enumerated twenty-eight species under his genus *Phormidium*, not designating any particular one as the type. Certain species (about three) are not retained by Gomont, but most of them are and there can be no question as to the main ideas of Kuetzing which have been retained by Gomont. These are, first, the possession of sheaths less firm than those of *Lyngbya*, and second, the diffluence and coalescing of these sheaths into a membrane or layer of jelly enclosing a large number of filaments. In some species admitted to the genus by Gomont there is little if any trace of a sheath. These species are approximate to *Oscillatoria*. During the period, or stages, of hormogonial production, the hormogonia, or even whole trichomes, escape from the sheaths and enclosing jelly. In this stage they are exactly like the trichomes of *Oscillatoria*.

The majority of the species of *Phormidium* are inhabitants of fresh water, but a few occur also in brackish water, while a very few are exclusively marine.

KEY TO THE SPECIES.

1. Trichomes distinctly torulose (**Moniliformia**)...2
1. Trichomes rarely or scarcely torulose (**Euphormidia**)...3
    2. Trichomes attenuate at the apices, terminal cell acutely conical.................
    ....1. **Ph. fragile** (p 69)
    2. Trichomes not attenuate at the apices, terminal cell swollen, subspherical
    ....2. **Ph. hormoides** (p 69)
3. Apices of trichomes neither attenuate nor capitate.........3. **Ph. ambiguum** (p 70)
3. Apices of trichomes both attenuate and capitate...4
    4. Trichomes slightly and abruptly tapering at the apices, terminal cell
    rounded, calyptrate..4. **Ph. lucidum** (p 71)
    4. Trichomes gradually tapering at the apices, terminal cell with depressed
    conical calyptra...5. **Ph. submembranaceum** (p 71)

Section 1. Moniliformia Gomont

Trichomes distinctly torulose, even moniliform in some cases, neither curved nor capitate at the apices.

Gomont, Monogr. des Oscill., 1892, p. 159 (1893, p. 179, Repr.).

1. *Ph. fragile* (Menegh.) Gomont
2. *Ph. hormoides* S. and G.

1. **Phormidium fragile** (Menegh.) Gomont

Filaments forming a thin yellowish or dark blue-green stratum, at times lamellate; trichomes more or less flexuous, bright blue-green, moniliform, 1.2–2.3μ diam.; apices attenuate, acute conical, without calyptra; protoplasm homogeneous.

San Francisco Bay, California.

Gomont, Monogr. des Oscill., 1892, p. 163, pl. 4, figs. 13–15 (1893, p. 183, Repr.); Collins, Holden and Setchell, Phyc. Bor.-Amer. (Exsicc.), no. 1609. *Anabaena fragilis* Meneghini, Conspectus Algol. Eugan., 1837, p. 8.

The type locality for this species is the thermal water of the Euganean springs, but, as seemingly often happens, the species reappears in shallow brackish waters. The specimens referred here agree with the description of Gomont so far as aggregation and diameter and torulosity of the trichome are concerned, but the trichomes are not so attenuate as that represented on plate 4, figure 14, but fully as much so as that represented by figure 15. It agrees also with a specimen from Kiel collected by Reinbold and determined by Gomont.

2. **Phormidium hormoides** S. and G.

Plate 6, fig. 23

Filaments forming a thin, expanded, gelatinous stratum; trichomes short, somewhat flexuous, moniliform, 2.4–2.7μ diam.; sheaths hyaline, ample gelatinous, confluent; cells quadrate or subquadrate, extremely constricted at the dissepiments, terminal cell larger, subspherical, end wall not thickened.

Forming a thin layer on glass aquaria of salt water from the Pacific Ocean. Physiological Laboratory, University of California, Berkeley, California, 1905.

Setchell and Gardner, *in* Gardner, New Pac. Coast Alg. III, 1918a, p. 467, pl. 40, fig. 23.

Phormidium hormoides is very closely related to *Ph. foveolarum* (Mont.) Gomont, which was found growing in small pits in calcareous rock on the coast of France, and from which it differs in habitat, in the size of the filament and in the shape of the terminal cell. The filaments as viewed under the microscope are not uniformly distributed in the stratum but seem to have a tendency to aggregate into fascicles which anastomose freely, giving the stratum somewhat the appearance of a very delicate net, although many short filaments crawl out into the interstices.

In the specimen collected, *Phormidium hormoides* is mixed with another undetermined filamentous species of Myxophyceae, about 0.8μ diam., with cylindrical cells longer than the diameter and with conspicuous cross walls.

Ph. hormoides has some resemblance to the young filaments of a very delicate species of *Anabaena* on account of the pronounced moniliform trichomes, but there is no indication of heterocysts or spores.

Section 2.   Euphormidia Gomont

Trichomes rarely or scarcely torulose, straight or curved at the apices, in many species capitate.

Gomont, Monogr. des. Oscill., 1892, p. 159 (1893, p. 179, Repr.).

> 3. *Ph. ambiguum* Gomont
> 4. *Ph. lucidum* (Ag.) Kuetz.
> 5. *Ph. submembranaceum* (Ard. and Straff.) Gomont

3. **Phormidium ambiguum** Gomont

Stratum more or less expanded, dark or yellowish green or at times aeruginous, filaments long and flexuous; sheath either firm or mucous and diffluent, somewhat thick and lamellate; trichomes aeruginous, slightly constricted at the dissepiments, 4–6μ diam.; cells one-fourth the diameter long, occasionally granular at the cross walls; apices neither attenuate nor capitate; terminal membrane slightly thickened.

Growing on wood and on stones in the upper littoral belt. Bridge across an estuary to Bay Farm Island, Alameda, California, September, 1903.

Gomont, Monogr. des Oscill., 1892, p. 178, pl. 5, fig. 10 (1893, p. 198, Repr.).

We are inclined to refer to *Phormidium ambiguum* a specimen collected in Alameda, California, which forms a distinct dark green layer, with sheaths fairly thick, firm and distinct, but only slightly, if at all, stratified. The cells are very short (only 1.5μ at times) and there are no granules on the dissepiments. The apical cell is rounded or even, at times, almost conical, with a very slightly thickened membrane.

4. Phormidium lucidum (Ag.) Kuetz.

Plate 1, fig. 4

Trichomes bright blue-green, floating separately while in the free motile stage, later forming dense entangled flocculent masses of dull blue-green color, attached to soil along the margin of a pond or floating, slightly but definitely constricted at the joints, 7–8µ diam.; sheaths quite mucous and more or less diffluent, apices straight, very slightly tapering; cells 2–2.5µ long; end cell rounded, calyptrate and with thickened end walls; protoplast with two layers of fine granules at the cross walls.

Growing in warm salt water. Oakland, California.

Kuetzing, Phyc. Gen., 1843, p. 194; Gomont, Monogr. des Oscill., 1892, p. 179, pl. 5, figs. 11, 12 (1893, p. 199, Repr.). *Oscillatoria lucida* Agardh, Aufzählung, *in* Flora, vol. 10, 1827, p. 633.

The type of *Phormidium lucidum* is credited to Carlsbad and as having grown on walls exposed, at least intermittently, to steam. The specimens referred to the species from our coast grew in warm salt water in the Key Route pool at Oakland, California, and are perhaps rather thermal than marine. The species also has been found in a salt marsh near the Key Route pool, in shallow ditches made warm by the sun. The description given above has been framed to fit our plant, but differs from that of Gomont only in the matter of color and thickness of the layer. In ours the color is bright blue and the layer thin. In both these respects no. 127 of Kuetzing's Decades (*Oscillaria lucens* Kuetz.), which Gomont has cited as genuine *Ph. lucidum*, agrees fairly well with our plants.

5. Phormidium submembranaceum (Ard. and Straff.) Gomont?

Stratum membranaceous, coriaceus, dark green, trichomes without sheath, embedded in an amorphous mucus, constricted at the joints, 5µ diam.; apices of trichomes straight, long and gradually tapering, capitate; cells 4–10µ long; terminal cell provided with a depressed conical calyptra; protoplasm homogeneous.

Upper littoral belt. San Francisco Bay, California.

Gomont, Monogr. des Oscill., 1892, p. 180, pl. 5, fig. 13 (1893, p. 200, Repr.); Collins, Holden and Setchell, Phyc. Bor.-Amer. (Exsicc.), no. 1162. *Oscillaria submembranacea* Ardissone and Strafforello, Enumer. d. Alg. di Lig., 1877, p. 66.

This species is one of the truly marine species, the type locality being Porto Maurizio in Liguria on the northwestern coast of Italy. The plants from Alameda, on San Francisco Bay, formed a layer on the wall of the buttress of a bridge. In habit, diameter of trichomes, and proportion of cells our plant corresponds with Gomont's description. The trichomes, however, are not so gradually attenuated, not quite so capitate, and not quite so constricted at the dissepiments as Gomont's figure indicates. In the absence of opportunity of consulting either the type or an authentic specimen, it seems best to refer our specimens provisionally to this species.

18. Lyngbya Ag.

Filaments with a sheath, simple, either free or aggregated into dense, floccose, caespitose or pannose masses, attached or free-floating; sheath thin and homogeneous or thick and at times lamellate, either hyaline or of a sordid yellow or brown color; trichome not constricted at the dissepiments, not tapering, or only slightly so at the apex, terminal cell wall often thickened.

Agardh, Syst. Alg., 1824, p. xxv.

The genus *Lyngbya* is usually given as having been described by C. A. Agardh in 1824 (Syst. Alg., p. xxv, p. 73), but Pfeiffer (1874, p. 184) gives the date 1820, viz., the Aphorismi botanici (p. 98). The name *Lyngbya* is given in the Aphorismi (D. VII, p. 98) as containing thirteen species but without description or synonymy. Pfeiffer also refers with doubt to C. A. Agardh's *Species Algarum,* I, giving the date as 1820 and not quoting a page number. The *Species Algarum* is dated 1820 to 1828, and in no part of it do we find reference to *Lyngbya.* In 1824 C. A. Agardh referred seven species and three varieties to *Lyngbya.* Of these Gomont identifies four as either wholly or in part *Lyngbya* in the proper sense, four as "species inquirendae," and the remaining two as wholly made up of other things. As now established, *Lyngbya* contains about sixty species, largely marine.

KEY TO THE SPECIES.

1. Filaments caespitose, fixed (epiphytic) at the center and free at both ends............
....(**Leibleinia**) 2
1. Filaments not caespitose, free or with attached base (**Eulyngbya**).................... 4
   2. Filaments spirally twisted about other filamentous algae.............................
....2. **L. epiphytica** (p 74)
   2. Filaments not spirally twisted.. 3
3. Trichomes 5–8µ thick..1. **L. gracilis** (p 73)

3. Trichomes 1.5–2μ thick.................................,..3. **L. Willei** (p 74)
    4. Sheaths finally yellow brown...4. **L. aestuarii** (p 75)
    4. Sheaths always hyaline.. 5
5. Trichomes 9–25μ thick, apices never attenuate-capitate...
....5. **L. confervoides** (p 77)
5. Trichomes 5–12μ thick, apices often attenuate-capitate....6. **L. semiplena** (p 78)

Section 1. **Leibleinia** (Kuetz.) Gomont

Filaments caespitose, fixed when submerged, attached by the central portion and intertwined, erect at both ends; sheaths thin, hyaline, not evidently lamellate; trichomes not attenuated at the apices; marine species of the aspect of *Calothrix*.

Gomont, Essai Class. Nost. Hom., 1890, p. 354; Kuetzing, Phyc. Gen., 1843, p. 221 (as genus, char. mut. et sp. excl.).

The species referred to the subgenus *Leibleinia* are epiphytic, attached at the middle and free at both ends. They seem to be wholly marine.

1. *L. gracilis* (Menegh.) Rab.
2. *L. epiphytica* Hieron.
3. *L. Willei* S. and G.

1. **Lyngbya gracilis** (Menegh.) Rab.

Tufts extended, dense, floccose, slippery, purple violet, often losing color on drying and becoming dirty yellow, up to 1.5 cm. high; filaments elongated, flexible, angularly flexuous; sheath thin, smooth, does not turn blue with chlor-zinc iodine; trichomes rose-red, torulose (in dried specimens), 5–8μ thick, not attenuated at the apices; cells quadrate or twice shorter than the diameter, 2.8–4.6μ long; protoplast finely granulate; apical cell rounded, with the membrane slightly thickened above.

On *Chaetomorpha aerea* in pool, Pacific Beach, southern California. Mrs. M. S. Snyder.

Rabenhorst, Fl. Eur. Alg. II, 1865, p. 145; Gomont, Monogr. des Oscill., 1892, p. 124 (1893, p. 144, Repr.); Collins, Holden and Setchell, Phyc. Bor.-Amer. (Exsicc.), no. 853. *Leibleinia gracilis* Meneghini, Alg. Spec. Nov., 1844, p. 304 (*fide* Gomont).

The only specimen known to us is that mentioned above, distributed as no. 853, in the Phycotheca Boreali-Americana. It is certainly of the subgenus *Leibleinia* and seems to be very close to *L. gracilis* so far as answering to the description is concerned. The determination is by Frank S. Collins. It is accompanied by other epiphytes.

2. **Lyngbya epiphytica** Hieron.

Filaments adhering firmly to the surface of other filamentous algae, often regularly, spirally twisted around them, commonly having the ends free; trichomes aeruginous; cells 1–1.5μ diam., 2μ long with non-granular protoplasts; sheath very thin, hyaline; terminal cell not attenuate and neither capitate nor calyptrate.

Growing on *Lyngbya confervoides* Ag. in tide pools along high-tide level. Carmel Bay, Monterey County, California.

Hieronymus, *in* Kirchner, *in* Engler and Prantl, Natürl. Pflanzenfam., 1898, p. 67 (nomen nudum); Lemmermann, Plankton, schwed. Gewässer., 1903, p. 103, pl. 1, fig. 10; Collins, Holden and Setchell, Phyc. Bor.-Amer. (Exsicc.), no. 2206.

This is one of the most curious members of the genus, growing tightly coiled about other filamentous algae, first found by Hieronymus at Berlin growing on *Oedogonium* and *Tolypothrix,* and later by Lemmermann who had it from a garden pond in Strömsberg in Sweden. It is also given as occurring in a lagoon on Chatham Island. It is curious to find it so far away from its type locality, but our plant seems to fit the description and figure most exactly.

3. **Lyngbya Willei** S. and G.

Filaments epiphytic on larger algae, either attached at one end with the other end free or attached in the middle with both ends free, solitary or aggregated into small caespitose masses; sheath very delicate, hyaline, adhering closely; trichomes pale blue-green or grayish green, torulose, 1.5–2μ diam.; cells quadrate or one-half the diameter long; apices not attenuate and neither capitate nor calyptrate; terminal cell wall convex, not thickened; protoplasm homogeneous.

Growing on *Rhizoclonium riparium* var. *polyrhizum* Rosenv. Near the mouth of Tomales Bay, Marin County, California. August, 1916.

Setchell and Gardner, *in* Gardner, New Pac. Coast Alg. III, 1918*a*, p. 468. *Lyngbya epiphytica* Wille, Algol Notizen XXII–XXIV, 1913, pp. 22–25, pl. 1, figs. 14–17 (not Hieronymus).

The type, *Lyngbya epiphytica* Wille, was discovered on the coast of Norway at Trondhjem, growing on *Rhizoclonium hieroglyphicum* (Ag.) Kuetz. What seems to be exactly the same plant has been detected on the coast of California growing on *Rhizoclonium riparium* var. *polyrhizum* Rosenv. The latter species was distributed under no. 2238*a* of the Phycotheca Boreali-Americana. The epiphyte is rather sparsely represented on our material of *Rhizoclonium*. The identification was made after the distribution of the *Rhizoclonium* and hence is not certainly known to be represented in all of the specimens of the distribution.

The specific name, *epiphytica,* having already been occupied by Hieronymus (1898, p. 67) to designate a small, unique species coiling around the filaments of larger Myxophyceae, it became necessary to rename the species of the *Leibleinia* type found later by Wille, and the species was consequently dedicated to him.

Subgenus 2. **Eulyngbya** Gomont

Filaments intertwined into a floccose or woolly expanded mass or tufted with attached base, or even floating free; sheaths, as they become older, often thick and lamellate, at times yellowish brown; trichomes sometimes attenuate at the apices; marine, fresh water or thermal species, found on rocks, but very rarely epiphytic.

Gomont, Essai Class. Nost. Hom., 1890, p. 354.

> 4. *L. aestuarii* (Mert.) Liebm.
> 5. *L. confervoides* Ag.
> 6. *L. semiplena* (Ag.) J. Ag.

4. **Lyngbya aestuarii** (Mert.) Liebm.

Filaments forming dense, dark olive green or blue-green strata on mud, or more or less floccose floating masses, elongated, very flexuous, densely interwoven, occasionally false-branched; sheath at first hyaline, thin, smooth, increasing in thickness with age, becoming rough, lamellate, or of yellowish or brownish color; trichomes blue-green or olive green, not constricted at the dissepiments; apices slightly attenuate, capitate, with slightly thickened end wall, rarely subacute, conical, 8–24μ diam., length of cells up to six times shorter than the diameter, 2.7–5.6μ long; protoplasm finely granular, usually with a layer of granules at the cross walls.

Floating in lagoons and pools of salt and brackish water, probably along the entire coast from Puget Sound southward.

Liebman, Bemärk. tilläg danske Algenflora, 1841, p. 492. *Conferva aestuarii* Mertens, *in* Jürgens, Algae aquaticae. Decas II, no. 8, 1816 (*fide* Gomont).

Lyngbya aestuarii, as commonly understood, is a widespread and, to some extent, polymorphous species, yet generally readily recognized in a normal form. It occurs in shallow ditches and pools of brackish water all over the world except in the polar regions. It varies considerably as to the color and stratification of the sheath as well as in habit.

Lyngbya aestuarii is properly a species of the tropical (or at least warmer) waters and is an excellent example of how such a species may invade zones of cooler waters, by finding in shallow pools and ditches waters warmed by the sun to the temperature favorable to their growth and reproduction.

The following forms have been detected on our coast:

Lyngbya aestuarii f. aeruginosa (Ag.) Wolle
Plate 1, fig. 16

Plant mass pale blue-green; sheaths moderately thin, hyaline.

Floating in lagoons, salt water. Whidbey Island, Washington, and probably common in similar habitats along the Pacific Coast to the south.

Wolle, *in* Wittrock and Nordstedt, Alg. Aq. Dulc., no. 282; Setchell and Gardner, Alg. N.W. Amer., 1903, p. 187; Gomont, Monogr. des Oscill., 1892, p. 130 (1893, p. 150, Repr.); Collins, Holden and Setchell, Phyc. Bor.-Amer. (Exsicc.), no. 902. *Lyngbya aeruginosa* Agardh, Syst. Alg., 1824, p. 74.

Lyngbya aestuarii f. ferruginea Gomont

Plants forming dense floating masses, of dark or ferruginous color; sheaths thick, lamellate, deep yellowish brown color.

Floating in pools, salt marsh, Whidbey Island, Washington.

Gomont, Monogr. des Oscill., 1892, p. 130 (1893, p. 150, Repr.); Setchell and Gardner, Alg. N.W. Amer., 1903, p. 187.

Lyngbya aestuarii f. limicola Gomont

Plants inhabiting moist soil, occasionally inundated, forming rather thin compact strata; filaments closely interwoven and very tortuous.

Growing on mud in salt marsh, Whidbey Island, Washington.

Gomont, Monogr. des Oscill., 1892, p. 129 (1893, p. 149, Repr.); Setchell and Gardner, Alg. N.W. Amer., 1903, p. 186; Collins, Holden and Setchell, Phyc. Bor.-Amer. (Exsicc.), no. 903.

Lyngbya aestuarii f. natans Gomont
Plate 1, fig. 15

Plants at first forming layers on the mud at the bottom of salt marsh pools, later floating; filaments elongated, flexuous or nearly straight and loosely intertwined.

Floating in pools, salt marsh, Whidbey Island, Washington.

Gomont, Monogr. des Oscill., 1892, p. 129 (1893, p. 149, Repr.), Setchell and Gardner, Alg. N.W. Amer., 1903, p. 187; Collins, Holden and Setchell, Phyc. Bor.-Amer. (Exsicc.), no. 904.

Lyngbya aestuarii f. spectabilis (Thur.) Gomont

Sheaths up to 14μ thick, hyaline on the exterior, changing to yellowish golden color on the interior.

Growing in salt marsh, Whidbey Island, Washington.

Gomont, Monogr. des Oscill., 1892, p. 130 (1893, p. 150, Repr.); Setchell and Gardner, Alg. N.W. Amer., 1903, p. 187. *Lyngbya spectabilis* Thuret, *in* Holmes and Batters, Rev. List Brit. Mar. Alg., 1890, p. 68 (*fide* Gomont, *loc. cit.*).

5. Lyngbya confervoides Ag.

Filaments attached, caespitose, at times twisted into small, rope-like fascicles, mucous, 1–5 cm. high, dark olive green or yellowish brown with age; sheath hyaline, homogeneous or slightly lamellate, more or less roughened on the surface, 3.5–5μ thick; trichomes olive green or aeruginous, not constricted at the dissepiments, not attenuate, 9–25μ, usually 10–16μ diam., cells 2–8 times shorter than the diameter; dissepiments generally granular; terminal cell wall convex, not thickened.

Growing in shallow rock pools along high-tide level. Central California.

Agardh, Sp. Alg., 1824, p. 73; Gomont, Monogr. des Oscill., 1892, p. 136 (1893, p. 156, Repr.); Collins, Holden and Setchell, Phyc. Bor.-Amer. (Exsicc.), no. 2206, as the host of *Lyngbya epiphytica* Hieron.

The type locality for the species is Cadiz, Spain, but it is generally recognized as occurring in the warmer water all over the globe. Thus far it has been found only on the central coast of California, but is probably of wider distribution. The specimen distributed under no. 2206 of the Phycotheca Boreali-Americana agrees with the description, but is twisted above into a rope-like strand. It is also infested with *Lyngbya epiphytica* Hieron.

6. Lyngbya semiplena (Ag.) J. Ag.

Plant mass caespitose, mucous, about 3 cm. high, dull yellowish green, sometimes becoming dark violet on drying; filaments ascending from a decumbent, entangled base, flexuous; sheaths hyaline, submucous, lamellate with age, 3μ thick; trichomes blue-green or yellowish green, slightly attenuate at the apices, capitate, not constricted at the joints, $5–12\mu$ diam. (commonly $7–10\mu$), cells $2–3\mu$ long, with granules at the cross walls; apical cell provided with a depressed, conical calyptra.

Growing on rocks and on mud in the littoral belt. Puget Sound, Washington and San Diego County, California.

J. Agardh, Alg. Mar. Medit. et Adriat., 1842, p. 11; Gomont, Monogr. des. Oscill., 1892, p. 138 (1893, p. 158, Repr.); Collins, Holden and Setchell, Phyc. Bor.-Amer. (Exsicc.), no. 1905; Setchell and Gardner, Alg. N.W. Amer., 1903, p. 187. *Calothrix semiplena* Agardh, Aufzählung, etc., *in* Flora, vol. 10, 1827, p. 634.

The type locality is Trieste at the upper end of the Adriatic Sea, where it was found on rocks at the upper limits of the littoral belt. As now recognized it is of wide distribution. It bears a close resemblance to *Lyngbya aestuarii* especially to the narrower forms with hyaline sheaths. Our specimens agree well with descriptions and published specimens.

19. Plectonema Thuret

Trichomes provided with sheaths, with false lateral branching in pairs or singly, caused by rapid growth in certain regions in the filaments rupturing the sheaths, frequently constricted at the joints; apices straight, rarely tapering, without calyptra; sheaths hyaline, or rarely colored.

Thuret, Essai Class. Nost., 1875, pp. 375, 379.

Thuret names *Plectonema mirabile* (Dillw.) Thuret first, and this may be taken as the type of the genus. He also includes *P. tenue* Thuret as the second member of the genus. Thuret (*loc. cit.*) places *Plectonema* between *Symploca* and *Lyngbya,* on the one side, and *Scytonema* and *Tolypothrix,* on the other. Gomont (1890, p. 5) places it in his subtribe Lyngbyeae, next to *Lyngbya*. Hansgirg (1892, p. 39) places it near *Tolypothrix* and *Scytonema*. In this he is followed by Kirchner (1898, pp. 77, 78) and Forti (1907, p. 488). West (1916, p. 43), however, follows Gomont. Since occasional false branching is known in *Symploca* and even rarely in *Lyngbya,* and since some species of *Plectonema* have few branches and none of them have heterocysts, it seems best to place this genus as has been done by Gomont and West.

KEY TO THE SPECIES.

1. Layer blackish or brownish green, trichomes 2–3.5μ thick...................................
   ....1. **P. Battersii** (p 79)
1. Layer rose colored or reddish brown, trichomes 1.2–2μ thick...........................
   ....2. **P. Golenkinianum** (p 80)

1. **Plectonema Battersii** Gomont
Plate 1, fig. 1

Filaments forming a black or dark green stratum, elongated, flexuous, abundantly and repeatedly false-branched; branches usually in pairs and smaller than the primary filaments; sheath hyaline, thickened slightly in the primary filaments; trichomes pale blue-green, torulose, 2–3.5μ diam., with long and gradually tapering apices; the older trichomes often very densely contorted within the sheath; cells up to four times shorter than the diameter, with homogeneous contents; apical cell rounded.

Growing on rocks near high-tide limit. Golden Gate, San Francisco, California.

Gomont, Sur quelq. Oscill. nouv., 1899, p. 36; Collins, Holden and Setchell, Phyc. Bor.-Amer. (Exsicc.), no. 1712.

The two members of the genus *Plectonema* credited to our coast were found in the same locality and it is difficult at times to separate them. *Plectonema Battersii,* as represented in no. 1712 of the Phycotheca Boreali-Americana, has slightly thicker trichomes and is described as forming a black or brownish green layer. Unfortunately the color of our specimens is not, at present, discernible. The branching in our specimens, while present, is not plentiful. But this species and the next twine about other filamentous Myxophyceae and then stream outward. The branching seems more common in younger specimens and toward the base of the cluster of entangled filaments. This is shown in specimens from the same locality but of another collection than that distributed.

2. **Plectonema Golenkinianum** Gomont

Rose colored or reddish yellow, forming an expanded layer on rocks, or distributed among other algae; filaments intricate, elongated, flexuous, repeatedly false branched; false branching in pairs, very evident, smaller than the primary filaments, subflagelliform, sheath hyaline, somewhat thickened; trichomes rose colored, with short articulations, torulose, 1.2–2μ diam., articulations up to three times shorter; protoplasm homogeneous; apical cell rounded.

Growing on rocks near high-tide limit. Golden Gate, San Francisco, California.

Gomont, Sur quelq. Oscill. nouv., 1899, p. 35, pl. 1, fig. 11; Collins, Holden and Setchell, Phyc. Bor.-Amer. (Exsicc.), no. 1713.

The specimen distributed under no. 1713 of the Phycotheca Boreali-Americana shows no trace of rose or rosy brown color, but the filaments are more slender than those of *P. Battersii* as issued under no. 1712 of the same distribution. No branching has been made out with certainty but in other respects the plant resembles closely the specimens under no. 603 of the Phycotheca Boreali-Americana, duplicates of which were determined by Gomont.

20. **Symploca** Kuetz.

Filaments loosely attached at the base, forming prostrate or erect, pointed, dense fascicles; trichomes single within a sheath, sparingly false branched, straight at the apices, branches solitary; sheaths thin, hyaline, firm or slightly mucous.

Kuetzing, Phyc. Gen., 1843, p. 201, Phyc. germ., 1845, p. 167 (as "*Synploca*," orthographic error); Sp. Alg., 1849*a*, p. 270 (corrected to "*Symploca*").

The original spelling of the name of this species was "*Synploca*" (cf. Kuetzing, 1843 and 1845, as noted above). This was undoubtedly an orthographic error and was changed by Kuetzing himself to "*Symploca*" in 1849 (*loc. cit.*) Originally seven species were referred to it, of which four still remain, two are doubtful, and one is now assigned to *Schizothrix* (*Sch. Friesii* (Ag.) Gomont). The type may be considered to be *Symploca muralis* Kuetz. There are twenty-six additional species referred to the genus, at present, by far the greater number being terrestrial or of fresh water. Certain species are distinctly and exclusively marine.

Species of *Symploca* may usually be readily recognized by their habit of occurring in more or less extended layers with the filaments at first prostrate, but soon forming erect fascicles which give the surface the effect of being covered by more or less crowded tooth-like tufts. In certain stages, however, and even under certain conditions, these fascicles are not formed. In such cases it may be difficult to distinguish the species from those of *Phormidium*. The symplocoid habit is also found in certain species of *Schizothrix, Scytonema,* and *Calothrix,* and one of us (W. A. S.) has found, particularly among thermal algae, undoubted species of *Phormidium* assuming a symplocoid habit under certain conditions.

KEY TO THE SPECIES.

1. Apical cell rounded, not calyptrate.. 2
1. Apical cell convex, calyptrate....................................3. **S. aeruginosa** (p 83)
   2. Apical cell slightly swollen, trichomes 6–8µ thick.......1. **S. hydnoides** (p 81)
   2. Apical cell not at all swollen, trichomes 3-3.5µ thick.....................................
....2. **S. funicularis** (p 82)

1. **Symploca hydnoides** Kuetz.

Plate 1, figs. 12, 13

Filaments forming erect, sordid, or dark blue-green, sharply pointed fascicles, 1–2 cm. high, in age often discolored at the base, sparingly false branched; sheaths thin, slightly mucous; trichomes blue-green, 6–8µ diam., slightly torulose near the apex; length of cells variable, 5–14µ long; protoplasm filled with large granules, particularly near the cross walls; apical cell slightly inflated, without calyptra.

Growing on logs in the littoral belt. Whidbey Island, Washington.

Kuetzing, Sp. Alg., 1849*a*, p. 272; Gomont, Monogr. des Oscill., 1892, p. 107, pl. 2, figs. 1–4 (1893, p. 126, Repr.); Setchell and Gardner, Alg. N.W. Amer., 1903, p. 188; Collins, Holden and Setchell, Phyc. Bor.-Amer. (Exsicc.), no. 905 (under var. *genuina* Gomont).

We have retained the name *Symploca hydnoides* for this species, since this was the final decision of Gomont. We note, however, that Gomont (*loc. cit.,* p. 108; p. 129 of Repr.) states that the first name given to this species was *Blennothrix elegans* Kuetz. (1845, p. 181) and that this was referred by Kuetzing later (1849*a*, p. 272) to *Symploca* as *S. elegans.* In the same account (1849*a*, p. 270), however, occurs also a description of the earlier *Symploca elegans* Kuetz. described in 1843 (p. 201). In the Tabulae Phycologicae (vol. 1,

p. 44, volume dated 1846–1849, this part certainly, then, of 1849) Kuetzing rectifies his error, bestowing on this species the new name of *Symploca pulchra*. It seems that this last name ought to have been chosen by Gomont for the species, since, according to Bornet and Flahault (1886, p. 356) the *Calothrix hydnoides* of Harvey (in Hooker, 1833, p. 368), upon which Kuetzing founded *Symploca hydnoides,* is referred to *Calothrix pulvinata* (Mert.) Ag., but without indication as to this opinion being borne out by examination of the type or other authentic specimen.

The inclusion of *Symploca hydnoides* among our Pacific Coast algae rests upon a single specimen distributed under no. 905 of the Phycotheca Boreali-Americana. This seems too close to the description and too nearly in agreement with other distributed specimens to be safely separated, but the trichomes are somewhat slender, even for var. *genuina* Gomont and the dissepiments are usually obscure. It may possibly prove to be an undescribed species.

2. **Symploca funicularis** S. and G.

Plate 7, fig. 29

Filaments twisted into fine anastomosing rope-like fascicles 3–4 mm. high; trichomes aeruginous, slightly torulose, 4.5–5.5μ diam., cells quadrate or up to two times as long as the diameter, 5–8μ long; sheath hyaline, gelatinous, and diffluent in the fascicles; terminal cell slightly longer, convex, with small convex calyptra.

Growing on moist soil and on other plants in a salt marsh. Bay Farm Island, near Alameda, California.

Setchell and Gardner, *in* Gardner, New Pac. Coast Alg. III, 1918*a*, p. 469, pl. 41, fig. 29. *Symploca atlantica* Collins, Holden and Setchell, Phyc. Bor.-Amer. (Exsicc.), no. 1356 (not of Gomont).

S. funicularis differs from *S. atlantica* in having a somewhat diffluent sheath, in having the filaments often tightly twisted into rope-like threads, in having smaller trichomes, less torulose, and with slightly longer cells; and particularly in the character of the terminal cell, which has a small, depressed, convex calyptra covering only about one-third of the end cell, instead of being "depressed conical" as is the case of *S. atlantica* as described and figured by Gomont (1892, p. 109, pl. 2, fig. 5; p. 129, Repr.).

3. **Symploca aeruginosa** S. and G.

Filaments 4–4.5μ diam., forming an aeruginous stratum of erect, loose fascicles, 1 mm. high; trichomes decidedly torulose, 3–3.5μ diam.; cells quadrate, terminal cell much rounded, neither capitate nor calyptrate; protoplast homogeneous, pale aeruginous, dissepiments obscure; sheath very thin, colorless, close, at first smooth, later becoming roughened on the outside.

Growing on mud covered rocks, near the upper-tide limit. St. Michael, Alaska.

Setchell and Gardner, *in* Gardner, New Pac. Coast Alg. III, 1918*a*, p. 469. *Symploca laeteviridis* Setchell and Gardner, Alg. N.W. Amer., 1903, p. 188 (not of Gomont).

In the Algae of Northwestern America the plant described above was referred to *Symploca laeteviridis* Gomont with the statement that "it certainly seems strange to find a plant, hitherto known only from the tropical locality of Key West, so far north." Since then we have been able, through the kindness of Professor W. G. Farlow, to examine an authentic specimen of Gomont's species. It is heavily incrusted with lime and has a decidedly conical apical cell, while the cells of the trichome are almost always decidedly longer than broad. The diameter of the cells is 1.5–2μ and the dissepiments are comparatively broad and transparent. In all of these respects it differs from our Alaskan plant, although that comes within the limits of thickness assigned by Gomont to his species. The Alaskan plant is more slender than *Symploca atlantica* Gomont, with less distinct sheath, and more rounded terminal cell, which is not at all thickened above. It differs also in having inconspicuous dissepiments. On account of these various differences it seemed necessary to separate the Alaskan plant and consider it an independent species.

SUBFAMILY 4. SCHIZOTRICHEAE FORTI

Trichomes in well developed filaments, two to many within a common sheath; sheaths in many species colored (yellowish brown, red or blue).

Forti, *in* De-Toni, Syll. Alg., vol. 5, 1907, p. 315. *Vaginarieae* Gomont, Essai Class. Nost. Hom., 1890, p. 351.

We have preferred to follow Forti in choosing the designation of this subfamily, since the genus *Vaginaria* has, justly and by common consent, been relegated to synonymy (cf., however, Otto Kuntze,

1891, p. 926, and below, under *Microcoleus*). With the exception of *Porphyrosiphon*, the genera referred to the subfamily Schizotricheae consist of species whose individuals always show a fair proportion of more than one trichome in a sheath.

KEY TO THE GENERA.

1. Trichomes few in a sheath..21. **Hydrocoleum** (p 84)
2. Trichomes many in a sheath...22. **Microcoleus** (p 85)

21. Hydrocoleum Kuetz.

Plant mass forming a compact, smooth or caespitose cushion on the substratum, at times encrusted with lime; filaments with hyaline, cylindrical, sublamellate sheaths, sparingly false branched, more or less mucous, at times completely diffluent; trichomes few and loosely aggregated in a sheath, apices straight, more or less attenuate and capitate, with terminal cell wall thickened.

Kuetzing, Phyc. Gen., 1843, p. 196.

It has seemed best to retain the original spelling of the name of this genus, although, of late, it is being written *"Hydrocoleus"* (cf. Forti, 1907, p. 315, and others). The assumption is, presumably, that *"Hydrocoleum"* is an orthographic error. The change may also be made to make it agree with *Microcoleus*. There seem to be two words in Greek, or even three, meaning sheath, viz., ὁ κολεός or τό κόλεον, the sheath or scabbard of a sword, and τό κολέον (alternative κουλεόν), a sheath; the two latter are Ionic forms to be sure, but no attempt has been made to limit the choice of Greek, in derivatives, to the Attic dialect alone. τό κόλεον, however, is Attic, so that it seems perfectly proper to retain the name in the form in which Kuetzing wrote it, since either *"Hydrocoleum"* or *"Hydrocoleus"* may be considered orthographically correct.

The genus *Hydrocoleum* was founded by Kuetzing on two fresh water species which he named *H. homoetrichum* and *H. heterotrichum* respectively. The first, which may be considered properly as the type of the genus, was found on stones in a small cataract near Trieste. The second species was apparently founded on a mixture of *H. homoetrichum* and another species, for which the name has been retained by Gomont. About twenty-eight species have been described as belonging to this genus of which one quarter are marine. The species are to be distinguished from those of *Microcoleus* and *Sirocoleum*, the only other marine genera, by having the trichomes less numerous in the sheath.

The trichomes of species of *Hydrocoleum* are always capitate while those of the other two genera are rarely so, the exception being one species of *Microcoleus*.

Hydrocoleum lyngbyaceum Kuetz.

Plate 1, fig. 10

Plant mass forming a smooth, compact, mucous stratum, or sometimes caespitose when young; filaments unbranched below, but with numerous false branches above; sheaths wide, irregular, apices acuminate, varying often to open, sometimes completely diffluent; trichomes dark blue-green, numerous at the base of the filaments, usually solitary in the branches, sometimes spirally twisted, 8–16μ diam., tapering and truncate at the apices; cells 2.5–4.5μ long, dissepiments granular.

Forming a thin layer of considerable extent on piles, in the middle littoral belt. Lands End, San Francisco, California.

Kuetzing, Sp. Alg., 1849*a*, p. 259; Gomont, Monogr. des Oscill., 1892, p. 337, pl. 12, figs. 8–10 (1893, p. 75, Repr.).

Thus far a single species of *Hydrocoleum* from a single locality has been found on our coast. It seems referable to the var. *rupestre* Kuetz. (*loc. cit.*), of which the type locality is the shores of Calvados in northern France. Our specimens seem to agree well with the descriptions and figures.

22. **Microcoleus** Desm.

Trichomes numerous, within a common, firm, hyaline, cylindrical sheath; sheath usually unbranched and tapering at both ends; filaments forming densely intertwined, more or less diffluent, masses on earth, in salt water or in fresh water; trichomes attenuate, with straight apices, tightly interwoven within the sheath; apical cells acute or rarely obtuse, conical or capitate.

Desmazières, Cat. des plantes, 1823, p. 7.

We have followed Gomont in retaining the name *Microcoleus* of Desmazières (1823) as against that of *Vaginaria* Bory (or Gray) 1821. The original *Vaginaria* Rich (in Persoon, *Synopsis*, 1805, p. 70) has passed, at least seemingly, into the synonymy of *Fuirena* Rootb. Bory later (1826, p. 524) adopted *Microcoleus* in place of his *Vaginaria*. The decision of Gomont, however, settles the case according to the rules of the Brussels Congress, and the existence of the earlier *Vaginaria,* according to the American Code. Otto Kuntze (1891, p. 926) has stated the case for the retention of *Vaginaria.*

Microcoleus contains between twenty and twenty-five species and is, for the most part, readily recognizable. The species are to be distinguished from those of *Sirococoleum* rather by habit than by any other character, while from *Hydrocoleum* they differ in the larger number of usually smaller trichomes in a sheath. There are species, however, and *Microcoleus confluens* S. and G. is one of them, which seem to be intermediate between the two genera.

KEY TO THE SPECIES.

1. Cells longer than broad.. 2
1. Cells shorter than broad.. 3
   2. Trichomes 2.5–6µ thick.....................................1. **M. chthonoplastes** (p 86)
   2. Trichomes 1.5–2µ thick......................................2. **M. tenerrimus** (p 87)
   3. Trichomes 7–8µ thick, apical cell acute conical....................3. **M. Weeksii** (p 87)
   3. Trichomes 4.4–5µ thick, apical cell rounded..............4. **M. confluens** (p 88)

1. Microcoleus chthonoplastes (Mert.) Thuret

Filaments forming dense, thick, coriaceous, extensive layers on damp mud or inundated ground, or at times intermingled with other species of algae, dark olive green; sheaths cylindrical, tapering at both ends, sometimes completely closed, or open and the trichomes projecting; trichomes blue-green, short, nearly straight, numerous, and closely packed within the sheath, constricted at the joints, 2.5–6µ diam., cells 3.6–10µ long; apices attenuate, acute, conical.

Growing on mud in salt marshes. Common along the coast from Puget Sound, Washington to Southern California.

Thuret, Essai Class. Nost., 1875, p. 378; Setchell and Gardner, Alg. N.W. Amer., 1903, p. 188; Collins, Holden and Setchell, Phyc. Bor.-Amer. (Exsicc.), no. 906; Gomont, Monogr. des Oscill., 1892, p. 353, pl. 14, figs. 5–8 (1893, p. 91, Repr.). *Conferva chthonplastes* Mertens, *in* Hornemann, Fl. Dan., 1813, vol. 9, fasc. 25, p. 6, pl. 1845.

This widespread species was found first ''in sinu Othiniensi'' in Denmark by Hofman Bang and described by Mertens in 1813. It is usually found in shallow pools in salt marshes and prefers the waters or moist mud warmed by the sun, whence it may readily be distributed through the agency of water birds in the mud adhering to their feet. Our specimens seem clearly of this species.

2. **Microcoleus tenerrimus** Gomont

Filaments aggregated into dense, blue-green layers, or dispersed among other species of algae, simple or sparsely branched; sheaths ample, somewhat irregular, apices acuminate or open and blunt; trichomes dark olive green, moderately numerous within the sheath, elongate, flexuous, somewhat loosely entwined, extremely constricted at the joints, apices frequently long attenuate, $1.5–2\mu$ thick; cells $2.2–6\mu$ long, with occasional granules at the cross walls; apical cell very acute conical, not capitate.

Growing on mud in salt marshes. Puget Sound, Washington, to central California.

Gomont, Monogr. des Oscill., 1892, p. 355, pl. 14, figs. 9–11 (1893, p. 93, Repr.); Setchell and Gardner, Alg. N.W. Amer., 1903, p. 188.

The specimens referred to *Microcoleus tenerrimus* are so placed with some doubt. The general characters agree but the cells are not distinct, the dissepiments being somewhat obscured instead of appearing hyaline and there are no evident constrictions. A further study of living material is very desirable.

3. **Microcoleus Weeksii** S. and G.

Plate 6, fig. 24

Sheath thin, hyaline, gelatinous, irregular; trichomes 20–40 in a sheath, loosely intertwined, aeruginous, not torulose, $7–8\mu$ diam., ends attenuated; protoplast homogeneous; cells $1.8–2.5\mu$ long; end cell acutely conical, neither capitate nor calyptrate.

Growing on *Griffithsia* sp. in the lower littoral belt. Near Pacific Grove, California. Collected by Mrs. J. M. Weeks, February 1, 1896.

Setchell and Gardner, *in* Gardner, New Pac. Coast Alg. III, 1918*a*, p. 470, pl. 40, fig. 24.

In 1896 Mrs. J. M. Weeks sent some specimens of *Griffithsia* for determination, and in making a microscopic examination of the material there were detected specimens of a *Microcoleus* which seemed to be undescribed. The material being scanty was not named and described at that time, but incorporated into the Herbarium awaiting further investigation, and with the hope that more material might be discovered. Since then, unfortunately, no more material has been reported, but the species has been described since it seems very distinct.

M. Weeksii is a species which seems at present to have its nearest affinities with *M. chthonoplastes* (Mert.) Thuret, on the one hand, and *M. acutirostris* (Crouan) Gomont, on the other. It differs from the first in the character of the sheath, being very much more indefinite, gelatinous, and amorphous, in having fewer trichomes within the sheath, and in the trichomes being larger as well as the cells much shorter. From the latter species it differs in having smaller trichomes with shorter cells, in not having the ends long, attenuated and very acute-conical as has *M. acutirostris,* according to the figures and description given by Crouan. The trichomes in *M. Weeksii* begin to taper about 6–10 cells back of the apices.

4. **Microcoleus confluens** S. and G.

Plate 6, fig. 25

Filaments erect, forming a blue-green, compact, spongy stratum of indefinite expansion on rocks; sheath gelatinous, moderately ample, irregular on the surface, hyaline, homogeneous, distinct at first or later diffluent; trichomes 1–8 in a sheath, separate or lossely entwined, very slightly constricted at the dissepiments, 4–4.5μ diam., straight and not attenuated at the apices; cells 2–4 times shorter than the diameter, protoplast homogeneous; apical cell moderately rounded, end wall slightly thickened.

Forming a spongy stratum in company with other species of Myxophyceae on rocks in the upper part of the littoral belt. Lands End, San Francisco, California.

Setchell and Gardner, *in* Gardner, New Pac. Coast Alg. III, 1918*a*, p. 471, pl. 40, fig. 25.

The species of the genera *Microcoleus* and *Hydrocoleum* intergrade, at least to a certain extent. Typical *Microcoleus* includes forms with numerous entwined trichomes within firm, fairly regular, unbranched sheaths. The trichomes are usually small, more or less pointed at the ends, mostly ranging below 7μ in diameter, except *M. subtorulosus* which is 6–10μ in diameter, and the cells are generally longer than the diameter.

The majority of the species which have thus far been placed in *Hydrocoleum* have fewer trichomes in a sheath, which is often more or less branched, and which is commonly less firm and in some cases even wholly diffluent. The trichomes are large, as a rule, with cells shorter than the diameter, the end cells usually blunt with thickened end cell walls. Borzi, however, has placed a species from New Guinea in this genus although it is but 2μ in diameter.

Microcoleus confluens does not conform wholly to the characters of either of the two above mentioned genera. It has the few trichomes and diffluent sheath of a typical *Hydrocoleum* but has the small trichomes of a typical *Microcoleus*.

SUBORDER 2.   HETEROCYSTINEAE nobis

Filaments usually attached, at least at first, but some floating free; trichomes made up of cells differing from one another either through heterocysts being present, or through trichomes ending in a hair, or through both, simple or branched; branching either true or false; sheath usually present, enclosing one or more trichomes, hyaline or colored, homogeneous or stratified; propagation chiefly by hormogonia; resting spores present in some species.

Heterocysteae, Hansgirg, Syst. einig. Süsswasseralg., 1884, p. 9.

The group of the heterocysted Myxophyceae is a fairly natural one and, in spite of some minor difficulties, it seems best to retain it. It was instituted by Hansgirg in 1884, and received its concrete form from Bornet and Flahault in 1886. Richter in 1896 (p. 274) urged certain objections but Gomont in 1899 brought forward certain facts tending to show the misconception under which Richter labored. Later writers have abandoned the group, but without substituting anything satisfactory. It is true that the heterocysts are lacking in certain species, but in nearly all cases, at least, there is, in such instances, a hair present. In other species, heterocysts may be scanty or rarely found, but all such cases show their affinities by other peculiarities of structure. We have, therefore, retained the group, simply changing the termination to conform with what seems to be the best usage.

FAMILY 4.   NOSTOCACEAE NAEG.

Thallus floating singly or gregariously, of definite, globular, ovoid, or regularly lobate form or irregularly expanded, solid or hollow, soft to firmly gelatinous; trichomes unbranched, never tapering to a hair, both extremities usually similar, heterocysts present; propagation by hormogonia and resting spores.

Naegeli, Neu. Algensyst., 1847, p. 132 (''Nostochaceae,'' lim, mut.). *Nostoceae* Kuetzing, Phyc. Gen., 1843, p. 203 (lim. mut.).

The family of the Nostocaceae is thoroughly distinct and its members easily recognizable. Naegeli used the term in an extended sense. Bornet and Flahault also used it in an extended sense in their title but their tribe Nostoceae (1887, p. 177) has the same limits as has gradually come to be customary and as we shall use it.

Most of the Nostocaceae are inhabitants of the fresh waters, but a few are marine or brackish water species. Only two species of this family have been detected on this coast and one of these is not strictly a marine or brackish water species. Probably several species remain to be detected.

Key to the Genera.

1. Trichomes contorted and twisted, united into a definite gelatinous thallus............
....24. **Nostoc** (p 92)
1. Trichomes straight, united into indefinite gelatinous masses................................
....23. **Anabaena** (p 90)

23. Anabaena Bory

Trichomes cylindrical or slightly tapering at the ends, without a sheath, or in some species with a thin mucous tegument, existing apart or united loosely by the mucous teguments into a stratum; heterocysts intercalary; spores single or seriate, spherical to cylindrical.

Bory, Dict. Class. d'hist. nat., vol. 1, 1822, p. 307.

Bory in 1822 established the genus *Anabaena* ("*Anabaina*") on five species: of which one, viz., *A. oscillarioides* Bory, remains to represent two of them; one species is uncertain; one is *Phormidium laminosum* (Ag.) Gomont, and the fifth is referred to "*Leptothrix lamellosa* Kuetz.'' The type of the genus is *Anabaena oscillarioides* Bory. The original spelling of the genus was "*Anabaina*," changed later into "*Anabaena*" to accord with usage in transliterating from Greek into Latin, the earlier diphthong *ai* becoming in classical Latin *ae*.

Key to the Species.

1. Trichomes 2.6–3μ in diam., spore wall hyaline................1. **A. propinqua** (p 90)
1. Trichomes 4–6μ in diam., spore wall brown.....................2. **A. variabilis** (p 91)

1. Anabaena propinqua S. and G.

Plate 8, fig. 9

Stratum very thin, light blue-green; trichomes flexuous, 2.6–3μ diam. without a distinct sheath; cells subquadrate, 1.8–3μ long, cylin-

drical, slightly constricted at the dissepiments which are very distinct;
terminal cell either blunt or acute conical; heterocysts subspherical to
cylindrical, 3.5–4.5μ diam., 5–7μ long; spores subspherical to broadly
ellipsoidal, 5.5–7μ diam., 7–9μ long, catenate, developing centrifugally,
membrane smooth, hyaline.

Forming a thin stratum, more or less continuous, on *Ruppia maritima*. In pools in a salt marsh, West Berkeley, California.

Setchell and Gardner, *in* Gardner, New Pac. Coast Alg. IV, 1919*b*,
p. 487. *Anabaena variabilis* Collins, Holden and Setchell, Phyc. Bor.-
Amer. (Exsicc.), no. 1209 (not of Kuetz.).

Anabaena propinqua appears to be closely related to *A. variabilis*
Kuetz., but comparison with material of that species from Herbarium
Thuret contributed by Dr. Bornet (Herb. Univ. Calif. no. 100300)
shows that it differs in the smaller dimensions of all parts of the plant,
vegetative cells, heterocysts, and spores, which are also more nearly
spherical than in *A. variabilis* and have hyaline walls instead of brown.

2. **Anabaena variabilis** Kuetz.

Plate 8, fig. 8

Filaments combined into a thin gelatinous stratum; trichomes
flexuous, mucous, 4–6μ diam., terminal cell obtuse-conical; cells doliiform, 2.5–6μ long, or at times subquadrate and deeply constricted at
the dissepiments; heterocysts spherical or oval, 6μ diam., 8μ long;
spores oval with truncate ends, usually seriate, 7–9μ diam., 8–14μ long,
developing centrifugally from midway between the heterocysts, with
smooth brownish walls.

Floating in quantity in bluish green gelatinous masses in brackish
pool. Presidio, San Francisco, December, 1895. Found in both fresh
and brackish waters.

Kuetzing, Phyc. Gen., 1843, p. 210; Bornet and Flahault, Rev. IV,
1888, p. 226.

This species has been found in brackish water only once with us,
but probably is to be found elsewhere than in San Francisco when
the various salt marshes along our coast are carefully explored. There
seems little doubt of the identity of the species although it is sterile.
It has been found in spore condition in fresh waters in our general
territory.

24. **Nostoc** Vaucher

Thallus mucous, gelatinous or coriaceous, usually globose or oblong at first, later either remaining so or assuming a variety of forms, solid or cavernous, usually with more or less firm periderm, floating or attached, composed of a mass of more or less densely intertwined and contorted filaments, with thin walls and thick mucous or gelatinous confluent teguments; trichomes usually torulose, cells cylindrical, subglobose or doliiform; heterocysts terminal and intercalary, spherical to oblong, seriate, beginning midway between the heterocysts and developing centrifugally.

Vaucher, Hist. des conferves, 1803, p. 203.

The genus *Nostoc* was founded by Vaucher in 1803 (p. 203) as a segregation from the old genus *Tremella* of Dillenius (1741, p. 41) which by common consent has been restricted to a group of species of basidiomycetous fungi. As founded by Vaucher, *Nostoc* is a very distinct genus, only one of five species described being now referred elsewhere and that is now referred to the lichen genus *Collema* whose algal constituent is *Nostoc*. *Nostoc* is adopted for the name of the genus by Bornet and Flahault (1888, p. 181), and this is the starting point of our nomenclature in this group. The type may be taken as *Nostoc commune* Vauch. and the species are practically entirely terrestrial or inhabitants of fresh water. The single species listed below is clearly a migrant into a salt water pool, but is worthy of inclusion because of its physiological significance.

Nostoc Linckia (Roth) Bornet

Thallus soft and gelatinous at maturity, of indefinite and irregular form, aeruginous or brownish with age; filaments abruptly contorted, intricately and closely intertwined, sheaths distinct in young colonies and on the surface of older ones, becoming confluent in the interior of the colonies, hyaline; trichomes $3.5–4\mu$ diam., pale grayish green, cells short, depressed-globose; heterocysts $5–6\mu$ diam., globose or subglobose; spores seriate, developing centrifugally, subglobose, $6–7\mu$ diam., $7–8\mu$ long; spore membrane smooth, hyaline or slightly brownish with age.

Floating in long, slender masses, 7–20 mm. diam., and up to 25 cm. long in a salt water pond, near the C. M. & St. Paul Railway station, Port Townsend, Washington. July, 1917.

Bornet, *in* Bornet and Thuret, Notes Algol., 1880, p. 86, pl. 28, figs. 1–12; Bornet and Flahault, Rev. IV, 1888, p. 192; Setchell and Gardner, Alg. N.W. Amer., 1903, p. 189 (growing in fresh water). *Rivularia Linckia* Roth, Neue Beiträge zur Botanik I, 1802, p. 265.

Our specimens form shapeless expansions, with much contorted trichomes and with excellent spores. They certainly seem to be *Nostoc Linckia* and their occurrence in salt water is worthy of note.

• FAMILY 5. RIVULARIACEAE RABENHORST

Filaments cohering by their gelatinous sheaths, forming more or less definite spherical or hemispherical colonies and radiating from the center outward, or parallel and forming more or less penicillate or caespitose strata, or rarely solitary, simple or branched, composed of sheaths and a single series of cells; trichomes usually much attenuated and terminated by a hair, one to several within a sheath; heterocysts basal and intercalary, rarely absent; branching false, usually beneath the heterocysts, more rarely between the heterocysts; multiplication by hormogonia. by resting spores or by both.

Rabenhorst, Fl. Eur. Alg., vol. 2, 1865, pp. 2, 200 (pro max. parte). *Rivularieae* Harvey, *in* W. J. Hooker, Brit. Fl., vol. 2, part 1, 1833, p. 391 (pro parte).

The Rivulariaceae form a very natural family distinguished by a combination of false branching and terminal hairs. Usually heterocysts are present, but occasionally they are lacking. In the Rivulariaceae as previously distinguished, the opposite ends of the trichomes are unlike, but in *Hammatoidea* (*Ammatoidea,* W. and G. S. West, 1916, p. 46) both ends of the trichome terminate in a hair while the central portion of the trichome is bent at nearly right angles to the end portions and is devoid of heterocysts. It seems to us that *Hammatoidea* bears the same relation to the non-heterocysted species of *Calothrix* that the subgenus *Leibleinia* does to the typical species of *Lyngbya*. Also there is a heterocysted species of *Calothrix* (*C. pilosa* Harv.) whose central portion is decumbent and both ends erect. We have removed the genus *Brachytrichia* from Rivulariaceae to Stigonemataceae on account of its true branching, although it possesses hairs.

The members of the Rivulariaceae are more largely marine than those of any other family of the Myxophyceae, yet there are many fresh water species in the family. The representation of the Rivulariaceae on the Pacific Coast of North America will undoubtedly be considerably increased as more thorough exploration and study is carried on.

25. Calothrix Ag.

Filaments simple or with false branching, and with cylindrical sheaths, usually aggregated into dense pulvinate masses; trichomes tapering from near the base upward or only at the upper end, frequently terminating in long hyaline hairs; spores or resting cells have been observed in some species; heterocysts basal or intercalary, single or several in a series.

Agardh, Syst. Alg., 1824, p. xxiv (diagnosis), 70 (list of species).

The genus *Calothrix* was founded by C. A. Agardh in 1824 (p. xxiv diag.) with *C. confervicola* (Roth) Ag. as the type. In his account Agardh (*loc. cit.*, p. 70) mentions twelve species, four of which still remain as true species of *Calothrix*. The genus is comparatively distinct, although some species are destitute of heterocysts and some lack a distinct terminal hair. Some species are branched and one species is decumbent in the middle and erect at both ends. Resting spores are known in some species. The species of *Calothrix* are the simplest of the Rivulariaceae, passing into those of *Dichothrix,* and thence through *Polythrix* to *Rivularia*. Species of *Calothrix* with undeveloped hairs bear a certain resemblance to species of *Microchaete,* which is usually placed among the Scytonemataceae.

The species of *Calothrix* are often troublesome to determine or to distinguish from one another, and the material from our coast is not so abundant as is desirable. We have attempted to place our species as accurately as possible, but the whole genus needs careful revision.

1. **Calothrix consociata** (Kuetz.) B. and F.

Filaments gregarious and stellately fasciculate, growing on other filamentous algae, dark green, rigid, 0.5 mm. high, 21–29μ diam., decumbent, curved, and slightly thickened at the base; sheaths close, membranaceous, dilated funnel-shaped, lamellate, inner layer brown, outer layer hyaline; trichomes 12μ diam., olive green; cells three times shorter than the diameter, heterocyst basal.

Growing on plants in a salt marsh. Whidbey Island, Washington.

Bornet and Flahault, Rev. I, 1886, p. 351; Setchell and Gardner, Alg. N.W. Amer., 1903, p. 197. *Schizosiphon consociatus* Kuetzing, Phyc. Gen., 1843, p. 234, Tab. Phyc. II, 1850–1852, p. 17, pl. 54, fig. 3.

Calothrix consociata was included by us in the Algae of Northwestern America on the evidence of a single collection which has, unfortunately, not been preserved or, at least, not placed in its proper place in the collections. It is, therefore, included in this account without further comment. The species is closely related to *C. confervicola* (Roth) Ag., from which it differs in color of trichome, greater diameter of filament, and thicker sheath, whose inner layers are colored. *C. confervicola*, although seemingly a widespread species, has not, as yet, been detected with certainty on our coast, although it is to be looked for.

2. **Calothrix robusta** S. and G.

Plate 6, fig. 22

Filaments caespitose, attached at the base and erect, or attached at the middle and both ends erect, somewhat flexuous, cylindrical, 1–2 mm. long, 30–40μ diam.; trichomes nearly cylindrical throughout, very abruptly pointed at the apex, rarely branching, not torulose, 16–20μ diam., bright blue-green, cells 2–3μ long, finely and uniformly granular; sheaths hyaline or slightly yellowish with age, lamellate, 7–8μ thick, closed at the apex when young; heterocysts basal, much compressed, 1–4μ seriate.

Forming a caespitose layer on rocks in small tide pools near high-tide level. Cypress Point, Monterey County, California.

Setchell and Gardner, *in* Gardner, New Pac. Coast Alg. III, 1918*a*, p. 473, pl. 40, fig. 22.

The material of this species shows but few heterocysts and these are all basal, usually much compressed, and occasionally 2–4 seriate. It has a thick, firm sheath, and this remains closed at the apex for some time. The branching, so far as observed, is through a rupture either on the side of the sheath, apparently caused by rapid growth of the trichome in that part, the closed sheath at the apex not expanding as rapidly as the trichome elongates, or more commonly by breaking through the sheath at the base of procumbent filaments.

Calothrix robusta is to be placed near *C. scopulorum* from which it differs decidedly in size, being, in fact, the largest species of the genus, so far as thickness of filament goes, with the exception of *C. pilosa* which also it resembles to a certain extent.

3. **Calothrix scopulorum** (Web. and Mohr) Ag.

Filaments contorted, erect, up to 1 mm. high, 10–18μ diam., moderately thickened at the base, forming an expanded, caespitose, olive green stratum; sheaths somewhat thick, hyaline, yellowish brown, or colored in zones, usually lamellate, repeatedly ocreate; trichomes olive green, 8–15μ diam., prolonged into a long, hyaline hair; heterocysts 1–3, basal, hormogonia numerous in a sheath, 4–5 times as long as the diameter.

Growing in rock pools and on wood along high-tide level and above. Puget Sound (Saunders, 1901, p. 399), and central and southern California.

Agardh, Syst. Alg., 1824, p. 70; Bornet and Flahault, Rev. I, 1886, p. 353; Bornet and Thuret, Notes Algol. II, 1880, p. 159, pl. 38; Setchell and Gardner, Alg. N.W. Amer., 1903, p. 197; Collins, Holden, and Setchell, Phyc. Bor.-Amer. (Exsicc.), no. 1720. *Conferva scopulorum* Weber and Mohr, Reise durch Schweden, 1804, p. 195, pl. 3, figs. 3*a*, 3*b*.

A fairly and readily recognizable species, possibly to be confused with *C. crustacea* when the intercalary heterocysts of the latter are scanty, growing usually on rocks but occasionally on wood, seldom as an epiphyte, but this condition not seen, as yet, in our territory.

4. Calothrix Contarenii (Zan.) B. and F.

Filaments forming a blackish green, smooth, glistening, compact, orbicular stratum, parallel, erect, moderately flexuous, up to 1 mm. high, 9–15μ diam., decumbent and thickened at the base; sheaths somewhat thick, hyaline or dark yellowish brown, homogeneous or lamellate, dilated above; trichomes 6–8μ diam., prolonged above into a long, delicate hair when actively growing; cells equaling the diameter or 2–3 times shorter; heterocysts 1–2, basal.

Growing sparsely on rocks near high-tide level. Southern Washington, central California, southern California.

Bornet and Flahault, Rev. I, 1886, p. 344. *Rivularia Contarenii* Zanardini, Sulle Alghe, 1839, p. 134.

This species is very close to the last (*C. scopulorum*) but is to be distinguished by its habit and its somewhat larger and more flexuous filaments. The specimens referred here have been subject to careful examination and seem rather to belong to *C. Contarenii* than to *C. scopulorum*. The number of basal heterocysts in *C. Contarenii* is given as "1–2," but there are three or more in most of our specimens.

5. Calothrix pulvinata (Mert.) Ag.

Filaments erect, flexuous, agglutinated into small irregular fascicles, 2–3 mm. high, 15–18μ diam., slightly enlarged at the base, sparingly false branched, forming expanded, spongy, dull green layers; sheaths thick, firm lamellate, hyaline or brownish; trichomes 8–12μ diam., olive green, attenuated into a short hair; cells 2–3 times shorter than the diameter.

Growing on woodwork, floating logs, etc. Along high-tide line and in salt marshes. Puget Sound, Washington, to central California.

Agardh, Syst. Alg., 1824, p. 71; Bornet and Flahault, Rev. I, 1886, p. 356; Setchell and Gardner, Alg. N.W. Amer., 1903, p. 197; Collins, Holden and Setchell, Phyc. Bor.-Amer. (Exsicc.), no. 957; Bornet and Thuret, Notes Algol. II, 1880, pp. 39, 161. *Ceramium pulvinatum* Mertens, *in* Jürgens, Alg. Aquat. Decas IV, 1817, no. 5.

This is a very distinct species in typical form because of its peculiar symplocoid habit. Even when this habit is obscure the filament and trichome are fairly typical. Our coast lacks its favorite habitat of piles and logs exposed to fairly pure and fairly warm salt water.

6. Calothrix parasitica (Chauv.) Thur.

Filaments aggregated among the cortical threads of *Nemalion,* blue-green, 0.5 mm. high, 9–10μ diam. (rarely 12–15μ diam.), slightly thickened in the middle, bulbous and curved at the base, up to 24μ thick; sheath thin, hyaline, often dilated at the apex; trichomes 7–8μ diam., cells short, prolonged into a very long, flexuous, hyaline thread above; heterocysts basal; several hormogonia in a sheath, 4–5 times as long as the diameter; color intense blue-green.

Growing in the fronds of *Nemalion lubricum.* Probably co-existent with the host along the Pacific Coast.

Thuret, Essai, 1875, p. 381 (*nomen nudum*); Bornet and Thuret, Notes Algol, II, 1880, p. 157, pl. 37, figs. 7–10; Bornet and Flahault, Rev. I, 1886, p. 357. *Rivularia parasitica* Chauvin, Rech. sur l'organ la fructification, 1842, p. 41.

The same species grows in the fronds of *Nemalion multifidum* on the Atlantic coasts of the United States and Europe. It seems to have no deleterious effect upon either species of *Nemalion,* and while it seems always to be associated with that genus, it has never been proven to be really parasitic upon it in the sense of obtaining nourishment.

There is some variation in the specimens found in *Nemalion* on both the Pacific Coast of North America and the Atlantic coasts of North America and Europe. The base is often swollen but not really bulbous. The sheath is sometimes strict and hyaline, but also often ample, brownish, and ocreate above. The trichomes are at times torulose, and again they show nothing of the sort. The heterocysts vary in different collections from one to four. These may be differences of age or of luxuriance of development, but also there may be more than one species confused under this name.

7. **Calothrix epiphytica** W. and G. S. West

Filaments· epiphytic upon larger algae, solitary or aggregated into small clusters, attached at the base and erect, or procumbent, 5–7.5μ diam., 250μ, rarely 350μ, long, gradually attenuated from the base to the apex, not branched; sheath transparent and colorless, inconspicuous; trichomes aeruginous, 3.5–4μ diam. at the base, tapering gradually into a delicate hair at the apex; cells quadrate, slightly longer above and shorter below, heterocysts small, solitary, basal.

Growing on *Rhizoclonium riparium* var. *polyrhizum* Rosenv., on sand rock along high-tide level. Near the mouth of Tomales Bay, Marin County, California. August, 1910. No. 3440, Gardner. Distributed in Collins, Holden, and Setchell, Phyc. Bor.-Amer. (Exsicc.), no. 2238*a*, on the above mentioned host.

West, W., and G. S., Welwitsch's African Alg., 1897, p. 240.

The type was found growing on *Oedogonium* sp. on the bank of the River Bero at Massamodes, Africa, August, 1859, no. 190, Welwitsch.

This is the first and only locality on our coast from which this species of *Calothrix* has been reported. The material is mixed with *Lyngbya Willei* S. and G. and is not very abundant. It seems strange to find a fresh water species on the edge of an ocean and it may be that our plant is different from that of the Wests. The agreement seems too close, however, to refer it elsewhere, but the material is not sufficiently abundant to make the determination thoroughly satisfactory.

8. **Calothrix crustacea** Thuret

Filaments caespitose, or velvety and widely expanded, dark green or brownish, erect, straight, dense, 1–2 mm. high, 12–20μ even up to 40μ diam., thickened at the base; sheaths somewhat thick, hyaline or yellowish brown, mostly lamellate, ocreate; trichomes olive green, 8–15μ diam., prolonged above into a very delicate acute hair; cells short, heterocysts 1–3, basal and intercalary; hormogonia numerous in a sheath, 4–5 times as long as the diameter; "spores seriate, oblong-cylindrical, smooth."

Attached to sticks, weeds, and stones in salt marshes, in the upper littoral belt. Washington to California.

Thuret, Notes Algol. I, 1878, pp. 13–16, pl. 4 (figures 5, 6 drawn from living material collected at Croisic) ; Setchell and Gardner, Alg.

N.W. Amer., 1903, p. 197; Collins, Holden, and Setchell, Phyc. Bor.-Amer. (Exsicc.) nos. 1212, 2104.

This species presents a great variety of variations. As interpreted by Thuret and by Bornet and Flahault the variation in the diameter of the filaments is from 12μ to 40μ. A variation so great as this has not been reported in any other species of the genus. It may be that we are dealing here with a heterogeneous group. The heterocysts are basal and intercalary, varying from one to several in each position. In its multiplicity of branching there is great diversity. Some collections have the filaments almost wholly without branching. This is considered to be the typical form. In other collections there are intermingled specimens without branches and those with profuse branching approximating to such species as *C. prolifera* Flah., *C. fasciculata* Ag., and even to the extreme represented by *C. vivipara* Harv.

9. **Calothrix rectangularis** S. and G.

Plate 6, fig. 21

Filaments simple or occasionally sparsely branched, attached at the base to various other algae, prostrate when young, soon becoming erect, 400–500μ long, 24–28μ diam. at the base; trichomes 16–18μ diam. at the base, tapering gradually upward nearly to the apex, then rather abruptly ending in a short hair; cells 2–4μ long, protoplasm finely granular, pale blue-green; sheath hyaline when young, changing to light yellow with age, strict at the apex; heterocysts basal and intercalary, the basal spherical to subconical, the intercalary cylindrical, usually single, near the base of the trichome, 18–25μ long, the same diameter as the trichome.

Epiphytic on various species of marine algae, near low-tide level. East Sound, Orcas Island, Washington.

Setchell and Gardner, *in* Gardner, New Pac. Coast Alg. III, 1918a, p. 472, pl. 40, fig. 21.

This species comes within the range of measurements stated by Bornet and Flahault (Rev. I, 1886, p. 359) for the diameter of the filaments of *C. crustacea* Thuret with which it seems to be most closely related, the filaments are, however, much shorter than the measurements given for that species. The small size, the habit of growth on other algae, the presence of cylindrical, intercalary heterocysts always near the base of the trichomes, the infrequency of branching, and the strict sheath seem sufficient to separate this species from any of the forms of *C. crustacea*. The cylindrical heterocysts above mentioned, appearing rectangular to the view, are sufficiently striking to suggest the specific name.

10. Calothrix prolifera Flahault

Layer expanded, velvety, brownish green; filaments 2 mm. high, 15–18μ thick, slightly flexuous, curved at the base and noticeably thickened, at times falsely branched, branches emerging from beneath a heterocyst after the fashion of *Tolypothrix;* sheath thick, lamellate, firm, many times ocreate, with the ocreae dilated and lacerate; above hyaline, below yellowish; trichomes 8–12μ thick, bright verdigris green, attenuated at the apices into hairs, cells 3–4 times shorter than the diameter; heterocysts basal (1–2) and intercalary (numerous).

Growing among other algae on boards in salt marshes. San Francisco Bay, California.

Flahault, *in* Bornet and Flahault, Rev. I, 1886, p. 361. *Calothrix crustacea* f. *prolifera* Collins, *in* Collins, Holden and Setchell, Phyc. Bor.-Amer. (Exsicc.), no. 1168.

This is one of the near relatives of *Calothrix crustacea* which helps to make the whole *crustacea* group of species (or varieties?) puzzling. At times this species is well developed and seemingly distinct, but the specimens referred to it here are too close to *C. crustacea* to be at all satisfactorily placed under a separate species. The same, however, may be said of specimens often placed under both *C. fasciculata* and *C. vivipara*. It has seemed best to continue the Bornet and Flahault arrangement for the present, or until culture experiments can be carried out to determine the polymorphism that may exist.

11. Calothrix vivipara Harv.

Filaments decumbent and intricate at the base, becoming erect and parallel, 3–5 mm. high, 12–24μ diam., somewhat flexuous, false branching numerous and usually in pairs as in *Scytonema,* forming broadly expanded, velvety coating on rocks; sheaths fairly thick, somewhat mucous, homogeneous, hyaline, or becoming yellowish brown; trichomes olive green, 9–15μ diam., gradually attenuate from the base upward, prolonged into a hair; cells quadrate or shorter than the diameter; heterocysts basal or intercalary, 2–4 seriate.

Growing in rock depressions of the breakwater, San Pedro, California. This is the only known locality on the Pacific Coast of North America.

Harvey, Ner. Bor.-Amer. III, 1858, p. 106; Bornet and Flahault, Rev. I, 1886, p. 362; Collins, Holden and Setchell, Phyc. Bor.-Amer. (Exsicc.), no. 2105.

The only specimen we have found sufficiently characteristic to refer confidently to this species has been compared with a fragment of the type specimen from Sakonnet Point, Rhode Island. The two agree very well indeed, except that our specimen shows more abundant and seemingly more characteristic branching.

12. Calothrix pilosa Harv.

Plate 8, fig. 4

Filaments forming a widely expanded, caespitose, pilose, dark blue-green stratum, decumbent, and densely interwoven at the base, both ends erect above, elongate, rigid, 2–10 mm. high, 10–40μ thick; sheaths firm, thick, golden, at times yellowish brown, subopaque, homogeneous; trichomes 10–20μ diam., olive brown to pale blue-green, briefly attenuated above; heterocysts basal and intercalary.

Forming a dense velvety coating on rocks in tide pools along high-tide level, or often above. Point Carmel, in the vicinity of Carmel Bay, California.

Harvey, Ner. Bor.-Amer. III, 1858, p. 106, pl. 48c; Bornet and Flahault, Rev. I, 1886, p. 363; Collins, Holden and Setchell, Phyc. Bor.-Amer. (Exsicc.), no. 859.

The plants referred here were all found in a single locality, viz., Point Carmel in Monterey County, California, but were collected in different years and months and independently by each of us. They form velvety linings to minute pools high up in the littoral region where the water is much warmed and concentrated by the sun. It is usual to find salt crystals in these pools, to such an extent has the evaporation proceeded. Harvey describes and figures the branches as single and after the fashion of *C. prolifera,* and his figures agree in all respects with a fragment of the type of *C. pilosa.* Certain of our specimens show the same method. Others with decumbent base show branching similar to *C. fasciculata,* and still others with the same sort of base show the branching of *C. vivipara.*

26. **Dichothrix** Zanard

Filaments aggregated into caespitose or cushion-shaped masses, subdichotomously and falsely branched; trichomes 1–6, rarely more, within a sheath, tapering above, often ending in a long, hyaline hair; heterocyst basal or intercalary, rarely absent, single or seriate; plants marine or fresh water.

Zanardini, Plant. Mar. Rubro., 1858, p. 297 (p. 89, Repr.).

The type of the genus is *Dichothrix penicillata* from the Red Sea, but the genus now contains about a dozen species distributed through both fresh and salt waters, one of which is devoid of heterocysts. It has been our experience to find two seemingly good and endemic species on our coast.

Key to the Species.

1. Heterocysts 2–6 seriate, trichomes 9–11μ thick......................1. **D. seriata** (p 103)
1. Heterocysts single, trichomes 4–5μ thick (at base).............2. **D. minima** (p 104)

1. **Dichothrix seriata** S. and G.

Plate 6, fig. 20

Filaments forming a caespitose stratum on rocks, 1–1.5 mm. high, 25–35μ diam., erect; repeatedly more or less fasciculately pseudobranched, ultimate branches mostly strict, acuminate; sheath homogeneous, ample, hyaline below, yellowish brown above, strict, not ocreate or only slightly so; trichomes aeruginous, 9–11μ diam., almost cylindrical, with acuminate apices; branches long, included in the common sheath; cells 2.5–3.5μ long; heterocysts basal, 2–6 seriate, diminishing in size downward, subspherical to disk-shaped.

Growing on rocks interspersed with *Rhodochorton Rothii*, in sheltered localities along high-tide level. Cape Flattery, Washington.

Setchell and Gardner, *in* Gardner, New Pac. Coast Alg. III, 1918*a*, p. 473, pl. 40, fig. 20.

Of all the known forms of *Dichothrix*, *D. seriata* seems most closely related to *D. rupicola* Collins (1901, p. 290), from which it differs in having longer filaments, with greater diameter, which are more strict at the apices, in having seriate, intercalary heterocysts as well as basal, and in the sheaths not being ocreate.

2. **Dichothrix minima** S. and G.

Filaments forming small, dense clusters 80–140μ high; trichomes few, 2–5, and almost wholly included within the common sheath, pale aeruginous, tapering gradually upward from the heterocyst, slightly torulose, ending in a long hyaline hair; basal cells 4–5μ diam., quadrate or slightly longer, homogeneous; hair cells 0.8–1μ diam.; heterocysts basal, single, spherical or subconical, slightly larger than the trichomes; sheath strict, at first hyaline, soon becoming brown and distinct.

Setchell and Gardner, *in* Gardner, New Pac. Coast Alg. III, 1918*a*, p. 474.

Growing in small tufts on *Enteromorpha* sp. along high-tide level and above. Chuckanut Quarry near Bellingham, Washington.

The type material of *Dichothrix minima* was collected in July, 1899. There was a very small quantity and it was placed in the Herbarium of the University of California under no. 100,341, awaiting the accumulation of other material. The locality has not been revisited and other material has not been reported from elsewhere. It is hoped that this description may stimulate further search for it.

27. **Isactis** Thuret

Filaments erect, parallel, densely crowded, and coalescent into a compact layer attached to the substratum, simple or sparsely branched; heterocysts basal; reproduction by spores unknown.

Thuret, Essai Class. Nost., 1875, pp. 376, 382; Bornet and Flahault, Rev. I, 1886, p. 343.

The diagonsis given by Thuret (*loc. cit.*, p. 376) was short, but yet to the point. *Isactis* is a habit genus differing from *Rivularia* in its more simple trichomes which are crowded and parallel. This gives rise to flattened, more or less orbicular layers instead of spherical, hemispherical, or more or less convex and lobed expansions.

Isactis plana (Harv.) Thuret

Fronds scarcely 0.5 mm. thick, spread out indefinitely on the surface of rock, dark green; filaments densely crowded, mostly simple; trichomes 7–9μ diam., light blue-green, tapering into a delicate hair above when young.

Thuret, Essai, 1875, p. 382; Bornet and Thuret, Notes Algol. II, 1880, p. 163, pl. 40; Bornet and Flahault, Rev. II, p. 344. *Rivularia plana* Harvey, *in* Hooker, Brit. Fl., 1833, vol. 2, part 1, p. 394.

The only two species thus far described are *Isactis plana* (Harv.) Thuret and *I. centrifuga* Bornet. Our plants seem to belong to the former and both varieties are represented on our coast.

Isactis plana var. plana B. and F.

Plate 1, figs. 8, 9

Layer not zonate as a rule; filaments unbranched or nearly so, cohering together.

On rocks. Puget Sound, Washington, to southern California.

Bornet and Flahault, Rev. II, 1886, p. 345.

Isactis plana var. fissurata B. and F.

Layer more or less zonate; filaments pseudoramose, tightly cohering.

On rocks. Unalaska, Alaska.

Bornet and Flahault, Rev. II, 1886, p. 345; Setchell and Gardner, Alg. N.W. Amer., 1903, p. 198.

28. Rivularia Ag.

Thallus hemispherical, globose or irregularly lobed, at times cavernous, or confluent into a solid expanded stratum; filaments radiating from the center, or from the base, repeatedly false branched, heterocysts basal, some species producing cylindrical spores contiguous to the heterocyst.

C. Agardh, Disp. Alg. Suec., 1812, p. 43, Syn. Alg. Scand., 1817, pp. xxxviii, 130, 131, Syst. Alg., 1824, p. xix; Roth, Neue Beitr. z. Bot., 1803, p. 261 (pro parte), Cat. Bot., vol. 3, 1806, p. 332 (proparte) (not of Roth, Cat. Bot., vol. 1, 1797, p. 212; or Tent. Flor. Germ., vol. 3, 1800, p. 543); Bornet and Flahault, Rev. Nost. Het. I, 1886, p. 345.

The status of the name *Rivularia* was made plain and current by the action of the Third International Botanical Congress at Brussels in 1910, when it fixed the starting point of the nomenclature of the heterocysted Nostocaceae with Bornet and Flahault's ''Nostocacées Hétérocystées.'' The history of the name, however, is both interesting and confusing. The first genus *Rivularia* was established by Roth in 1797 (p. 212). The two species assigned to it are both of the earlier genus *Chaetophora* Schrank (1783, p. 124). In 1800 (pp. 544–546), Roth listed three species, all of the present *Chaetophora*. In 1803 (pp. 261–285), however, Roth increased the genus as to the number of species, adding some true *Rivularias,* and in 1806 (pp. 332–341) he still further added to it additional *Rivularia* species as now reckoned. In 1812 (p. 42) C. Agardh separated the species of *Chaetophora* from the others and also refounded *Rivularia* (*loc. cit.,* p. 43) in very nearly the sense in which we now use it. He repeated his account later in 1817 (p. xxxviii) and even again in 1824 (p. 24). The genus, then, had been refounded and well established even before the *Portacus* of Rafinesque (1819, p. 107) had been proposed.

KEY TO THE SPECIES.

1. Filaments separating from one another readily under pressure...........................
....1. **R. Biasolettiana** (p 106)
1. Filaments separating from one another with difficulty under pressure................... 2
   2. Fronds spherical or nearly so, at times more or less confluent into a layer.... 3
   2. Fronds globose to plicate-corrugate, expanded, often hollow........................
   ....3. **R. nitida** (p 108)
3. Fronds spherical or irregular, 0.5–1.0 mm. in diam., often gregarious and partially
   confluent into a pulverulent layer.................. .... 4. **R. mamillata** (p 109)
3. Fronds spherical or hemispherical, 3–5 mm. in diam., confluent (in var. **confluens**)
   into a flat, smooth layer...2. **R. atra** (p 107)

1. Rivularia Biasolettiana Menegh.

Thallus gelatinous, at first hemispherical, later becoming diffluent and broadly expanded, warty and cushion-shaped, often becoming encrusted with lime, dark olive green, becoming yellowish or brownish on the surface; filaments densely crowded, 18μ diam.; sheaths ample, mucous, lamellate, funnelform, hyaline or yellowish, or alternating hyaline and yellowish giving the thallus a zonate appearance; trichomes 5–9μ diam., prolonged into a delicate, flexuous hair; cells in basal portion slightly shorter than the diameter, one-third the diameter above, heterocyst oblong, basal, 1–3 seriate, or rarely intercalary.

Growing on dripping rocks in both fresh and salt water along high-tide line and above. Ranging from Unalaska, Alaska, to southern California.

Meneghini, *in* Zanardini, Syn. Alg. Mar. Adr., 1841, p. 42; Setchell and Gardner, Alg. N.W. Amer., 1903, p. 198; Saunders, Alg. Harriman Exped., 1901, p. 399; Bornet and Flahault, Rev. II, 1886, p. 352; Collins, Holden and Setchell, Phyc. Bor.-Amer. (Exsicc.), no. 358; Tilden, Amer. Alg. (Exsicc.), no. 570. *Rivularia nitida* Tilden, Amer. Alg. (Exsicc.), no. 571 (not of Agardh). *Rivularia atra* var. *coadunata* (Sommerf.) Bornet, Les Nost. Het. Syst. Alg., 1889, p. 6. *Linckia atra* var. *coadunata* Sommerfelt, Suppl. Fl. Lapp., 1826, p. 201. *Rivularia coadunata* Foslie, Contrib. Knowl. Mar. Alg. Norway II, 1891, p. 56 (p. 21, Repr.); Forti, *in* De-Toni, Syll. Alg., vol. 5, 1907, p. 667.

This species is not uncommon and is more often growing in strictly fresh water than in salt water. It grows on rocks and on hard soil, among mosses, etc., where fresh water from springs or streams trickles over it, but frequently the layers extend down into the salt water where it seems to thrive.

It seems best to retain the name of *Rivularia Biasolettiana* as chosen by Bornet and Flahault (*loc. cit.*) as against *R. coadunata* (Sommerfelt) Foslie, and an additional reason exists for this choice, since the name *coadunata,* as used by Sommerfelt, was simply varietal and was raised to specific rank subsequent to the proposal of *Biasolettiana* as specific (cf. Regl. Intern. de la Nom. Bot., Vienna, 1906, Art. 49).

2. **Rivularia atra** Roth

Thallus spherical, solitary or confluent, blackish green, up to 4 mm. diam.; filaments thickly compacted, not separating readily under pressure; sheaths thin, scarcely distinct, ampliate above, hyaline or yellow; trichomes 2.5–5μ thick, verdigris green, produced above into a slender hair; lower cells scarcely longer than broad, upper shorter.

On various solid living and non-living substrata. California. Dr. C. L. Anderson (fide. Bornet and Flahault, *loc. cit.,* p. 355).

Roth, Cat. Bot., vol. 3, 1806, p. 340; Bornet and Flahault, Rev. II, 1886, p. 353.

This species is given by Bornet and Flahault as occurring on the coast of California. The specimens we have seen belong to one or another of the two following varieties.

Rivularia atra var. **hemisphaerica** (Kuetz.) B. and F.

Plate 8, figs. 1, 2

Thallus hemispherical, firm, solitary, or occasionally slightly confluent by close proximity to others, of very dark green color, black when dry, 3–5 mm. diam.; filaments crowded; sheaths narrow, firm, inconspicuous below, spreading and slightly colored above; trichomes 2.5–5μ diam., blue-green, prolonged into a long, delicate, hyaline thread above; cells of equal diameter below, shorter above.

Growing on rocks in the upper littoral belt. Ranging from central to southern California.

Bornet and Flahault, Rev. II, 1886, p. 355; Collins, Holden and Setchell, Phyc. Bor.-Amer. (Exsicc.), no. 1561. *Rivularia hemisphaerica* Kuetzing, Actien, 1836 (cf. note under Bibliography).

Rivularia atra var. confluens (Kuetz.) Bornet

Thallus a flat confluent mass, 2–5 cm. across, adhering firmly to rock, deep blue-green, glossy; filaments closely adhering; sheaths hyaline when young, colored in zones with age; trichomes 5–7μ diam., 170μ long; heterocysts 7–10μ diam., nearly spherical, very numerous.

Growing on dripping rocks at high-tide level and above. Carmel Bay, Monterey County, California.

Bornet, Alg. de Schousb., 1892, p. 189 (p. 29, Repr.); Collins, Holden and Setchell, Phyc. Bor.-Amer. (Exsicc.), no. 2107. *Euactis confluens* Kuetzing, Sp. Alg., 1849a, p. 341, Tab. phyc., vol. 2, 1850–52, p. 24, pl. 77, fig. 1.

3. Rivularia nitida Ag.

Thallus variable in shape, spherical, and sometimes hollow, or expanded and plicate corrugate; up to 3 cm. diam.; olive green; filaments crowded; sheath narrow, close fitting, indistinct below, slightly expanding upward, hyaline or yellowish brown; trichomes 2–5μ diam., olive green, prolonged into a very narrow long hair; cells 3–4 times longer than the diameter below, but the upper cells shorter than the lower.

Growing on mud near high-water mark. St. Michael, Alaska.

Agardh, Dispositio Alg. Suec., 1812, p. 44; Bornet and Flahault, Rev. II, 1886, p. 357; Setchell and Gardner, Alg. N.W. Amer., 1903, p. 198.

The specimen from Alaska credited by us to this species seems to agree in microscopic details with the descriptions and in habit very well with figure 1, on plate 2518 of the Flora Danica (1849).

4. Rivularia mamillata S. and G.

Plate 6, fig. 19

Thalli 0.5–1 mm. diam., bright blue-green or brownish; spherical or irregular in outline, more or less confluent, forming congested, pulverulent layers, or scattered among other Myxophyceae; filaments repeatedly false branched, spreading widely above, separating only under considerable pressure; sheath hyaline below, yellowish brown above, distinct below but very thin above, ocreate, trichomes tapering gradually from the base upward, terminating in a long hyaline hair, 4–5μ diam. at base; cells 2–5μ long, slightly torulose, blue-green, decidedly granular; heterocysts spherical to bluntly conical, 5.5–8μ diam.

Growing on decaying logs along high-tide level in somewhat shaded localities. Cape Flattery, Washington.

Setchell and Gardner, in Gardner, New Pac. Coast Alg. III, 1918a, p. 475, pl. 40, fig. 19.

This species closely resembles R. nitida Ag. but has more distinct, ocreate sheaths, larger trichomes with shorter cells, and a less ample thallus.

FAMILY 6. STIGONEMATACEAE KIRCHNER

Filaments free, or forming a pannose or pulvinate stratum, or united into definite, more or less gelatinous colonies, composed of a single series of cells or of two or more series within a sheath, resulting from cell divisions in two or more planes; branches connately joined or in some genera both true and false branches; cells cylindrical, subspherical, or irregularly angular; heterocysts intercalary or terminal on short lateral branches; multiplication by hormogonia and by resting spores.

Kirchner, in Engler and Prantl, Natürl. Pflanzenfam. I, Th. 1, 1a Abt., 1898, p. 80. Stigonemeae Hassall, Brit. F. W. Alg., vol. 1, 1845 (also editions of 1852 and 1857), p. 227. Sirosiphoniaceae Rabenhorst, Fl. Eur. Alg., vol. 2, 1865, p. 2; Bornet and Flahault, Rev. III, 1887, p. 51. Stigonemaceae Forti, in De-Toni, Syll. Alg., vol. 5, 1907, p. 562.

Bornet and Flahault (1887, p. 51) use the name Sirosiphoniaceae of Rabenhorst (1865, p. 2), but Sirosiphon has been dropped from the list of valid genera. It seems necessary to reject the name (cf. Art. 52, Regl. Intern de la Nom. Bot., 1906) and to adopt Hassall's name of 1845 (p. 227) giving it the form adopted by Forti. This form

"Stigonemataceae" is orthographically correct and is more according to custom than "Stigonemaceae" adopted by Forti 1907, p. 562.

We have increased the scope of the family by removing from the Rivulariaceae the genus *Brachytrichia* and placing it here. Although the species of this genus have trichomes with terminal hairs, yet so do the trichomes of *Mastigocoleus,* which is assigned to this family by Bornet and Flahault (*loc. cit.,* pp. 53, 54). The branching in *Mastigocoleus* is mainly true branching but the piliferous branches appear to be false. From our point of view the branching in *Brachytrichia* is always true branching and we place it in Stigonemataceae near *Mastigocoleus* and *Nostochopsis.*

The majority of the Stigonemataceae are terrestrial or inhabitants of the fresh waters. Only two genera are made up of marine species.

KEY TO THE GENERA.

1. Sheaths not diffluent, shell or lime rock borers............29. **Mastigocoleus** (p 110)
1. Sheaths diffluent into a definite gelatinous thallus, not boring................................
....30. **Brachytrichia** (p 111)

29. **Mastigocoleus** Lagerh.

Filaments associated into tangled masses, branching freely and boring into the shells of various mollusks or into lime rock, mostly single within a sheath and of uniform diameter, or at times two or more within the same sheath and attenuated into long hyaline hairs, generally extending beyond the surface; branching mostly true, but double piliferous branches false; heterocysts single, terminal, or on short lateral branches; hormogonia formed in the piliferous branches.

Lagerheim, Note sur le Mastig., 1886, p. 65.

The single species thus far described for the genus *Mastigocoleus* inhabits dead shells of bivalves, boring its way into the calcareous material or sometimes boring into calcareous cliffs. Two sorts of branches occur, some bearing hairs and some devoid of hairs. The former are hormogonial and are often formed two in the same sheath and are false, after the fashion of the branching in *Scytonema.* The hair bearing, or piliferous branches reach to the surface of the shell. At the bases the true branches arise from a V-shaped structure similar to that of *Brachytrichia.*

The multiplication in *Mastigocoleus* is, according to Bornet and Flahault (1889, p. 17) entirely by hormogonia, the coccogonioid multiplication attributed to it by Lagerheim being founded on certain states of the associated species of *Hyella.*

Mastigocoleus testarum Lagerh.

Plants forming thin membranaceous layers at first, soon boring into the shells of mollusks, pale blue-green; filaments variously contorted; 6–10μ thick; sheaths very delicate, hyaline; trichomes 3.5–6μ thick, cells cylindrical or subcylindrical; heterocysts elliptical or spherical, 6–18μ diam.

Growing in oyster shells. San Francisco Bay, California.

Lagerheim, Note sur le Mastig., 1886, p. 65, pl. 1; Bornet and Fla-hault, Rev. III, 1887, p. 54; Setchell, Notes on Cyanophyceae III, 1899, p. 47.

Mastigocoleus testarum is abundant in oyster shells on the New England coast. These mollusks have been introduced into San Francisco Bay and plants growing on these are the only reported ones in our region. They are thus probably not indigenous. Something similar has also been found in the thin shell of some bivalves at San Pedro, California. The material, however, is scanty and fragmentary. Further search and comparison is necessary to establish the true status of the genus as regards our coast.

30. Brachytrichia Zanard.

Colonies more or less spherical, composed of a mass of filaments embedded in a firm, gelatinous matrix, solid when young, later becoming more or less inflated, somewhat cavernous, and usually plicate and rugose; filaments intertwined below, giving rise to numerous erect, true branches at the ends of loops, becoming parallel and attenuated toward the surface of the colony; heterocysts intercalary; spores unknown.

Zanardini, Phyc. Ind. Pug., 1872, p. 24.

The type species is *Brachytrichia rivularioides* Zanard. (1872, p. 24), which is identified with the *Nostoc Quoyi* Ag. (1824, p. 22), the type of which was collected at the Mariana Islands by Gaudichaud. Zanardini's type came from Sarawak, and this species seems more or less common through the Indo-Malayan region. The *Brachytrichia Quoyi* is large, firmly gelatinous, bullate, and dark green, while certain other species are flattened and brownish.

The branching of filaments in *Brachytrichia* is unique. The filaments at first are more or less horizontal. At certain points in them lateral loops are formed, sometimes only a few cells long, at other times many cells long (pl. 7, fig. 27). One cell at the end of the loop divides lengthwise, cutting off a cell which becomes the basal cell of a branch which develops toward the surface, giving the appearance of dichotomous branching, although virtually a single filament gives rise to a single lateral branch at certain intervals. Occasionally after the cortical portion of the thallus becomes dense, branches develop without the formation of loops (pl. 7, fig. 28). This is in reality true branching, for the erect filament is connately joined to the parent filament. Soon a heterocyst is formed near the base of the erect branch and eventually the upper part develops into a *Calothrix*-like filament tapering to a very delicate, hair-like point. Other intercalary heterocysts may appear. These masses of erect parallel filaments form the cortical portion of the thallus.

The branching in *Brachytrichia* is similar to that of *Mastigocoleus*, *Nostochopsis* and *Herpyxonema*. The trichomes in the last mentioned genus have branches (cf. Weber-van Bosse, 1913, pp. 36, 37) with V-shaped bases, and seemingly must arise exactly as do those in *Brachytrichia*. We feel certain that these are true branches and that the genus is closely related to *Brachytrichia*, differing in lack of hairs and in not forming a gelatinous thallus.

Brachytrichia affinis S. and G.

Plate 7, figs. 27, 28

Thallus small, 0.5–2 cm. diam., deeply plicate, bullate, cavernous, cartilaginous, light blue-green, or brownish with age; filaments intricate, loosely intertwined below, very densely crowded and parallel above, and tapering gradually into delicate hairs; cells of the lower interior filaments slightly ventricose, 4–5μ diam., 1.5–2 times as long as broad; cells of the erect filaments spherical to doliiform toward the base, 7–9μ diam., terminal cells 1μ diam.; heterocysts spherical or slightly compressed.

Growing in rock pools in the middle littoral belt. Laguna Beach, Orange County, California.

Setchell and Gardner, *in* Gardner, New Pac. Coast Alg. III, 1918*a*, p. 475, pl. 41, figs. 27, 28. *Brachytrichia Quoyi* Guernsey, Notes on Mar. Alg., 1912, p. 195; Collins, Holden and Setchell, Phyc. Bor.-Amer. (Exsicc.), no. 2106 type (not of Bornet and Flahault).

Brachytrichia Quoyi was accredited to our coast by Bornet and Flahault (Rev. II, 1886, p. 373) on the authority of Grunow, but remained unknown to workers on this coast until recently. It has not been possible to compare our plant with the type of Agardh's *Nostoc Quoyi* from the Mariana Islands, nor with Zanardini's *Brachytrichia rivularioides* from Sarawak, Borneo, but comparison has been made with *B. Quoyi* from Woods Hole, Massachusetts, which, as has been suggested by Collins (1890, p. 176), probably was introduced some years ago from some of the southern islands through the instrumentality of guano ships or other human agencies. It has also been compared with material of what seems even more likely to be the same species from the Philippine Islands.

Our plant is much smaller than those of *B. Quoyi* from Woods Hole and from the Philippines, the specimens from Woods Hole sometimes measuring up to 7 cm. in diameter. The largest specimen yet seen of *B. affinis* measures about 2 cm. in diameter, the majority being less than 1 cm. It is also much more firm and cartilaginous, more profusely lobed and saccate. The cells are in general slightly larger and the filaments more profusely branched. It may also be said that the Massachusetts plants differ somewhat from those of the Philippines.

LITERATURE CITED

AGARDH, C. A.
  1812. Dispositio algarum Sueciae, D. IV. Lund.
  1817. Synopsis algarum Scandinaviae. Lund.
  1817–1825. Aphorismi botanici. Decades I–XVI. Lund.
  1821. Decas VII.
  1820–1828. Species algarum rite cognitae cum synonymis, differentiis specificis et descriptionibus succinctis. Lund.
    1820. Vol. 1, part 1. (The Lund edition has 1820, while the ''Gryphiswald'' edition has 1821.)
    1822. Vol. 1, part 2.
    1828. Vol. 2, part 1.
  1824. Systema algarum. Lund.
  1827. Aufzählung einiger in den östreichischen Landen gefundenen neuen Gattungen und Arten von Algen, nebst ihrer Diagnostik und beigefugten Bemerkungen. Flora, vol. 10, p. 625.
AGARDH, J. G.
  1842. Algae maris Mediterranei et Adriatici. Paris.
ARDISSONE, F., and STRAFFORELLO, I.
  1877. Enumerazione delle alghe di Liguria. Milano.
ATKINSON, F. A.
  1905. A college text-book of botany, ed. 2. Henry Holt & Co., New York.
BÖRGESEN, F.
  1902. The marine algae of the Fäeröes. Botany of the Fäeröes, II, p. 339, Copenhagen.

Bornet, E.

1889. Les Nostocacées Hétérocystées du systema algarum de C. A. Agardh (1824) et leur synonymie actuelle (1889). Bull. Soc. Bot. de France, vol. 36, p. 144, Paris.

1892. Les algues de P.-K.-A. Schousboe, récoltées au Maroc et dans la Mediterranée de 1815 à 1829. Mém. Soc. Nat. Sci. Nat. et Math., vol. 28, p. 165.

Bornet, E., and Flahault, C.

1885. Tableau synoptique des Nostocacées filamenteuses Hétérocystées. Mém. Soc. Nat. Sci. Nat. et Math. de Cherbourg, vol. 25, p. 195.

1886–1888. Revision des Nostocacées Hétérocystées. Ann. des Sci. Nat., 7 sér., Bot.

    1886. Part   I, vol. 3, pp. 323–380.

    1886. Part II, vol. 4, pp. 343–373.

    1887. Part III, vol. 5, pp. 51–129.

    1888. Part IV, vol. 7, pp. 177–262.

1888. Note sur deux nouveaux genres d'algues perforantes. Jour. de Bot., vol. 2, p. 161.

1889. Sur quelques plantes vivant dans le teste calcaire des mollusques. Bull. Soc. Bot. de France, vol. 36, p. cxlvii, pls. 6–12.

Bornet, E., and Thuret, G.

1876–1880. Notes algologiques; recueil d'observation sur les algues. Paris.

    1876. Fascicle I.

    1880. Fascicle II.

Bory de Saint Vincent, J. B. (M. A. G.).

1822. *Anabaena. In* Dictionnaire classique d'histoire naturelle, vol. 1, p. 307.

Borzi, A.

1882. Note alla morfologia e biologia delle Alghe ficocromacee. Nuovo Giorn. Bot. Ital., vol. 14, p. 272, pls. 16, 17.

Brand, F.

1901. Bemerkungen über Grenzzellen und über spontan rothe Inhaltskörper der Cyanophyceen. Ber. deut. bot. Ges., vol. 19, p. 152.

Brébisson, A. de.

1839. De quelques nouveau genres d'algues (pp. 5, Planche, Mai, 1839, Falaise, Imp. Levasseur).

1841. *Agmenellum* (p. 187), *Anacystis* (p. 417) ... *In* D'Orbigny's Dictionnaire universel d'histoire naturelle, vol. 1.

    We have quoted 1841 as the date of this volume although the titlepage of this and of all the other volumes of the set bear the date 1849. Pritzel (1872, p. 361) gives as the date of the set, 1842–1849. Brunet's dates are 1841–1849, but in neither case is any date assigned to any particular volume.

Briquet, J.

1906. International rules of botanical nomenclature adopted by the International Congress of Vienna, 1905.

1912. *Ibid.*, and Brussels, 1910.

Brunet, J. C.

1860–1865. Manuel du Libraire et de l'amateur de livres. Vols. 1–6.

CHAUVIN, J. F.
  1842. Recherches sur l'organisation la fructification, et la classification de plusiers genres d'algues avec la description de quelques especes inédites. Caen.

CLEMENTE, S. DE ROXAS.
  1807. Essayo sobre las variedades de la vid comun que vegetan en Andalucia, con un indice etimológico y tres listas de plantes con two laminas. Madrid.

COHN, F.
  1879. Ueber sein 1871 aufgestelltete Thallophytensystem. Jahresber. Schles. Ges. f. vaterl. Cultur, 1879, p. 279.

COLLINS, F. S.
  1890. *Brachytrichia Quoyi* (Ag.) Bornet and Flahault. Bull. Torr. Bot. Club, vol. 17, p. 175.
  1901. Notes on algae. IV. Rhodora, vol. 3, pp. 289–293.
  1909. Green algae of North America. Tufts College studies, vol. 2, no. 3, Scientific series.

COLLINS, F. S., and HERVEY, A. B.
  1917. The algae of Bermuda. Proc. Amer. Acad. Arts and Sci., vol. 53, no. 1.

COLLINS, F. S., HOLDEN, I., and SETCHELL, W. A.
  1895–1917. Phycotheca Boreali-Americana. Fascicles I–XLV, A–E, Malden, Mass. (Exsicc.).

CROUAN, P. L., and H. M.
  1858. Note sur quelques algues marines nouvelle de la rade de Brest. Ann. des Sci. Nat., 4 sér., Bot., vol. 9, p. 69, pl. 3.
  1860. Liste des algues marines découvertes dans le Finistère, depuis la publication des algues de ce department en 1852. Bull. Soc. Bot. de France, vol. 7, pp. 367–373.
  1867. Florule du Finistère, contenant les descriptions de 360 èspeces nouvelles de sporogames de nombreuses observationes et une synonymie des plantes cellulaires et vasculares qui croissent spontanément dans ce département. Paris.

DESMAZIÈRES, J. B. H. J.
  1823. Catalogue des plantes omises dans la botanographie belgique, et dans les Floures du nord de la France. Lille.
  1825. Plantes cryptogames du nord de la France II. Lille. (Exsicc.)

DE-TONI, J. B. (or G. B.).
  1907. Sylloge algarum omnium hucusque cognitarium. Vol. 5. Patavii.

DILLENIUS, J. J.
  1741. Historia muscorum, in qua circiter sex centae species veteres et novae ad sua genera relatae describunter, et iconibus genuinis illustrantur; cum appendice et indice synonymorum. Oxonii.

EHRENBERG, C. G.
  1838. Die Infusionsthierchen als vollkommene Organismen. Leipzig.

ENDLICHER, S.
  1843. Mantissa botanica altera sistens generum plantarum. Supp. III.

ENGLISH BOTANY, or coloured figures of British plants with their essential char-
  1790–1814. acters, synonymes, and places of growth, to which will be added occasional remarks by James Edward Smith. The figures, by James Sowerby. Vols. 1–36, London.

FARLOW, W. G.
    1881. Marine algae of New England and adjacent coast. Rep. U. S. Fish
        Comm., 1879, Washington.

FORTI, A.
    1907. Sylloge Myxophycearum. *In* De-Toni, Sylloge algarum, vol. 5.

FOSLIE, M.
    1891. Contribution to the knowledge of the marine algae of Norway II.
        Tromsö Mus. Aarshefter, vol. 14, p. 36, Tromsö.

GARDNER, N. L.
    1906. Cytological studies in Cyanophyceae. Univ. Calif. Publ. Bot., vol. 2,
        pp. 237–296, pls. 21–26.
    1917. New Pacific coast marine algae I. *Ibid.*, vol. 6, pp. 377–416, pls. 31–35.
    1918. New Pacific coast marine algae II. *Ibid.*, pp. 429–454, pls. 36–37.
    1818a. New Pacific coast marine algae III. *Ibid.*, pp. 455–486, pls. 38–41.
    1919. New Pacific coast marine algae IV. *Ibid.*, pp. 487–496, pl. 42.

GOMONT, M.
    1890. Essai de classification des Nostocacées Homocystées. Jour. de Bot.,
        vol. 4, pp. 349–357.
    1891. Faut-il dire *Oscillatoria* ou *Oscillaria*. *Ibid.*, vol. 5, p. 273.
    1892–1893. Monographie des Oscillariées. (Nostocacées Homocystées.)
        1892. Ann. des Sci. Nat., 7 sér., Bot., vol. 15, part 1, pp. 263–368, pls.
            6–14; vol. 16, part 2, pp. 91–264, pls. 1–7.
        1893. Reprint, pp. 1–302, plates and figures same as original.
    1899. Sur quelques Oscillariées nouvelles. Bull. Soc. Bot. de France, vol. 46,
        part 1, p. 25.
    1901. Myxophyceae hormogoneae. *In* Schmidt, Flora Koh Chang IV, Botanisk
        Tidsskrift, vol. 24, part 2, pp. 202–211, pl. 5.

GRAY, J. E.   (*See* S. F. Gray.)

GRAY, S. F.   (*See also* J. E. Gray.)
    1821. A natural arrangement of British plants. Vol. 1, London.
        J. E. Gray, son of S. F. Gray, did the systematic work in the two
        volumes which came out under the above title, with S. F. Gray as the
        author. J. E. Gray is sometimes cited as the author.

GUERNSEY, J.
    1912. Notes on the marine algae of Laguna Beach. 1st Ann. Rep., Laguna
        Mar. Lab., Pomona College, California, pp. 195–218.

HANSGIRG, A.
    1884. Bemerkungen zur Systematik einiger Süsswasseralgen. Oester. bot.
        Zeitschr., vol. 34, pp. 313, 318, 351, 358, 389, 394.
    1887. Physiologische und algologische Studien. Prag.
    1889. Beiträge zur Kenntniss der quarnerischen und dalmatinischen Meeres-
        algen. Oester. bot. Zeitschr., vol. 39, p. 4.
    1890. Physiologische und algologische Mittheilungen. S.–B. d. k. böhm. Ges.
        Wiss. II, pp. 83–140, pl. 3. Cf. Just's Bot. Jahresber., 1890, p. 248.
    1892. Neue Beiträge sur Kenntniss der Meeresalgen-und Bacterien-flora der
        österreichisch-ungarischen Küstenländer. *Ibid.*, pp. 212–249.
    1892a. Prodromus der Algenflora von Böhmen II, Die blaugrünen Algen
        (Myxophyceen, Cyanophyceen) nebst Nachträgen zum erstem Theile
        und einer systematischen Bearbeitung der in Böhmen verbreiteten
        saprophytischen Bakterien und Euglenen. Arch. d. naturw. Landes-
        durchf. v. Böhmen, vol. 8, no. 4. The date 1892 appears on the title
        page, but 1893 appears on the cover.)

HARVEY, W. H.
1846–1851. Phycologia Britannica. London.

According to a memorandum from E. M. Holmes to F. S. Collins, this work was issued in parts of six plates each, part 1 having been issued January, 1846, then monthly parts to part 42 issued June 1, 1849. After that the issues were irregular, the last part (60) having been issued August, 1851. The issues may be summarized, so far as we have the information, as follows:

1846, plates 1–72.
1847, plates 73–144.
1848, plates 145–216.
1849, plates 217–258.
1849–1851, plates 259–354.
1851, plates 355–360.

The title pages of the volumes are dated as follows:
1846, vol. 1 (plates 1–120).
1849, vol. 2 (plates 121–240).
1851, vol. 3 (plates 241–360).

1852–1858. Nereis Boreali-Americana. Washington, D. C.
1852. Part I, Melanospermeae.
1853. Part II, Rhodospermeae.
1858. Part III, Chlorospermeae.

HASSALL, A. H.
1845. A history of the British fresh water algae including descriptions of the Desmideae and Diatomaceae with upwards of one hundred plates illustrating the various species. Vols. 1, 2, London.
1845. Ed. 1.
1852. Ed. 2.
1857. Ed. 3.

HAUCK, F.
1885. Die Meeresalgen Deutschlands und Oesterreichs. *In* Rabenhorst's Kryptogamen-flora von Deutschland, Oesterreich und der Schweiz. Vol. 2, part 10, Schizophyceae. (Vol. 2 was issued in parts between 1833 [1 Lief.] to 1885 [10 Lief.].

HOLMES, E. M., and BATTERS, E. A. L.
1890. A revised list of the British marine algae. Annals of Botany, vol. 5, pp. 63–107.

HOOKER, W. J.
1833. The English flora of Sir James Edward Smith, class XXIV, Cryptogamia, vol. 5 (vol. 2 of Dr. Hooker's British flora), part 1, comprising the Mosses, Hepaticae, Lichens, Characea and Algae. London.

HORNEMANN, J. W.
1818. Icones plantarum sponte nascentium in regno Daniae, et in ducatibus Slesvici, Holsatiae et Lauenburgiae ad illustrandum opus de iisdem plantis, regio jussu exarandum, Florae Danicae nomine inscriptum. Vol. 9, fasc. 25–27, pls. 1441–1620.
Work usually cited as Flora Danica.
1813. Fasc. 25.
1816. Fasc. 26.
1818. Fasc. 27.

HOWE, M. A.
1914. The marine algae of Peru. Mem. Torr. Bot. Club, vol. 15, pp. 1–185, pls. 1–66.

JANCZEWSKI, E. DE.
1883. *Godlewskia*, nouveau genre d'algues de l'ordre des Cryptophycées. Ann. des Sci. Nat., 6 sér. Bot., vol. 16, pp. 227–230, pl. 14.

JÓNSSON, H.
1901. The marine algae of Iceland. I, Rhodophyceae. Botan. Tidsskrift, vol. 24, part 2, p. 127, November.
1903. II, Phaeophyceae. *Ibid.*, vol. 25, part 2, p. 141.
1903. III, Chlorophyceae; IV, Cyanophyceae. *Ibid.*, vol. 25, part 3, p. 337.

JÜRGENS, G. H. B.
1816–1822. Algae aquaticae quas in littore maris dynastiam Javaranam et Friseam orientalens alluentis rejectas et in harum terrarum aquis habitantes collegit ... Decades I–XX, Jever. (Exsicc.)

KIRCHNER, O.
1898. Schizophyceae. *In* Engler and Prantl, Natürl. Pflanzenfam. I Th. I Abt., 177. Lief.

KUETZING, F. T.
1833–1836. Algarum aquae dulcis germanicarum. Decades I–XVI, Halis saxon. (Exsicc.)
1833a. Ueber ein neue Gattung der Confervaceen. Flora, vol. 1, 1833, p. 517.
1836. Actien
    Actien refers to a collection of algae made by Kuetzing in Dalmatia, upper Italy and Switzerland apparently the first reference to which was made in Flora, 1836, vol. 19, part 1, Intelligenzblatt no. 1, p. 4, and again in the same volume, part 2, Intelligenzblatt no. 1, p. 13. In the latter a list of the species is given with a star to indicate the new species proposed.
1843. Phycologia generalis, order Anatomie, Physiologie und Systemkunde der Tange. Leipzig.
1845. Phycologia germanica d. i. Deutschlands Algen. Nordhausen.
1849a. Species Algarum. Leipzig.
1846–1871. Tabulae Phycologicae.

1846–49. Vol. 1.	1856. Vol. 6.	1861. Vol. 11.	1866. Vol. 16.
1850–52. Vol. 2.	1857. Vol. 7.	1862. Vol. 12.	1867. Vol. 17.
1853. Vol. 3.	1858. Vol. 8.	1863. Vol. 13.	1868. Vol. 18.
1854. Vol. 4.	1859. Vol. 9.	1864. Vol. 14.	1869. Vol. 19.
1855. Vol. 5.	1860. Vol. 10.	1865. Vol. 15.	1871. Index.

Vol. 1. 1845–1849.
    The title page which was issued in 1849, gives the dates as above. According to the Botanische Zeitung of the dates, 1846, pp. 303, 903, 1847, p. 854, and 1850, p. 566, the various "Lieferungen" were issued as follows:
    Lief. 1, pls. 1–10, pp. 1–8, Jan., 1846.
    Lief. 2, pls. 11–20, pp. 9–16, later in 1846.
    Lief. 3–5, pls. 21–50, pp. 17–36, 1847.
    Lief. 6–10, pls. 51–100, pp. 37–54, with title page and introduction, in 1849.

Vol. 2. 1850–1852.
   Lief. 11–13, pls. 1–30, pp. 1–8, 1850.
   Lief. 14–15, pls. 31–50, pp. 9–16, 1851.
   Lief. 16–20, pls. 51–100, pp. 17–37, with title page and introduction, in 1852.
   From ''Lieferungen'' 16–20 on through the rest of the volume the initial page of each series of ''Lieferungen'' is labelled and dated at the lower left hand corner. F. S. Collins has called our attention to these aids towards the proper dates of issue.

KUNTZE, O.
   1891. Revisio generum plantarum, I. Würzburg.

LAGERHEIM, G.
   1886. Note sur le *Mastigocoleus,* nouveau genre des algues marines de l'ordre des Phycochromacées. Notarisia, vol. 1, pp. 65–96, pl. 1.

LE JOLIS, A.
   1863. Liste des Algues marines de Cherbourg. Mém. Soc. Imp. Sci. Nat. de Cherbourg, vol. 10, p. 1, Paris. Reprinted in 1880.
      The paper was published as a separate in 1863, before it appeared in the Mémoires, 1864.

LEMMERMAN, E.
   1903. Das Plankton schwedischer Gewässer. Arkiv för Botanik., vol. 2, no. 2, pl. 1.

LIEBMAN, F.
   1841. Bemärkninger og tilläg til den danske Algenflora. Kröyers Tidskrift, p. 592, Copenhagen.

LINK, H. F.
   1809. Nova plantarum genera e classe Lichenum, Algarum, Fungorum. *In* Schrader, Neues Jour. f. d. Bot., vol. 3, part 1, p. 1.

MENEGHINI, G.
   1837. Conspectus Algologiae Euganeae, germanicis naturalium rerum scrutatoribus Pragae anno 1837 convenientibus oblatus. Patavii typ. Minervae. Seorsim impr. ex Comment. di medicina del Dr. Spongia, fasc. Sept., 1837. Cf. Pritzel, 1872, p. 212.
   1838. Cenni sulla organografia e fisiologia delle alghe. Estratti del vol. 4 dei nuovi saggi dell I. R. Accademia di scienze littere ed arti di Padova.
   1840. Synopsis desmidiearum hucusque cognitarum.
   1842. Monographia Nostochinearum italicarum addito specimine de Rivulariis. Estratto del ser. 2, vol. 5, Memoire delle R. Accademia della Science di Torino. Cf. Pritzel, 1851, p. 189.
   1844. Algarum species novae vel minus notae a Professor J. Meneghini propositae. Giorn. Bot. Ital. da Filippo Parlatore, vol. 1, p. 296.

MEYEN, J.
   1839. Jahresbericht über die Resultate der Arbeiten im Felde der physiologischen Botanik von dem Jahre, 1838. Wiegmann's Archiv für Naturgeschichte, vol. 2. Berlin.

Naegeli, C.
   1847.   Die neue Algensysteme und Versuch zur Begründung eines eigenen Systems der Algen und Florideen. Zürich.
   1849.   Gattungen einzelliger Algen, physiologisch und systematisch. Denkschriften allg. schweiz. Natur. Ges. vol. 8. Zürich.

Nordstedt, O.
   1911.   Algological notes V–VII. Botaniske notiser, p. 263. Lund.

Persoon, C. H.
   1805.   Synopsis plantarum, seu Enchiridium botanicum, complectens enumerationem systematicam specierum hucusque cognitarum. Vol. 1, Paris.

Pfeiffer, L.
   1873–1875.   Nomenclator botanicus nominum ad finem anni 1858 publici juris factorum, classes, ordines, tribus, familias, divisiones, genera, subgenera vel sectiones designantium enumeratio alphabetica. Adjectis auctoribus, temporibus, locis systematicus apud varios, notis literariis atque etymologicis et synonymis. Casselis.
       1873.   Vol. 1, part 1.
       1875.   Vol. 1, part 2.
       1874.   Vol. 2, part 1.
       1874.   Vol. 2, part 2.

Pritzel, G. A.
   1851.   Thesaurus literaturae botanicae omnium gentium inde a rerum botanicarum initiis ad nostra usque tempora, quindecim millia operum recensens. Leipzig.
       1851.   Ed. 1.
       1872.   Ed. 2.

Rabenhorst, L.
   1864–1868.   Flora Europaea algarum aquae dulcis et submarinae. Leipzig.
       1864.   Vol. 1.
       1865.   Vol. 2.
       1868.   Vol. 3.

Rafinesque, C. S.
   1819.   Prodromus des nouveau genres de plantes observes en 1817 et 1818 dans l'interieur des Etats-Unis d'Amerique. Journ. de Phys., vol. 89, part II, pp. 96–107.

Reichenbach, H. G. L.
   1841.   Der deutsche Botaniker. Erster Band: Das Herbariumbuch.
           Repertorium herbarii, sive nomenclator generum plantarum systematicus, synonymicus et alphabeticus, in usum practicum accommodatus, quo affinitas naturalis et locus cujusque generis in herbario citissime explorentur. Dresdae et Lipsiae, Arnold. Zwei Abteilungen. Cf. Pritzel, 1851, p. 243; 1872, p. 260.

Reinsch, P. F.
   1874, 1875.   Contributiones ad algologiam et fungologiam. Leipzig.

Rosenvinge, L. K.
   1893.   Grönlands Havalger. Meddelelser om Grönland, vol. 3, pp. 765–981.

Roth, A. G. (or A. W.).

1802. Neue Beiträge zur Botanik I. Frankfurt.

1797–1806. Catalecta botanica quibus plantae novae et minus cognitae descri-
buntur atque illustrantur. Leipzig.

       1797. Fasc. 1.

       1800. Fasc. 2.

       1806. Fasc. 3.

1788–1800. Tentamen florae germanicae.

       1788. Vol. 1.

       1789. Vol. 2, part 1.

       1793. Vol. 2, part 2.

       1800. Vol. 3.

Sachs, J.

1874. Lehrbuch der Botanik. Ed. 4.

Saunders, D.

1901. Papers from the Harriman Alaska Expedition. XXV, The Algae.
Proc. Wash. Acad. Sci., vol. 3, pp. 391–486, pls. 43–62, Washington.

1904. Harriman Alaska Expedition with coöperation of Washington Academy
of Sciences. Alaska, vol. 5, Cryptogamic botany. The Algae of the
Expedition, pp. 153–212, pls. 10–29, New York.

This is a reprint of the above paper, 1901, with change in title,
volume, pages and plates, otherwise it is the same.

Sauvageau, C.

1892. Sur les algues d'eau douce récoltées en Algérie pendant la session de la
Société botanique en 1892. Bull. Soc. Bot. de France, vol. 39, p. civ.

1895. Sur le *Radaisia*. Jour. de Bot., vol. 9, pp. 372–376, pl. 7. In Reprint,
see pp. 1–4.

Schmidle, W.

1901. Ueber drei Algengenera. Ber. deut. bot. Ges., vol. 19, p. 10, pl. 1.

Schrank, F. von P.

1783. Botanische Rhapsodien. Der Naturf., part 19, p. 116.

Setchell, W. A.

1899. Notes on Cyanophyceae III. Erythea, vol. 7, no. 5.

1912. Algae novae et minus cognitae I. Univ. Calif. Publ. Bot., vol. 4, pp.
229–263, pls. 25–31.

Setchell, W. A., and Gardner, N. L.

1903. Algae of Northwestern America. Univ. Calif. Publ. Bot., vol. 1, pp.
165–418, pls. 17–27.

Sommerfelt, C.

1826. Supplementum florae lapponicae quam edidit D. G. Wahlenberg. Chris-
tianiae.

Stizenberger, E.

1852. *Spirulina* und *Arthrospira*. Hedwigia, vol. 1, p. 32.

1860. Dr. Ludwig Rabenhorst's Algen Sachsens, resp. Mittel-Europas.
Dekaden I–C. Systematische geordnet mit Zugrundelegung eines
neuen Systems. Dresden.

Thuret, G.

1854. Note sur la synonymie des *Ulva Lactuca* et *Ulva latissima*. Mém. Soc.
Imp. Sci. Nat. de Cherbourg, vol. 2, p. 13.

1875. Essai de classification des Nostochinées. Ann. des Sci. Nat., 6 sér.,
Bot., vol. 1, p. 372.

TILDEN, J. E.
    1894–1909. American Algae. Centuries I–VI, fasc. 1. (Exsicc.) Minneapolis, Minn.
    1910. Minnesota Algae, vol. 1. The Myxophyceae of North America and adjacent regions, etc. ''Report of the Survey, Botanical Series, VIII.''

TREVISAN, V.
    1842. Prospetto della Flora Euganea. Cf. Flora, oder Allgemeine botanische Zeitung, vol. 26, 1843, p. 464 for further information.

TURPIN, P. J. F.
    1827. Spiruline oscillarioïde. *In* Dictionnaire des sciences naturelles (de Levrault), vol. 50, p. 309, fig. 3, 3*a*, 3*b*, 3*c*.

VAUCHER, J. P.
    1803. Histoire des conferves d'eau douce, contenant leurs differens modes de reproduction, et la description de leurs differens modes de reproduction, et la description de leurs principales espèces, suivie de l'histoire des tremelles et des ulves d'eau douce. Geneva.

WALLROTH, C. F. W.
    1833. Flora Cryptogamica Germaniae. Vol. 2, Norimberg.

WEBER, F., and MOHR, E. M. H.
    1804. Naturhistorische Reise durch einen Theil Schweden. Göttingen.

WEBER-VAN BOSSE, A.
    1913. Liste des Algues du Siboga. I, Myxophyceae, Chlorophyceae, Phaeophyceae avec le concours de M. Th. Reinbold. Monographie LIX a de: Uitkomsten op zoologisch, botanisch, oceanographisch en geologisch Gebied verzameld in Nederlandsch Oost-Indië 1899–1900 aan boord H. M. ''Siboga'' onder commando van Luitenant ter Zee, 1ᵉ kl. G. F. Tydeman.

WEST, G. S.
    1916. Algae. I, Myxophyceae, Peridinieae, Bacillarieae, Chlorophyceae. Cambridge botanical handbooks.

WEST, W., and G. S.
    1797. Welwitsch's African fresh-water algae. Jour. of Bot., British and Foreign, vol. 35, pp. 235–243.

WILLE, N.
    1900. Algologische Notizen I–VI. Nyt Magazin for Naturvidensk, vol. 38, part 1.
    1903. Ueber einige von J. Menyhardt in Südafrika gesammelte Süsswasseralgen. Oester. bot. Zeitschr., vol. 53, p. 89.
    1906. Algologische Untersuchungen an der biologischen Station in Drontheim I–VII. Kgl. norske Vidensk. Selsk. Skrifter, no. 3.
    1913. Algologische Notizen XXII–XXIV. Nyt Magazin for Naturvidensk, vol. 51, part 1, p. 1.

WITTROCK, V. B., and NORDSTEDT, O.
    1877–1903. Algae aquae dulcis exsiccatae praecipue Scandinavicae quas adjectis algis marinis chlorophyllaceis et phycochromaceis. Upsala.
        1877–1889. Fasc. 1–21.
        1893–1903. Fasc. 22–35.
    In 1896 G. Lagerheim was added as one of the distributors.

ZANARDINI, G.

1839. Sulle alghe. Bibliotheca Italiano, vol. 96, p. 131.

1841. Synopsis algarum in mari Adriatico hucusque cognitarum, cui accedent monographia siphonearum nec non generales de algarum vita et structura disquisitiones cum tabulis auctoris manu ad vivum depictis. Mem. della Reale Accademia delle Scienze, ser. 2, vol. 4.

1858. Plantarum in mari rubro hucusque collectarum enumeratio. Mem. R. Instit. Veneto II, vol. 7, p. 297.

1872. Phycearum indicarum pugillus. Mem. R. Instit. Veneto, vol. 17.

ZUKAL, H.

1894. Neue Beobachtungen über einige Cyanophyceen. Ber. deut. bot. Ges., vol. 12, p. 256, pl. 19.

EXPLANATION OF PLATES

All of the figures of the following plates were drawn under the direction of the authors by Dr. Helen M. Gilkey, with the exception of figures 5, 11, plate 1, 5, 6, plate 8, and 30, plate 7, which were drawn by H. N. Bagley, and figures 8, 9, plate 8, figs. 15, 16, plate 1, drawn by Miss Almeda H. Nordyke.

PLATE 1

Plectonema Battersii Gomont

Fig. 1. Plants with hormogonia just beginning false branching. × 500.

Gomphosphaeria aponina Kuetz.

Fig. 2. Surface view of a colony showing the ends of radiating filaments. × 500.

Fig. 3. A small fragment showing the branching stalks of the cells. × 500.

Phormidium lucidum (Ag.) Kuetz.

Fig. 4. Terminal portion of a single filament. × 1000.

Spirulina major Kuetz.

Fig. 5. Two filaments, one coiled loosely throughout, the other with one-half very tightly coiled. × 1200.

Synechococcus curtus Setchell

Fig. 6. Showing various stages of division. × 1000.

Synechocystis aquatilis Sauvageau

Fig. 7. Small group of cells showing various stages of division. × 1000.

Isactis plana var. *plana* B. and F.

Fig. 8. Diagrammatic section showing perpendicular, parallel arrangement of the filaments. × 12.

Fig. 9. Section of the thallus showing the character of mature filaments. × 250.

Hydrocoleum lyngbyaceum Kuetz.

Fig. 10. The end of a filament showing trichomes within the sheath. × 250.

Oscillatoria acuminata Gomont

Fig. 11. Three terminal portions of trichomes showing different degrees of attenuation and uncination. × 500.

Symploca hydnoides Kuetz.

Fig. 12. A habit sketch showing characteristic, erect, attenuate fascicles. × 10.

Fig. 13. A single filament. × 500.

Chroococcus turgidus f. *submarinus* Hansg.

Fig. 14. Showing one-, two-, and four-celled stages. × 250.

Fig. 15. *Lyngbya aestuarii* f. *natans* Gomont. × 500.

Fig. 16. *Lyngbya aestuarii* f. *aeruginosa* (Ag.) Wolle. × 500.

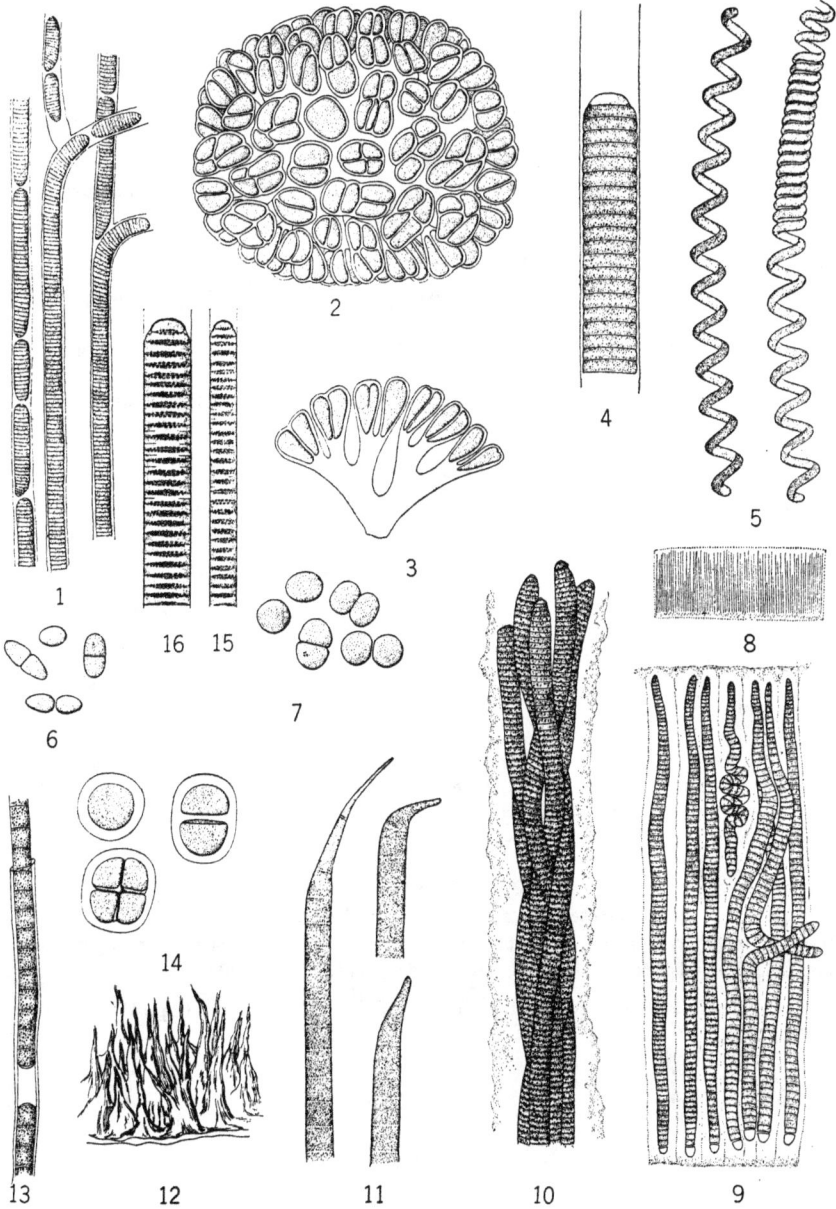

PLATE 2

Chlorogloea lutea S. and G.

Fig. 1. A section through the host plant showing the general character of the endophyte. × 500.

Xenococcus Chaetomorphae S. and G.

Fig. 2. A few elongated vegetative cells with interspersed gonidangia. × 125.

Fig. 3. Vegetative cells on different parts of the cells of the host, *Chaetomorpha.* × 125.

Fig. 4. A group of cells near the cross wall of the host plant showing the formation of gonidia. × 250.

Hyella socialis S. and G.

Fig. 5. A section of the host plant showing the habit of growth of the endophyte. × 250.

Chlorogloea conferta (Kuetz.) S. and G.

Fig. 6. Habit sketch of a colony growing on *Rhodochorton Rothii.* × 500.

Dermocarpa sphaeroidea S. and G.

Fig. 7. Habit sketch of vegetative cells and two gonidangia showing shapes and relative sizes. × 250.

Hyella linearis S. and G.

Fig. 8. Section of the host *Prionitis*, diagrammatic, showing an affected portion inhabited by the endophyte which is shown as radiating groups of cells, and with two gonidangia on the surface. × 125.

Dermocarpa suffulta S. and G.

Fig. 9. Plants selected to show variation in shapes and sizes. The two empty cells are vegetative, the other three are gonidangia, one showing all of the contents changed into gonidia, the exceptional condition. × 500.

This plate is a reprint of Gardner, Univ. Calif. Publ. Bot., vol. 6, p. 452, pl. 36, 1918.

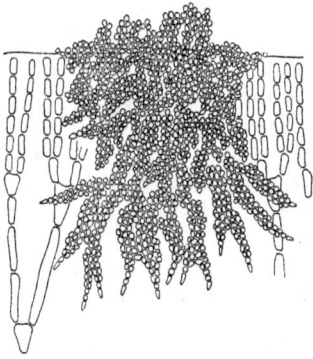

1

2

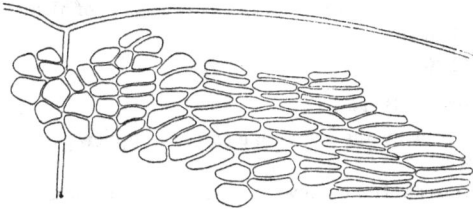

3

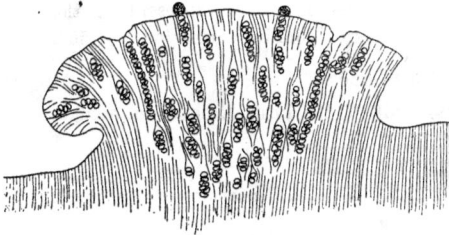

4

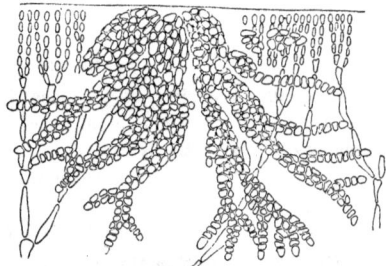

5

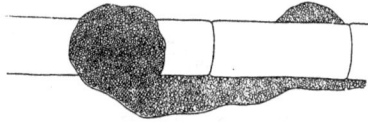

6

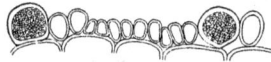

7

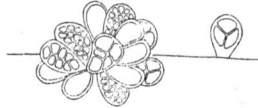

8

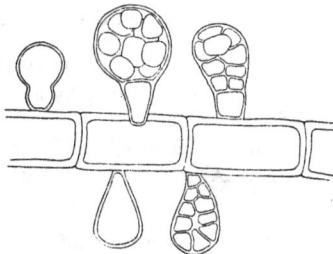

9

PLATE 3

Radaisia epiphytica S. and G.

Fig. 10. A portion of the basal layer of a plant, as seen from above, showing a few rows of cells one layer deep at the right and beginning to develop erect filaments at the left. × 250.

Fig. 11. A few erect filaments showing complex gonidangia at the outer ends. × 250.

Radaisia subimmersa S. and G.

Fig. 12. A section through the host perpendicular to the flat surface, showing various stages of development of Radaisia on the surface at the left, and variously embedded groups to the right. × 250.

Fig. 13. Various stages in the development of the basal layer. × 250.

Radaisia Laminariae S. and G.

Fig. 14. A few highly magnified erect filaments, three of them bearing terminal gonidangia. × 500.

Fig. 15. The same as figure 14 but less magnified, showing the margin of the colony at the left. × 250.

Fig. 16. A portion of the basal layer showing the method of its development. × 375.

Radaisia clavata S. and G.

Fig. 17. A perpendicular section through a plant showing the character of the erect filaments and one gonidangium. × 250.

Fig. 18. A portion of the basal layer. × 250.

Hyella Littorinae S. and G.

Fig. 19. Showing a few forms assumed by the erect filaments, the upper portions representing the outer ends. × 250.

Fig. 20. A few groups of cells as seen from above, representing the outer ends of the filaments. × 250.

Dermocarpa hemisphaerica S. and G.

Fig. 21. Showing one cell in the lower left-hand corner representing the vegetative condition, and the other in various stages of gonidia formation × 500.

Dermocarpa pacifica S. and G.

Fig. 22. A group of cells at the cross wall of the host, a common, wedge-shaped form. × 250.

Fig. 23. A group of spherical gonidangia. × 250.

Fig. 24. A group of vegetative cells from the dead portion of the host. × 250.

This plate is a reprint of Gardner, Univ. Calif. Publ. Bot., vol. 6, p. 454, pl. 37, 1918.

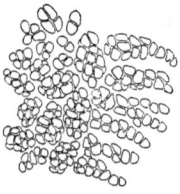
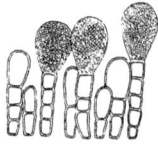
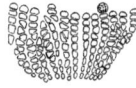

18

14

17

10

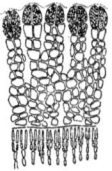
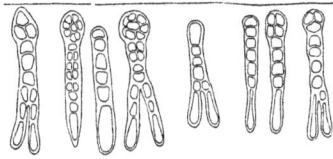

19

20

11

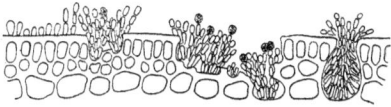
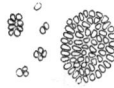

12   13   22

23

16   15   24

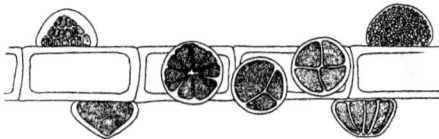

21

PLATE 4

Placoma violacea S. and G.

Fig. 1. Habit sketch of the thallus. × 75.

Fig. 2. Perpendicular section through young confluent thallus, showing the more or less radial arrangement of the cells. × 250.

Fig. 3. Surface view of a group of cells. × 375.

Dermocarpa protea S. and G.

Fig. 4. A series of gonidangia illustrating different shapes and sizes. × 250.

Fig. 5. A group of cells showing various stages of gonidial formation. × 250.

Anacystis elabens (Kuetz.) S. and G.

Fig. 6. Surface view of a colony, showing tendency to break up into groups of cells. × 125.

Fig. 7. A single group of cells from the colony. × 500.

Xenococcus Cladophorae (Tilden) S. and G.

Fig. 8. A group of vegetative cells in various stages of growth and interspersed gonidangia. × 250.

Pleurocapsa entophysaloides S. and G.

Fig. 9. *a–g*, A series of vegetative cells in various stages of development; *h–i*, two gonidangia. × 250.

Fig. 10. Section view through a large colony showing the somewhat entophysaloid arrangement of groups of cells. × 250.

This plate is a reprint of Gardner, Univ. Calif. Publ. Bot., vol. 6, p. 480, pl. 38. 1918.

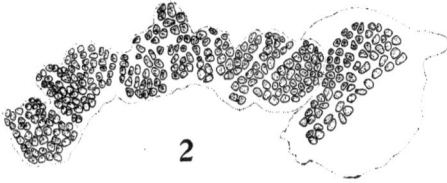

2

3

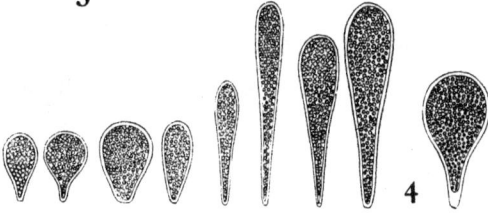

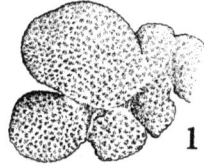

4

1

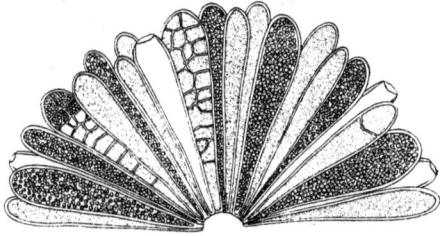

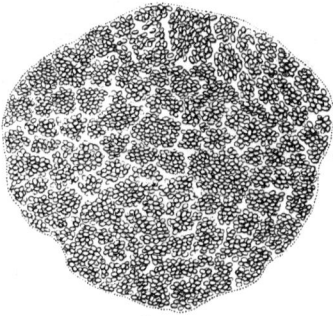

6

5

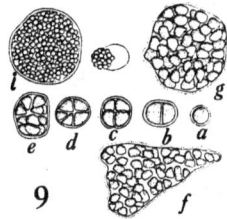

7

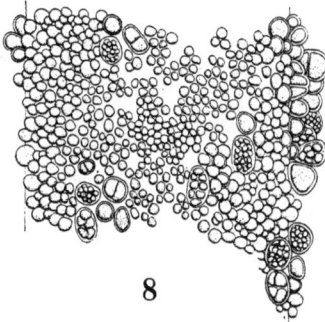

9

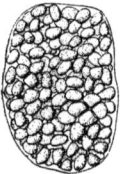

8

10

PLATE 5

Xenococcus Gilkeyae S. and G.

Fig. 11. Group of vegetative cells and gonidangia in various stages of development. × 500.

Xenococcus pyriformis S. and G.

Fig. 12. Groups of vegetative cells as viewed from different angles, and three gonidangia. × 500.

Xenococcus acervatus S. and G.

Fig. 13. Vegetative cells, some showing cell divisions, others showing a heaped-up condition as seen in median plane view of the host. × 500.

Dermocarpa sphaerica S. and G.

Fig. 14. Vegetative cells and gonidangia, *a–b*, with the wall still in position, and *c–d*, showing where the walls have dissolved in position. × 500.

Pleurocapsa gloeocapsoides S. and G.

Figs. 15, 16. Groups of vegetative cells in various stages of development. × 500.

Fig. 17. Showing formation of gonidia. × 500.

Arthrospira breviarticulata S. and G.

Fig. 18. An individual showing the loose spiral doubled and entwined upon itself.

This plate is a reprint of Gardner, Univ. Calif. Publ. Bot., vol. 6, pl. 39, p. 482, 1918.

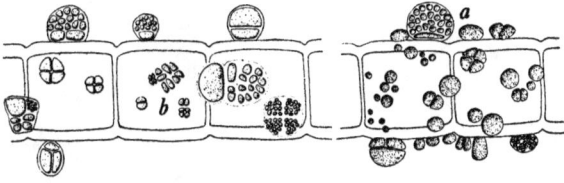

11

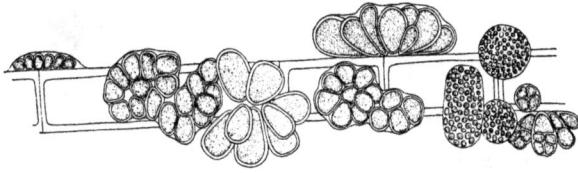

12

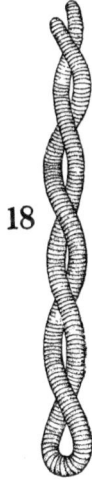

18

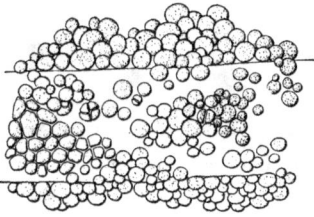

13

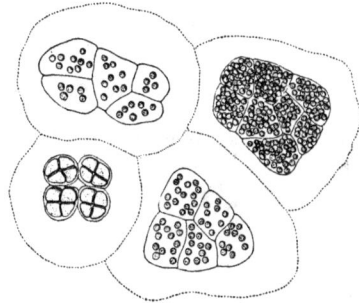

17

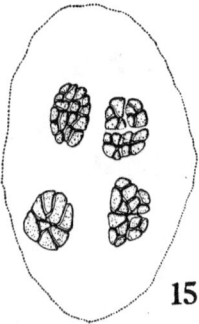

15

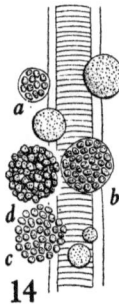

14

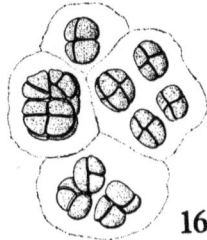

16

PLATE 6

Fig. 19.　*Rivularia mamillata* S. and G.　× 250.

Fig. 20.　*Dichothrix seriata* S. and G.　× 125.

Fig. 21.　*Calothrix rectangularis* S. and G.　× 125.

Fig. 22.　*Calothrix robusta* S. and G.　× 125.

Fig. 23.　*Phormidium hormoides* S. and G.　× 500.

Fig. 24.　*Microcoleus Weeksii* S. and G.　× 500.

Fig. 25.　*Microcoleus confluens* S. and G.　× 500.

This plate is a reprint of Gardner, Univ. Calif. Publ. Bot., vol. 6, p. 484, pl. 40, 1918.

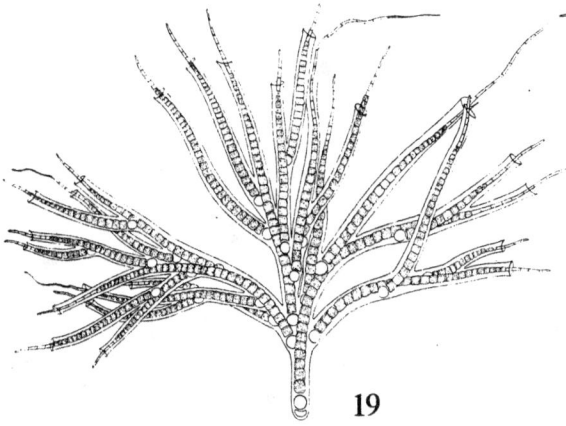

19

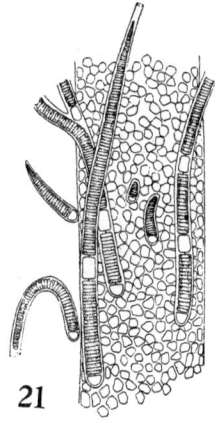

21

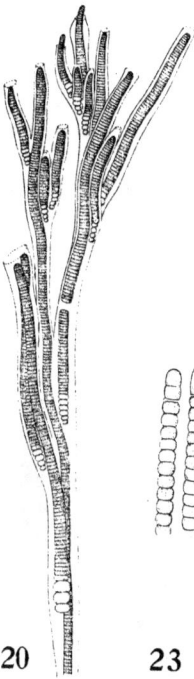

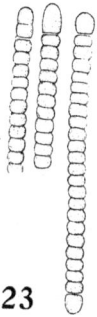

20

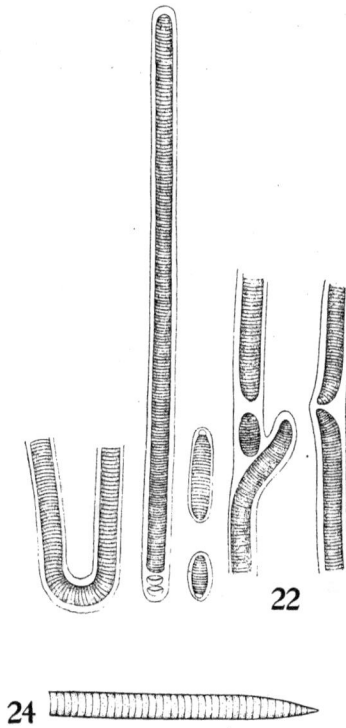

22

23

24

25

PLATE 7

Arthrospira breviarticulata S. and G.

Fig. 26.   An individual showing the general characters of a plant taken at random.

Brachytrichia affinis S. and G.

Fig. 27.   A section view through the cortex of a young thallus showing the characteristic method of branching, and young hairs. $\times$ 330.

Fig. 28.   A section view through an old thallus showing true branches spring- ing directly from the main filaments without the formation of loops. $\times$ 330.

Symploca funicularis S. and G.

Fig. 29.   Portion of a single filament showing the characters particularly of the terminal cell. $\times$ 330.

Pleurocapsa entophysaloides S. and G.

Fig. 30.   Showing the characteristic manner in which the groups of colonies arrange themselves in the stratum, after the manner of *Entophysalis*.

This plate is a reprint of Gardner, Univ. Calif. Publ. Bot., vol. 6, p. 486, pl. 41, 1918.

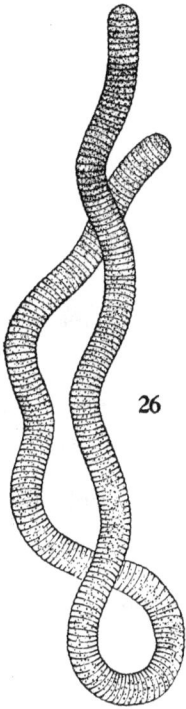

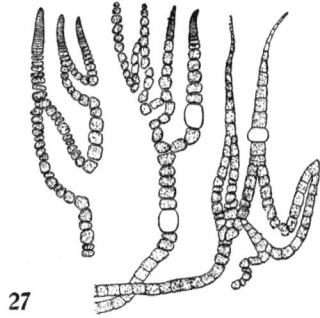

27

26

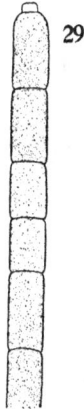

29

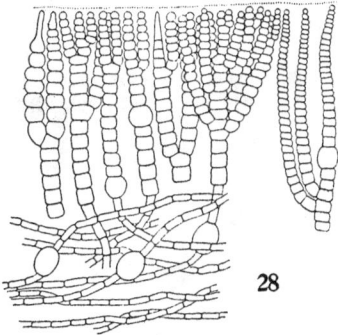

28

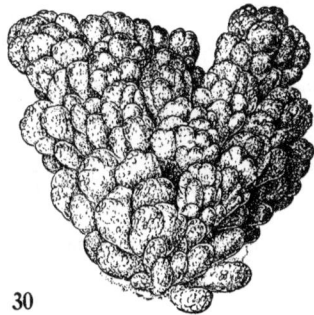

30

PLATE 8

Rivularia atra var. *hemisphaerica* (Kuetz.) B. and F.

Fig. 1. Diagrammatic section of thallus perpendicular to the base showing the radiating filaments and concentric zones. × 5.

Fig. 2. Portion of radial section of thallus, somewhat diagrammatic. × 20.

Arthrospira maxima S. and G.

Fig. 3. An entire mature plant. Reprinted from Gardner, Univ. Calif. Publ. Bot., vol. 6, p. 412, pl. 33, fig. 3, 1917.

Calothrix pilosa Harvey

Fif. 4. A small group of filaments showing the position of heterocysts and some of the methods of false branching. Semidiagrammatic.

Dermocarpa fucicola Saunders

Fig. 5. Diagram of host plant, showing the position and distribution of the epiphyte.

Fig. 6. Section through a pustule showing the shapes of the vegetative cells and one gonidangium. × 1400.

Merismopedia Gardneri (Collins) Setchell

Fig. 7. Surface view of a portion of the thallus. × 350.

Anabaena variabilis Kuetz.

Fig. 8. Two portions of plants, showing typical vegetative cells, heterocysts and mature spores. × 360.

Anabaena propinqua S. and G.

Fig. 9. A group of plants showing typical vegetative cells, heterocysts, and mature spores. × 400. Reprinted from Gardner, Univ. Calif. Publ. Bot., vol. 6, p. 496, pl. 42, fig. 1, 1919.

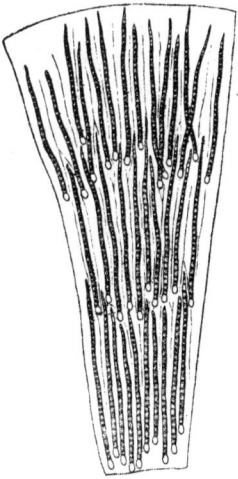

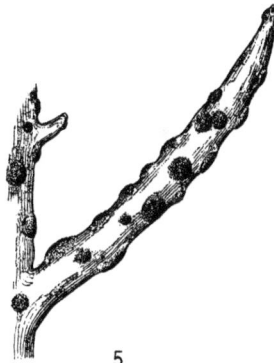

2

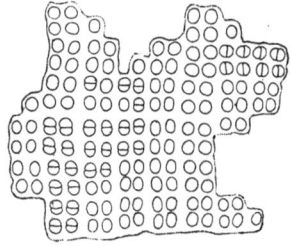

5

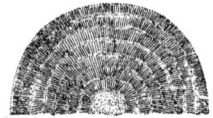

7

1

8

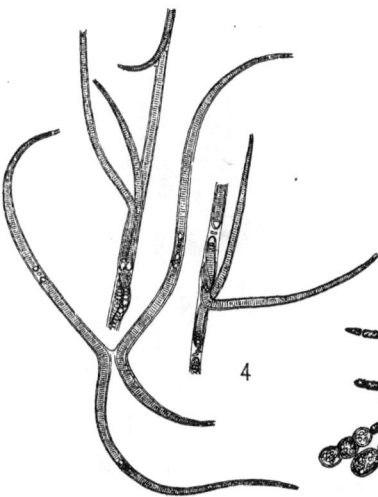

4

6

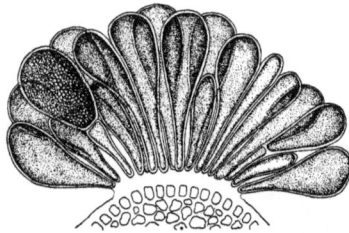

9

3

UNIVERSITY OF CALIFORNIA PUBLICATIONS

IN

BOTANY

Vol. 8, No. 2, pp. 139-374, plates 9-33   July 14, 1920

THE MARINE ALGAE OF THE PACIFIC COAST OF NORTH AMERICA

PART II

CHLOROPHYCEAE

BY

WILLIAM ALBERT SETCHELL

AND

NATHANIEL LYON GARDNER

SUBCLASS 2. CHLOROPHYCEAE KUETZING

Thallophytes containing only the pigments, chlorophyll and xanthophyll; thallus varying from strictly single cells (uninucleate), or more or less simple or complex colonies (as in the Protococcales), to multicellular individuals which are either made up of cells (i.e., uninucleate divisions, as in the Ulvales, Schizogoniales, and the Ulotricales) or are coenocytes (i.e., of multinucleate divisions), the latter being either septate (i.e., provided with partitions, as in the Siphonocladiales) or destitute of septa (as in the Siphonales), and ranging in size from microscopic forms to individuals of more than a meter in at least two dimensions; cell walls varying in structure and composition, mostly of cellulose but sometimes largely of pectose, occasionally more or less externally mucilaginous, generally simple, moderately thick and structureless, but at times thick and variously stratified, occasionally incrusted with lime; nuclei well developed; chromatophores usually distinctly differentiated, of varying shape and number, often containing starch centers, or pyrenoids, and colored by chlorophyll and xanthophyll, the former usually in excess; reproduction vegetative, by non-sexual spores, and by zygotes; vegetative reproduction by cell division, fragmentation, and by gemmae (or cysts); non-sexual spores either motile (zoospores, zoogonidia, or planospores) or non-motile (aplanospores, formed inside the wall of the original cell, and akinetes where the outer wall of the original cell is included in the spore formation); zygotes formed either from isogametes (either isoplanogametes, i.e., both equal and motile, or isoaplanogametes, i.e., both equal and non-motile) or from aniso-gametes (or heterogametes, i.e., unlike gametes); germination of the zygote various.

Chlorophyceae Kuetzing, Phyc. Germ., 1845, p. 118; Wille, *in* Engler and Prantl, Natürl Pflanzenfam., 1 Th., 2 Abt., 1897, p. 24; Oltmanns, Morph. und Biol. der Alg., vol. 1, 1904, p. 133; West, Algae, vol. 1, 1916, p. 126. *Chlorospermeae* Harvey, *in* Mackay, Fl. Hibern., part III, 1836, p. 163, Genera So. African Plants, 1838, p. 403.

Harvey seems to have been the first to propose the classification into green algae, brown or olive-green algae and red algae, as now usually adopted, and coined the names Chlorospermeae, Melano-spermeae and Rhodospermeae. In his Chlorospermeae, he included also what are now separated under Myxophyceae, very few of which, however, were known to him. Kuetzing was the first to use the ending phyceae, but his Chlorophyceae included both the Chloro-spermeae and Melanospermeae of Harvey and is consequently very different in its content from that of more recent writers.

There is some difference of opinion, even at present, as to the exact content of the Chlorophyceae. It has seemed best to follow West (1916, p. 126) in including also the Conjugatae as well as the groups with zoospores, although the decision assumes no practical importance in the present account, since none of the Conjugatae is properly marine.

The great majority of the marine Chlorophyceae are inhabitants of the littoral belt, a few descending to the upper portion of the sublittoral belt, while those occurring in deeper waters are, so far as the extra tropical portions of the Pacific Coast of North America are concerned, very few indeed.

The great majority of the Chlorophyceae are either subaerial or inhabitants of strictly fresh waters, so that a number of the important groups are not found in the salt waters. Unfortunately for this account, also, the strictly tropical waters of the Pacific Coast of North America are, as yet, unexplored, and little is known as to the occur-rence, or non-occurrence, of the more complex forms of the Codiaceae,

Valoniaceae and the Caulerpaceae. Undoubtedly many species of these families will be added later when the algal flora of the Pacific coasts of Mexico and Central America is collected and made known.

CHLOROPHYCEAE

SERIES 1. ISOKONTAE BLACKMAN AND TANSLEY

Unicellular, multicellular and coenocytic Chlorophyceae, reproducing by means of zoospores or planogametes provided with two or four equal cilia or, when multinucleate, with the cilia arranged in pairs (Vaucheriaceae) or exceptionally in a circle (Derbesiaceae).

Isokontae Blackman and Tansley, Revis. Class. Green Algae, 1902, p. 20; West, Algae, vol. 1, 1916, p. 156.

It has seemed best to adopt the classification as outlined by West (1916, p. 153), dividing the Chlorophyceae into four series according to the character and arrangement of the cilia of the zoospores and the planogametes. West adopts four series, viz., Isokontae, Akontae, Stephanokontae and Heterokontae. Of these, our marine flora has to deal only with the first, the Isokontae, where the cilia are equal and arranged in twos or fours. The other groups are not represented. Only two seeming exceptions need explanation. In the Vaucheriaceae the zoospore is large and covered with cilia. They are, however, arranged over the surface in pairs. The other, and only real, exception is the Derbesiaceae, where the cilia are arranged in a circle as in the Stephanokontae. It has seemed best to leave the Derbesiaceae with the Isokontae in this account, since this is the usual arrangement, but it is done with some reservation of opinion.

KEY TO THE ORDERS

1. Thallus of true cells (uninucleate segments)... 2
1. Thallus coenocytic (of multinucleate segments)... 6
    2. Cells solitary or in non-filamentous colonies..........1. **Protococcales** (p. 143)
    2. Cells in filaments or membranes.. 3
3. Thallus filamentous... 4
3. Thallus membranaceous... 5
    4. Filaments simple or, more often, branched; chromatophore parietal..........
    ....6. **Ulotrichales** (p. 281)
    4. Filaments simple; chromatophore axile.................5. **Schizogoniales** (p. 275)
5. Chromatophores parietal...4. **Ulvales** (p. 233)
5. Chromatophores axile..5. **Schizogoniales** (p. 275)
    6. Thallus septate...3. **Siphonocladiales** (p. 179)
    6. Thallus non-septate......................................2. **Siphonales** (p. 153)

Order 1. PROTOCOCCALES (meneghini) oltmanns

Unicellular isokontae, motile or nonmotile, often occurring singly or in larger or smaller, definite or indefinite colonies or coenobia, or even simple coenocytes, often provided with mucilaginous teguments of more or less ample dimensions and of various shapes, never properly multicellular; number and shape of chromatophores (chloroplasts) various; pyrenoids often present; reproductive methods various as outlined for the series Isokontae.

Protococcales Oltmanns, Morph. und Biol. der Algen, vol. 1, 1904, p. 169; West, Algae, vol. 1, 1916, p. 160. *Protococcoideae* Meneghini, Cenni sulla organ. e. fisiol. delle Alghe, 1838, p. 4 (of reprint); Black-man and Tansley, Revis, Class. Green Algae, 1902, p. 21; Wille, *in* Engler and Prantl, Natürl. Pflanzenfam., Nachtr. zum 1 Th., 2 Abt., 1909, p. 3.

The Protococcales, or Protococcoideae as they have long been designated, form a rather large and seemingly heterogeneous order. They are mostly inhabitants of the fresh waters and, although unicel-lular in the broad sense, are varied in their form, aggregation and methods of reproduction. Our marine species are few so far as known, but undoubtedly a considerable number yet remain to be detected. This is particularly true of the endophytic species of our coast whose presence and development are very little understood at present.

Key to the Families

1. Thallus of larger or smaller colonies (or aggregations of cells)..................................
....1. **Palmellaceae** (p. 143)
1. Thallus strictly unicellular.......................................2. **Chlorochytriaceae** (p. 146)

Family 1. PALMELLACEAE (decaisne) naegeli

Cells united into larger or smaller colonies by mucilaginous modi-fication of the outer walls, usually provided with a single parietal chromatophore (chloroplast) containing a single pyrenoid; repro-duction by biciliated zoospores or by isoplanogametes; fragmentation of colonies often takes place.

Palmellaceae Naegeli, Die neuern Algensyst., 1847, p. 123; West, Algae, vol. 1, 1916, p. 183. *Palmelleae* Decaisne, Essai sur une classe des Algues, etc., 1842, p. 327.

The names Palmellaceae and Palmelleae are used with different intent and to different extent by various writers and are interchangeable wholly or in part with other family designations. It has seemed best to follow West as to the usage in this account. The fundamental idea is that of the colony the cells of which are held together by means of the mucilaginous material produced by the transformation of the outer walls. Some of the colonies are microscopic while others are of considerable size reaching a length (e.g., in some species of *Tetraspora*) of 15 to 20 centimeters. The mucilaginous modification may be general or it may be localized on each cell wall, and the shape, as well as the size, of the colony may thereby be influenced. The cells possess a single parietal chromatophore with a single pyrenoid. Colonies may split up and multiply the plant. Reproduction by zoospores and planogametes is the rule. In zoospore formation several (4 or 8) may arise in a zoosporangium or an ordinary cell may be transformed directly into a zoospore. This latter method, as well as the general cell structure and colony formation, points directly toward relationship with the Volvocaceae.

1. **Collinsiella** S. and G.

Frond gelatinous, solid or later hollow, composed of pyriform cells, on dichotomous, gelatinous stalks tapering downward from the cells; all enclosed in the general gelatine; chromatophore band-shaped, with one large pyrenoid; the terminal cells become the zoosporangia(?)

Setchell and Gardner, Alg. N.W. Amer., 1903, p. 204.

There is a reason for difference of opinion as to whether *Collinsiella* is to be retained as an independent genus or merged with the genus *Ecballocystis* Bohlin. It has seemed best to place the discussion under the single species known from our coast.

Collinsiella tuberculata S. and G.
Plate 10, figs. 4–10

Colonies rugose-tuberculate, 2–4 mm. diam., dark green, gelatinous, firm, attached by a broad base; cells pyriform, 5–12μ diam., 12–20μ long; branches repeatedly dichotomous, proceeding in two planes perpendicular to each other and to the surface of the colony, some of the cells of the dichotomies pushing forward, forming the cortex, leaving behind the translucent, stalklike, gelatinous cell walls, while growth of other cells of the dichotomies is suppressed and they remain within the colony; the cell stalks show strong cellulose reaction to Chloriodide of Zinc; the cortical cells are changed into zoosporangia(?) containing 8–16 or occasionally more zoospores.

Growing on rocks and pebbles in tide pools in the middle and upper littoral belts. West coast of Whidbey Island, Washington, Port Renfrew, Vancouver Island, Farallones Islands and Point Carmel, California.

Setchell and Gardner, Alg. N.W. Amer., 1903, p. 204, pl. 17, f. 1–7; Collins, Green Alg. N. A., 1909, p. 141; West, Algae, vol. 1, 1916, p. 188; Collins, Holden and Setchell, Phyc. Bor.-Amer. (Exsicc.), no. 909. *Ecballocystis Willeana* Yendo, Three species of marine *Ecballocystis*, 1903, p. 199. *Ecballocystis tuberculata* (S. and G.) Wille, Nachträge, 1909, p. 28 (in part).

Wille (1909, p. 27) places *Collinsiella* as a synonym under *Ecballocystis* Bohlin, but Collins (1909, p. 141) and West (1916, p. 188) keep it distinct. It seems to differ from *Ecballocystis* in forming an extended and definite gelatinous thallus, in the more vertical and regular division of the cells, and in the longer gelatinous stalks to the cells. Because of the first of these differences, West (*loc. cit.*) places it in the subfamily Palmophylleae of the Palmellaceae and next to *Palmophyllum*. This disposition of the genus seems to be the most satisfactory and is adopted here. Collins (1909, p. 141) assigns the *Ecballocystis Willeana* Yendo, from Port Renfrew, British Columbia, to *Collinsiella tuberculata* S. and G. as a synonym, and draws his description from both those of Yendo and of Setchell and Gardner. Yendo, however, in his remarks (1903, p. 204) states that it seems to him highly probable that *Collinsiella tuberculata* may be a young and sterile form of a plant closely related to his *Ecballocystis Willeana*, if not the same species. We have not had the opportunity of examining a plant of *Ecballocystis Willeana*, but judging from Yendo's description and plates, there are some differences. In the Whidbey Island plant there is a sort of basement membrane from which bullate swellings rise as indicated in the habit figure of Setchell and Gardner (1903, pl. 17, f. 1). Yendo figures isolated, much folded thalli (1903, pl. 8, f. 1) attached by rhizoidal outgrowths on the underside (*loc. cit.*, pl. 8, f. 2, 6, 12). No such outgrowths have been detected in the plant from Whidbey Island or in any others of our collections. These are, perhaps, minor and unessential differences, but they indicate that there is a reason to feel uncertain as to the absolute identity of the two plants. The specimens from the central Californian coast, on the other hand, resemble more closely the Yendo plant in habit but show no attaching fibrils.

FAMILY 2. CHLOROCHYTRIACEAE NOM. NOV.

Thallus unicellular, not united into colonies, or single unseptate coenocytes, reproducing solely by zoospores and by isoplanogametes.

Planosporaceae West, Algae, vol. 1, 1916, p. 209. *Endosphaeraceae* Klebs, Organ. einig. Flagellaten-Gruppe, 1883, p. 344; Hansgirg, Prodr. d. Algenfl. v. Böhmen, II, 1888, p. 124; Blackman and Tansley, Rev. Class. Green Algae, 1902, p. 95.

This is a small but fairly natural family including a number of genera which are, however, mostly inhabitants of the fresh waters. West (1916, p. 212) has reduced a number of genera under *Chlorochytrium*, as indicated elsewhere, among them being *Endosphaera*. West has, consequently, set aside the family name Endosphaeraceae, as adopted by Blackman and Tansley, and substituted the designation "Planosporaceae." Since this does not embody one of the genera of the family as reorganized it seems best to consider it inapplicable under the present rules of nomenclature and to adopt "Chlorochytriaceae" as a fitting family name.

As stated by West, "this family is established to include all those non-coenobic members of the Protocaccales which are reproduced solely by zoogonidia or isoplanogametes." The great majority of the members of this family are either epiphytic or endophytic.

KEY TO THE GENERA

1. Cells spherical to more or less ovoid..............................2. **Chlorochytrium** (p. 146)
1. Cells cylindrico-oblong with a more or less elongated stipe......3. **Codiolum** (p. 151)

2. **Chlorochytrium** Cohn

Thallus unicellular, rounded, with chromatophore covering more or less of the outer wall and continuous or with radial prolongations, containing one to several pyrenoids; asexual reproduction by akinetes or by 2- or 4-ciliated zoospores escaping singly or enclosed in a gelatinous mass; sexual reproduction by 2-ciliated isogametes escaping in a gelatinous utricle mass and conjugating before separation, or escaping singly and conjugating in the water; zygote, 4-ciliated, at first motile, later coming to rest and penetrating the host plant.

Cohn, Ueber parasitische Algen, 1872, p. 102, diagnosis.

West reduces *Endosphaera* Klebs, *Scotinosphaera* Klebs, *Chloro-cystis* Reinhardt and *Stomatochytrium* Cunningham to one genus and unites that with *Chlorochytrium* Cohn. "They all agree in being holophytic, unicellular, spherical or nearly so, wholly or partly endo-phytic plants with a single chromatophore, covering the wall more or less completely and containing one or more pyrenoids. Reproduction is by plano-gametes or by zoospores or by both" (cf. Gardner, 1917, p. 383). It seems best to follow West in his conception of the genus and our species are, consequently, assigned to *Chlorochytrium* in this extended sense. The three species thus far credited to our territory are immersed in the tissues of various membranous or expanded red algae.

KEY TO THE SPECIES

1. Cells with apiculate tips..1. **C. inclusum** (p. 147)
1. Cells without apiculate tips.. 2
    2. Cells spherical, chromatophores with one pyrenoid..
    ....3. **C. Porphyrae** (p. 150)
    2. Cells clavate or ovoid, chromatophores with two or more pyrenoids..........
    ....2. **C. Schmitzii** (p. 149)

1. **Chlorochytrium inclusum** Kjellm.

Plate 13, fig. 1

Cells in the vegetative condition, spherical or subspherical, entirely included within the host plant, at the time of the formation of the zoospores, slightly elongated, depressed conical, ampullaeform, ovoid or ellipsoid, at length exposed through the penetration of the cortical layer of the host by the apiculate tip, emitting the zoospores through an ostiole.

Endophytic in the fronds of various membranaceous algae, e.g., *Iridaea, Weeksia, Constantinea*, etc. Probably common along the coast from Sitka, Alaska, to Puget Sound, Washington.

Kjellman, Alg. Arctic Sea, 1883, p. 320, pl. 31, f. 8–17; Setchell and Gardner, Alg. N.W. Amer., 1903, p. 206?; Collins, Green Alg. N. A., 1909, p. 147 (in part); Collins, Holden and Setchell, Phyc. Bor.-Amer. (Exsicc.), no. 514?; Tilden, Amer. Alg. (Exsicc.), no. 389 ?.

The description, given above, is a fairly literal translation of the Latin diagnosis of Kjellman, who adds certain details in his remarks. The original host is *Dilsea integra* (Kjellm.) Rosenv. (*Sarcophyllis arctica* Kjellm.). The cells of the *Chlorochytrium* are placed, in most cases, near the surface of the host plant but sometimes occur in the middle layer. In vegetative condition the cells are from 80μ to 100μ in diameter, the cell wall is thin and of equal thickness throughout, while the chromatophore is thin and spread over the whole wall. The wall becomes thicker and apiculate at the outer end as the cell passes into the reproductive stages, the apiculate wall piercing the outer cortical tissues of the host. Kjellman states that the contents divide into a large number of closely packed zoospores which escape through an opening formed by the dissolution of the wall at the tip of the cell. These latter statements are evidently inferences because he distinctly says that he had only dried specimens for examination.

In an authentic specimen of the host plant distributed by Kjellman, young cells of the *Chlorochytrium* were found nearly spherical in shape, with uniformly thin walls, and with a chromatophore thin and dotted with numerous large pyrenoids. These cells are about 80μ in diameter.

Upon examining various specimens referred to this species, the conclusion has been forced upon us that there is some variety of species and possibly even of genera among the Pacific Coast plants referred to *Chlorochytrium inclusum* and it seems practically demonstrated that no one of those accessible to us is clearly the plant of Kjellman.

Very little can be accomplished from the study of dried specimens, but living specimens should be studied to obtain more exact information as to structure and development. Our present knowledge, even of the type, is so slight as to admit of little certainty, and Kjellman's statements as to the formation and emission of ''zoospores'' need to be carefully verified.

On reëxamining the various specimens referred to this species from our coast, we are able to make only a few general statements.

Freeman (1899, p. 186) describes a plant which he provisionally refers to *Chlorochytrium inclusum*, but he found only vegetative stages. It was endophytic in the blades of *Constantinea subulifera* Setchell. In the Algae of Northwestern America (1903, p. 206), we referred several specimens to the same species. Of these we may distinguish, at least, two very different kinds of endophytes. The first kind includes what are probably species of *Chlorochytrium*, possessing a single chromatophore with numerous starch centers, while the second is made up of plants seemingly possessing neither chromatophores nor chlorophyll and certainly devoid of starch. No. 290, N. L. Gardner, on *Iridaea* from the west coast of Whidbey Island, Washington, shows small plants $(40\mu \times 80\mu)$, broadly pyriform and with thick walls. It is to be referred provisionally to *Chlorochytrium*, but does not agree with Kjellman's description. No. 514, of Collins, Holden and Setchell's Phycotheca Boreali-Americana, shows large, thin walled cells, depressed vertically and measuring about 160μ by 240μ, seemingly a *Chlorochytrium*, but not in accord with the descriptions of either Kjellman or Freeman. The other references given by us, with the exception of Tilden's no. 389, which is Freeman's plant, are to be rejected. They are found to be based upon plants of the second type, which is probably *Chytridiaceous*, possibly being near to *Rhodochytrium*. They are probably the so-called gland cells mentioned by Schmitz as occurring in *Turnerella Mertensiana* (P. and R.) Schmitz (1896, p. 372) and figured as occurring in *Iridaea affinis* P. and R. (Postels and Ruprecht, 1840, pl. 40, f. 93). We have selected for illustration (pl. 13, f. 1) plants occurring endophytic in *Weeksia Fryeana* Setchell collected by Gardner near Sitka, Alaska. These seem to correspond more nearly than any of our other specimens with the description and figures of Kjellman.

2. Chlorochytrium Schmitzii Rosenv.

Cells clavate or ovoid, with rounded apex, without cone-shaped thickening of the cell wall, and with pointed base; up to 370μ long by 90μ diam.; chromatophore single, occupying the greater part of the cell wall, and with two pyrenoids.

Growing in various incrusting marine algae, e.g., *Petrocelis*. Alaska.

Rosenvinge, Groenl. Havalg., 1893, p. 964, f. 56; Collins, Green Alg. N. A., 1909, p. 147; Setchell and Gardner, Alg. N.W. Amer., 1903, p. 206 (in part).

In our Algae of Northwestern America (1903, p. 206), we referred two specimens to this species, one from Harvester Island in Uyak Bay, on the Island of Kadiak, Alaska, and another from the west coast of Whidbey Island, Washington. On reëxamination of these specimens it seems best to retain the former under this name, in spite of certain differences between it and the figures and descriptions of the Greenland plant as given by Rosenvinge. Certain of the cells in our specimens are rounded above and pointed below, seemingly in vegetative condition. Other cells have papillate swellings at one or both ends and are probably reproducing since the contents seem more or less broken up. There is no trace of a stalk (or tail) as in *Codiolum*, nor is the shape that of the cell (or clava) of that genus. Therefore it seems best to retain our Alaskan plant in *Chlorochytrium* and to refer it to *C. Schmitzii* pending further investigation of living material.

The Whidbey Island plant referred here has also been carefully reëxamined. In the shape of the "clava" and in the occasional possession of a stalk (or tail), it shows itself to be a *Codiolum* and is discussed below under that genus.

The Alaskan plant referred to this species varies in height from 123μ up to 220μ, and in width from 54μ to 66μ, thus coming within the measurements as given by Rosenvinge for *Chlorochytrium Schmitzii*.

3. Chlorochytrium Porphyrae S. and G.

Plate 15, fig. 1

Cells spherical, 40–60μ diam., embedded within the host on both sides; chromatophore, single, at first small, covering the upper part of the young plant, then increasing in size by sending out several radiating arms and finally covering the cell wall; pyrenoid, single, large, embedded within the chromatophore toward the upper part of the cell; cell wall 2–3μ diam., hyaline, not laminated; color, grass green; sexual reproduction by 2-ciliated isogametes, 3–4μ diam., fusiform to almost spherical, escaping singly through the oval opening in the outer wall; asexual reproduction by zoospores and akinetes unknown.

Growing completely embedded within the outer membrane of *Porphyra perforata* f. *segregata* Setchell and Hus. Washington (Cape Flattery) to central California (San Francisco).

Setchell and Gardner, *in* Gardner, New Pac. Coast Mar. Alg. I, 1917, pp. 379–384, pl. 32, f. 6; Collins, Holden and Setchell, Phyc. Bor.-Amer. (Exsicc.), no. 2280.

This species was discovered at Lands End, San Francisco, California, but since the first publication of its discovery it has been observed in the vicinity of Cape Flattery, Washington, where it grows in abundance on the same host as at San Francisco. It probably extends along our coast wherever the host plant grows.

A full account of the morphology and development of this species, as well as an extended discussion of the status of *Chlorochytrium* and the various plants referred to *C. Cohnii*, will be found under the reference to Gardner given above.

3. **Codiolum** A. Braun

Frond unicellular, ovoid to clavate or subcylindrical, the cell wall prolonged below into a longer or shorter stipe, attached by a simple or forked expansion; chromatophore covering the cell wall or more or less broken, with several pyrenoids; asexual reproduction by 4-ciliated zoospores, many in a cell.

A. Braun, Algarum Unic., 1855, p. 19.

This genus was first mentioned in 1852 by Braun before the 29th Congress of naturalists and physicians at Wiesbaden (cf. Flora, 1852, p. 755) and was excellently described and illustrated in full in 1855 in his ''Algarum Unicellularum Genera nova et minus cognita'' (p. 19). The type species is *Codiolum gregarium* A. Braun, and the type locality is Helgoland.

The species of *Codiolum* are all very similar and consist of a colorless stipe of longer or shorter dimensions bearing above a swollen cell which is elongated ovoid in shape and which is termed the ''clava.'' The dimensions of both stipe and clava differ somewhat even in the same species, but in the endophytic species the stipe may be abbreviated or even, most commonly, wanting.

KEY TO THE SPECIES

1. Cells with a long stipe not endophytic.............................1. **C. gregarium** (p. 151)
1. Cells with stipe short or wanting, endophytic.................2. **C. Petrocelidis** (p. 152)

1. **Codiolum gregarium** A. Braun

Plate 15, fig. 2

Clava narrowly elliptical in median section, definitely delimited from the long narrow stipe, up to 500μ long, and 100μ wide; stipe hyaline, unbranched, nearly cylindrical but slightly enlarging upward, $600–1000\mu$ long, $20–30\mu$ wide, somewhat disk-shaped at the base.

Reported from a single locality in our region, growing on an iron buoy near Friday Harbor, San Juan County, Washington.

A. Braun, Alg. Unic., 1855, p. 20, pl. I, f. 1–17; Collins, Green Alg. N. A., 1909, p. 152.

There have been described several species of *Codiolum* beside the endophytic species and these species have been dependent largely upon differences in various dimensions, but particularly on length of stipe. Börgesen, however, in his ''Marine Algae of the Faeröes'' (1902,

p. 517) comes to the conclusion that two species, or groups of species, stand out with fair distinctness, viz., *Codiolum gregarium* A. Braun, in which species (or group) the clava is definitely constricted at the line of union with the stipe, and *C. pusillum* (Lyng.) Kjellman, where the stipe passes insensibly into the clava. Our specimens are to be arranged with *C. gregarium* A. Braun and while their dimensions differ from those given by various authors for this species, yet it seems best not to attempt any separation at present. Our specimens vary in length of clava from 160μ to 240μ, and in width from 32μ to 64μ, while the stipe varies from 250μ to 550μ in length and from 16μ to 28μ in diameter.

2. **Codiolum Petrocelidis** Kuckuck

Clava ovoid or obovoid, 65–90μ long, 20–30μ wide; stipe relatively short or sometimes absent, often tapering abruptly below into a sharp point.

Growing within the thallus of *Petrocelis franciscana*, central California, and of *P. Middendorffii*, Whidbey Island, Washington.

Kuckuck, Bemerk. zur mar. Algenveg. Wiss. Meeres., vol. 1, 1894, p. 259, f. 27; Collins, Green Alg. N. A., 1909, p. 152; Collins, Holden and Setchell, Phyc. Bor.-Amer. (Exsicc.), no. 2281. *Chlorochytrium Schmitzii* Setchell and Gardner, Alg. N. W. Amer., 1903, p. 206 (in part).

Codiolum Petrocelidis was described by Kuckuck from specimens growing in *Petrocelis Hennedyi* at Helgoland, where it had first been detected many years previously by Ferdinand Cohn. It has also been described as growing in *Petrocelis* on the coast of New England. Two specimens of *Codiolum* growing in species of *Petrocelis* have been collected on the Pacific Coast of North America, one in *P. Middendorffii* on the west coast of Whidbey Island, Washington, and the other in *P. franciscana* on the coast of central California at Fort Point, San Francisco. These two sets of plants differ somewhat from one another and also both differ in dimensions from *C. Petrocelidis* as described by Kuckuck. Kuckuck gives (as in description above) 65μ to 90μ long and 20μ to 30μ wide as the dimensions of his type. The Washington plant varies from 136μ to 180μ long and from 20μ to 44μ wide for the clava, while the Californian plant shows clavae from 80μ to 140μ long and 28μ to 42μ wide. It seems best, however, to refer them both to *Codiolum Petrocelidis* Kuckuck, at least for the present.

As to the stalk (or stipe) it is very distinct in some specimens while absolutely wanting in most of the others, but the shape of the cell is, in general, sufficiently distinctive to permit of the ready separation of these specimens from those usually referred to *Chlorochytrium*.

Order 2. SIPHONALES (Grev.) Oltmanns

Fronds filamentous, either simple or variously entangled or interwoven, sometimes producing complex individuals, devoid of septa (or very nearly so) in the actively vegetative portions, but septa appearing in the reproductive portions, multinucleate and with many small chromatophores; multiplication vegetative, by non-sexual spores, and by zygotes; vegetative by abscission of proliferous shoots or fragmentation; non-sexual spores, either aplanospores or zoospores, produced usually in specialized zoosporangia; zygotes from either isogametes or heterogametes.

Siphonales Oltmanns, Morph. und Biol. d. Algen, vol. 1, 1904, pp. 134, 291; Blackman and Tansley, Revis. Class. Green Algae, 1902, p. 114; Collins, Green Alg. N. A., 1909, p. 385; West, Algae, vol. 1, 1916, p. 222. *Siphoneae* Greville, Alg. Brit., 1830, p. 183.

There is a very considerable variety both in the structure of the frond and in the methods of multiplication to be found among the Siphonales. From the simple globular but pedicellate *Halicystis* or the dichotomously filamentous *Vaucheria* or *Derbesia*, through the more or sometimes less specialized species of *Bryopsis*, to the elaborately constructed fronds of the Codiaceae, which are, however, made up of interwoven filaments with or without calcareous incrustation, there is a series of increasing complexities. In sexual multiplication there is also a series of increasing complexities from the isoplanogametes of *Bryopsis* through the heteroplanogametes of *Codium* to the condition in *Vaucheria*, where the female gamete is large and motionless and the male gamete is small and motile. Complexity of form and differentiation of gametes do not proceed along the same lines, e.g., *Vaucheria* has a frond of simple structure, while its male and female gametes are most widely different from one another.

3. Filaments pinnately branched..4. **Bryopsidaceae** (p. 156)
3 Filaments irregularly or dichotomously branched..................................... 4
    4. Sexual reproduction anisogamous; zoospore single, covered with cilia........
    ....7. **Vaucheriaceae** (p. 177)
    4. Sexual reproduction unknown; zoospores several in a sporangium, provided
    with a crown of cilia.............................5. **Derbesiaceae** (p. 163)

FAMILY 3.   PROTOSIPHONACEAE BLACKMAN AND TANSLEY

Thallus small, more or less globular, with or without colorless rhizoids or pedicels, unseptate in vegetative condition; nuclei several; chromatophore single and reticulate, or several, with or without pyrenoids; multiplication vegetative, by aplanospores, through micro- and macro-zoospores, and possibly also through isoplanogametes.

Blackman and Tansley, Rev. Class. Green Alg., 1902, p. 115; Collins, Green Alg. N. A., 1909, p. 153; West, Algae, vol. 1, 1916, p. 223.

A small family separated by Blackman and Tansley from the Botrydiaceae to contain particularly *Protosiphon* Klebs as a segregate from *Botrydium* Wallroth which, in turn, was removed from the Isokontae to the Heterokontae. Besides *Protosiphon*, it is usually made to include *Blastophysa* Reinke and *Halicystis* Aresch. The result is a not over homogeneous assemblage and one not readily or satisfactorily to be defined. Our only representative is *Halicystis ovalis* (Lyngb.) Aresch, which is described further on.

4. Halicystis Aresch.

Thallus globular to ovoid, unseptate, multinucleate, with penetrating rhizoidal portion; chromatophores small, disk-like, destitute of pyrenoids; asexual reproduction by 2-ciliated zoospores without stigma, escaping through one or more openings; similar but smaller zoospores or zoogametes(?) formed in separate individuals; after the emission of the spores the openings close and several new generations of spores can be similarly produced.

Areschoug, Phyc. Scand., part II, 1850, p. 446; Collins, Green Alg. N. A., 1909, p. 372.

A genus of two marine species of the northern oceans, seemingly occurring in both the north Atlantic and the north Pacific. It has been separated from *Valonia* because the vegetative body consists of an unseptate coenocyte of an ovoid shape with short pedicel and rhizoidal portion. It is now placed near to *Protosiphon,* which it more closely resembles than any other genus. The description given above is adapted from Collins (*loc. cit.*) and expresses well the general characters.

Halicystis ovalis (Lyngb.) Aresch.

Plate 14, fig. 3

Thallus solitary or gregarious, obovate-ovoid, 0.5–1 cm. high, about half as wide; membrane tough, 10–12μ thick; basal prolongation penetrating the substratum; zoospores 12–14μ long, 7–8μ wide; gametes (?) 7–8μ long, 2–3μ wide.

Growing on *Lithothamnion* and on other crustaceous corallines adhering by means of the rhizoidal portion penetrating deeply into the host. Vancouver, British Columbia, to Monterey, California.

Areschoug, Phyc. Scand., part II, 1850, p. 447; Kuckuck, Abhandl. Meeresalg., 1907, p. 139, pl. III; Collins, Green Alg. N. A., 1909, p. 372. *Valonia ovalis* (Lyngb.) Agardh, Sp., vol. 1, part 2, 1822, p. 431; Saunders, Four Siphon. Alg., 1899, p. 2, pl. 350, f. 2 a, b; Setchell and Gardner, Alg. N. W. Amer., 1903, p. 232. *Gastridium ovale* Lyngbye, Hydr. Dan., 1819, p. 72, pl. 18 B.

This very curious and interesting species has been most carefully studied, described and illustrated by Kuckuck (1907) and seems to be the same as that found on our own coast. Very little material is available for study, neither of the present writers has had the opportunity of collecting it, and only one of them (Setchell) has had the privilege even of examining a living plant. In general appearance and structure, however, the Pacific Coast plant agrees thoroughly with the descriptions of the European writers.

It was first credited to our coast by Saunders (1899) who found it at "Point Lobos" (or Point Carmel) in Monterey County, California. Later it was found in successive years at a locality near Point Cypress, only a few miles north of "Point Lobos" by Professor Harold Heath of Stanford University, and the third and last locality is Port Renfrew in British Columbia, where it was collected by Misses Butler and Polley. It is always found growing on living crustaceous corallines into the thallus of which it bores its way. It will probably be found at other points along the coast, since it undoubtedly escapes observation as it is small and grows at, or just below, the lowest tide mark.

FAMILY 4. BRYOPSIDACEAE (BORY) DE-TONI

Thallus a more or less branched, unseptate coenocyte, arising from rootlike, creeping, often rhizome-like filaments which originate as lowermost branches; branching more or less regularly or irregularly pinnate and lateral pinnules of definite growth arranged pinnately, and either distichous or polystichous, never interwoven; wall thin, neither incrusted nor provided with trabeculae (as in the Cauler-paceae); chromatophores and nuclei numerous and small, the former elliptically discoid and provided with a single pyrenoid each; vege-tative reproduction by a detachment of pinnules or breaking off of proliferations or the creeping rhizome; zoospores unknown; sexual reproduction by 2-ciliated anisogametes produced in gametangia which are slightly modified pinnules (*Bryopsis*) or ovoid or pyriform out-growths from the pinnules (*Pseudobryopsis*); monoecious or dio-ecious; female gamete the larger, with large posterior chromatophore, male gamete smaller, brownish-red with reduced chromatophore; zygote germinating at once.

Bryopsidaceae De-Toni, Syll. Alg., vol. 1, 1889, p. 427; Collins, Green Alg. N. A., 1909, p. 402; West, Algae, vol. 1, 1916, p. 225. *Bryopsideae* Bory, Voyage Coquille (Duperrey), Bot., 1828, p. 203; Thuret, Rech. sur les zoosp. et les antherid. des Crypt., 1850, p. 217 (*sub* "Bryopsidees").

The genus *Bryopsis* is the only representative of the family Bryopsidaceae on our coast. The family closely resembles the Der-besiaceae, from which it is distinguished by its method of branching and the possession of 2-ciliated, motile, reproductive bodies, the Codiaceae, from which it is distinguished by not having its branches interwoven to form a complex frond, and the Caulerpaceae, from which it is distinguished readily by the thin wall and lack of internal reënforcing plates or trabeculae. The fernlike fronds of our species distinguish them at a glance from all our other filamentous Chloro-phyceae.

The name Bryopsidaceae, as first used by Bory, included other Siphonales as well as *Bryopsis,* particularly species of *Caulerpa* and *Vaucheria.* The present concept of the family dates from about 1850 when Thuret published his classic paper on zoospores and anthe-ridia.

The account of the reproduction is adopted from Oltmanns (1904, p. 304 *et seq.*) and has not been verified, as yet, in our species.

5. **Bryopsis** Lamour.

Thallus unseptately coenocytic, much branched; chromatophores numerous small disks, each with one pyrenoid; the axis producing rhizoids below and branches above both of unlimited and limited growth; in the latter large, 2-ciliated, green, female gametes, and on separate individuals, smaller, brown, 2-ciliated male gametes are formed; by the union of the two a zygote is formed germinating immediately.

Lamouroux, Observ. sur la physiol. des alg. mar., 1809, p. 333, Mém. sur trois nouv. gen. de la famille des alg. mar., 1809a, p. 133; Collins, Green Alg. N. A., 1909, p. 402.

The above description, taken largely from Collins, expresses the technical characteristics of the family and of the genus *Bryopsis*. In this genus the gametes are produced in the branchlets of limited growth which are little changed, but are shut off from the axis on which they are borne when they are transformed into gametangia. The genus contains about twenty-five species and inhabits warmer waters, but a few species proceed northward into decidedly cold water, e.g., *B. plumosa* (Huds.) J. Ag. being credited even to the icy waters of Baffin Bay. Most of the species have wonderfully symmetrical fern-like fronds of a beautiful dark green which, when spread out on paper, adhere closely to it and produce a very pleasing picture. Our Pacific Coast species are nowhere common, and are in need of more careful study to determine their habits of growth and reproduction, as well as their specific differences and identities.

The species of *Bryopsis* present problems of determination of exceeding complexity and difficulty. The specific limits do not seem to be at all well ascertained and the actual identity and limits of the described species must remain uncertain until some monographer, with ample facilities and patience, shall have unusual opportunities for study and illustration. Much remains to be determined as to the stability of the various characters of these plants. A preliminary study leads us to believe that many characters, even of minor morphological importance, may prove stable and suitable for use in distinguishing species. The general habit, the number of orders of branching, the distinctness or lack of it of the "plumes," or feather-like divisions, the distichous, tetrastichous, or polystichous arrangement of the ultimate branchlets, or "pinnules," are characters now generally employed. We suggest also comparison of the exact shape and proportions of the pinnules and, especially, the shape of the bases of the older pinnules, as important characters. The bases of branches and branchlets, especially below on the main or secondary axes, may produce rhizoid-like, almost corticating, structures, and these seem to present differences, possibly of diagnostic value. M. A. Howe has particularly called attention to this in his "Marine Algae of Peru" (1914, p. 38 *et seq.* and pl. 7, f. 6–9). They were also made part of the distinction in *B. corticulans* Setchell, but they exist in many, or possibly all, species, varying in their shape and distribution.

Key to the Species

1. Thallus small, more or less simple................................1. **B. pennatula** (p. 158)
1. Thallus 8–14 cm. high, much branched.. 2
   2. Pinnules arising on all sides of the branches..............2. **B. hypnoides** (p. 159)
   2. Pinnules distichous.. 3
3. Base of pinnules abruptly constricted and unequally rounded.............................
....3. **B. corticulans** (p. 160)
3. Base of pinnules gradually tapering and not appreciably rounded.........................
....4. **B. plumosa** (p. 161)

1. Bryopsis pennatula J. Ag.

Thallus more or less simple, sublinear in outline, distichously pinnate; pinnules nearly equally long, cylindrical, obtuse.

Known only from the type locality "St. Augustin," on the Pacific coast of Mexico, where it was collected by Professor Liebmann of Copenhagen.

J. G. Agardh, Nya alger från Mexico ("Algae Liebmannianae"), 1847, p. 6; Kuetzing, Spec. Alg., 1849, p. 492, Tab. Phyc. vol. 6, 1856, p. 27, pl. 76, f. II. *Bryopsis pennata* var. *minor* J. G. Agardh, Till Alg. Syst., 1886, part 5, p. 23. *Bryopsis pennata* Collins, Green Alg. N. A., 1909, p. 405 (in part).

The only information available concerning this species is derived from Agardh's description and Kuetzing's figure. The latter seems to have been drawn from a specimen of the type collection. Agardh, later, as may be seen from the references above, reduced this species to a form of *Bryopsis pennata* Lamour. Comparing figures and specimens of *B. pennata*, there seems to be a close resemblance in habit and even in the shape of the pinnules, but the Mexican plant is very small and slender, as compared with typical *B. pennata*, and it seems best to us to keep it distinct until additional collections throw further light upon the relationships of the two plants. Concerning the type locality, which is also the type locality for other species published by J. G. Agardh in the same paper, it seems probable it is on the coast of the state of Oajaca in the vicinity of Pochutla and Pt. de Huatulco. (See Oersted in Liebmann, Chênes de l'Amerique Tropicale, 1869, p. viii.)

2. **Bryopsis hypnoides** Lamour.

Thallus 5–10 cm. high, flaccid, rather pale green, profusely and variously branched; branches in no definite order, growing smaller in the successive series, and with no sharp division between the lesser branches and the pinnules that clothe them on all sides, the latter themselves being frequently more or less branched; pinnules usually long and slender, gradually attenuate at the apices, suddenly constricted and symmetrically rounded at the bases.

Growing on logs, floats, shells, stones, etc. Ranging from Victoria, British Columbia, to San Pedro, California.

Lamouroux, Mém. sur trois nouv. genres., 1809a, vol. 2, p. 135, pl. 1, f. 2 a, b; Setchell and Gardner, Alg. N.W. Amer., 1903, p. 230; Harvey, Phyc. Brit., 1846, pl. 119; Vickers, Phyc. Barb., 1908, p. 30, pl. 53, f. 1, 2; Collins, Holden and Setchell, Phyc. Bor.-Amer. (Exsicc.), no. 1028.

Only three illustrations of this species are available to convey an idea of the characteristics of the species. The first, and only strictly authentic illustration, is that of Lamouroux (1809a, pl. 1, f. 2 a, b). Figure 2a represents the habit, while figure 2b shows the enlarged tip of a pinnule. The plant is evidently polystichous with the ultimate branchlets gradually attenuated in the lower third, or even half. The second is that of Harvey in the Phycologia Britannica (pl. 119), which does not represent the plant with sufficient detail to make as certain as desirable the shape of the base and apex of the pinnules. It is very evident, however, that the conception of Harvey was of a much branched, polystichous plant with long slender pinnules which are more or less constricted at the base and with the base itself unsymmetrical. The third illustration is that given by Anna Vickers in her Phycologia Barbadensis (pl. 53). The pinnules of this species, both as to proportions and as to branching, seem very different from those of both the others. The illustrations of the plumules (*loc. cit.*, f. 2) seem also to indicate that the gametes (?) are formed in restricted basal segments of the pinnules.

The branching of *Bryopsis elegans* Menegh. figured by Zanardini (1860–76, pl. 72) referred to by J. G. Agardh (1886, p. 28) as being possibly of *B. hypnoides* is different in detail, at least, from all the others.

The specimens from the Pacific Coast, referred here until more study and careful comparison with the type specimens can be made, seem reasonably uniform. They are much branched plants, polystichous, with less definite distinction between axes and with less regular plumes than *B. corticulans* shows. The pinnules are comparatively long and slender, long attenuate at the apex, but suddenly contracted to a broad, rounded base, and attached to the axis by a narrow neck. The older pinnules are very symmetrically rounded at the base and without any appearance of the production of rhizoidal outgrowths above, but possessing stout, rather long and branched rhizoids at the bases of the main branches (cf. M. A. Howe, 1914, p. 40, and Phyc. Bor.-Amer., no. 1028). The Pacific Coast plants referred here vary somewhat in coarseness and may ultimately be found to belong to more than one species.

3. **Bryopsis corticulans** Setchell
Plate 15, figs. 4, 5, and plate 27

Thallus rather stout and coarse, 8–14 cm. high, main stem 1 mm. diam.; dark green in the growing parts, glossy throughout; main stems not much divided, lower part naked, upper part, usually about half of the whole length, with abundant, patent, generally opposite branches constricted at the bases, naked below, above with rather stout, distichous pinnules, decreasing in length towards the tip of the branch and abruptly contracted at the unequal base; general outline of frond of individual branches pyramidal; conspicuous tufts of coarse, descending, slightly branched, rhizoidal filaments found at the bases of the branches and branchlets.

Growing on rocks in the lower littoral belt, from Vancouver, British Columbia, to southern California. Observed at Vancouver Island, British Columbia, Puget Sound, Washington, and also at Santa Cruz, Pacific Grove, Carmel, and San Pedro in California.

Setchell, *in* Collins, Holden and Setchell, Phyc. Bor.-Amer. (Exsicc.), 1899, no. 626; Collins, Green Alg. N. A., 1909, p. 404; Setchell and Gardner, Alg. N.W. Amer., 1903, p. 230. *Bryopsis plumosa* Tilden, Amer. Alg. (Exsicc.), no. 371 (not of C. Agardh).

Bryopsis corticulans is a coarse, dark green species, fairly regularly distichous and with pinnules little reduced in size from the axis whence they spring. It has, in the older plants at least, small clusters of short rhizoidal outgrowths at the bases of the lower (or even of the upper) branches. Thus far it has been observed only in winter and spring on the coast of California, but in Puget Sound it seems to occur also in mid-summer. It is closely related to *B. plumosa*, but it is coarser, with more regularly occurring corticulating rhizoids, and with the pinnae more elongated lanceolate. It is still a question whether we have true *B. plumosa* or not, and it is not absolutely certain that *B. corticulans* differs sufficiently from it to be always distinguishable. The pinnules of *B. corticulans*, however, are coarser, more robust, and more abruptly and unequally rounded at the base than those of any of the plants usually referred to *B. plumosa*. The lower plumules are not only abruptly and extremely constricted at the base, but possess bases which bulge out on the lower side where the rhizoidal growths issue. Thus far the great majority of the strictly distichous *Bryopsis* from our coast seem referable to *B. corticulans* rather than to *B. plumosa*.

4. **Bryopsis plumosa** (Huds.) Ag.

Plate 14, figs. 1, 2

Thallus not more than 10 cm. high, deep green and shining, more or less branched once or twice, seldom more, the ultimate branches forming plumes with distichous, slender pinnules gradually narrowed above and to a base which is slightly, if at all, rounded; bases of the lower branches showing several short lobes.

On floats, Puget Sound Marine Station, Friday Harbor, San Juan Island, Washington, collected by Annie M. Hurd.

Agardh, Sp. Alg., vol. 1, part 2, 1822, p. 448; Collins, Green Alg. N. A., 1909, p. 403. *Ulva plumosa* Hudson, Flora Anglica (2nd Ed.), 1778, p. 571.

The type locality of *Bryopsis plumosa* is Exmouth in Devon on the south coast of England, and no type specimen seems to be available. The species is widespread, as far as report goes, but it is very doubtful whether by any means all the plants, even of Europe, assigned to it, really are properly referred. It is distichous, as generally defined, and has broad triangular plumes. More investigation is needed to determine exactly the original application of the name, if possible, and also, as to how many varieties or even species are to be properly referred under it. The illustrations by Kuetzing (1856) both under *B. plumosa* (pl. 83, f. I) and under *B. abietina* (p. 80, f. I) are usually referred to *B. plumosa*. On comparing these illustrations with those of Harvey (1846, pl. 3), Greville (1830, pl. 19) and of *Bryopsis Lyngbyei* Hornemann (1818, pl. 1603) and its reproduction by Lyngbye (1819, pl. 19, f. B), it seems evident that there is considerable variety among the European plants referred to *B. plumosa*. Unfortunately the figure of *B. arbuscula* Lamouroux (1809, pl. 5, f. 1) is only of the habit of the plant.

Without opportunity of examining a distinct type, or any specimens from the type locality, it is impossible to determine with any certainty just what the *Ulva plumosa* Hudson may be. However, judging by specimens from Debray from the neighboring coast of France, it seems likely to prove to be a plant very similar to that figured by Kuetzing (1856, pl. 83, f. II), viz., a distichous plant with pinnules gradually tapering to both base and apex. This gradually tapering base of the pinnule is not found in any of our specimens of *B. corticulans* or *B. hypnoides*. It may possibly be that this is characteristic of only younger stages of the European plant, but this does not seem likely. An illustration which seems to be reasonably authoritative, and one that may serve as a basis for discussion is that of *Ulva plumosa* of the English Botany (vol. 33, 1814, pl. 2375). This seems also to be the basis for the figures of plate 19 of Greville's *Algae Britannicae* (1830). The agreement of these figures with that of Kuetzing is very close.

As to whether *Bryopsis plumosa* is represented on the Pacific coast of North America, or not, there is little to be said. The name appears in certain local lists and Tilden (Amer. Alg., no. 371) has distributed a plant from Tracyton, Washington, in the Puget Sound region which seems rather to be *B. corticulans* Setchell. There is a single specimen collected by Miss Hurd (Herb. Univ. Calif., no. 200726) which seems to agree fairly closely with the illustration in the "English Botany" and we refer this to *B. plumosa* with some hesitation. The shape of the pinnules seems to be fairly characteristic. In this specimen the bases of the lower branches show several short, blunt lobes. Older specimens might show that these grow out into short rhizoids such as are found abundantly in Atlantic Coast and European plants referred to this species. The description given above was drawn up with especial reference to our plant.

FAMILY 5. DERBESIACEAE (THURET) KJELLM.

Thallus of erect, simple or sparingly branched coenocytic filaments, arising from more slender creeping filaments which are attached to the substratum by short, branched, rhizoid-like holdfasts; chromatophores small, discoid, oval or elliptical in shape without or with one or two pyrenoids; non-sexual reproduction by zoospores provided with a crown, or circlet, of cilia, produced in special, lateral, globose to pyriform zoosporangia, sexual reproduction unknown.

Derbesiaceae Kjellman, Algae Arctic Sea, 1883, p. 316. *Derbésiées* Thuret, Rech. sur les zoospores des algues, etc., 1850, p. 231 (p. 22 Repr.). *Derbesieae* Thuret, *in* Le Jolis, Liste des alg. mar. de Cherbourg, 1863, p. 14.

The family of the Derbesiaceae differs from all others of the Isokontae in the possession of zoospores with a circlet of cilia similar to those of the Stephanokontae. In spite of this seemingly fundamental difference, all writers have placed it among the Isokontae rather than among the Stephanokontae. Davis, in his paper on ''Spore formation in *Derbesia*'' (1908), has followed out the nuclear behavior during zoospore formation and its relation to the development of a blepharoplast as well as the resulting circle of cilia. Unfortunately the development of the zoospore and of the circlet of cilia is not as yet known for *Oedogonium* or any other of the Stephanokontae. Davis (*loc. cit.*, p. 16) states that the zoospores of *Derbesia* and of *Oedogonium* are of similar structure and ventures to predict that those of *Oedogonium* will be found to develop a blepharoplast closely similar to that of *Derbesia*. Nevertheless, he warns against the danger of classifying the algae on the basis of the structure of zoospores and gametes and expresses as his idea that *Derbesia* should not be removed from the Siphonales. Davis also expresses the opinion that no one will be bold enough to suggest a relationship between *Derbesia* and *Oedogonium* on account of the resemblances of the zoospores.

It seems to us, however, that in the Stephanokontae, there exists a peculiar type which may be as early, or as primitive, as any of those under the Isokontae. Possibly there may have existed many forms of Stephanokontae, now lost, or possibly not yet discovered. We may assume then that as the Isokontae have advanced along several lines from multicellular to septate and then to unseptate coenocytic condition, the Stephanokontae may have done the same. It seems to us neither impossible, nor wholly inconsistent with what we find among the Isokontae, to consider *Derbesia* as a coenocytic genus of the Stephanokontae. For general convenience, however, connected with the fact that this account deals with marine species only, we leave it in the place usually assigned to it.

There are only two genera to represent this family, *Derbesia* Solier and *Bryobesia* Weber-van Bosse. Of these *Derbesia* alone has been found, thus far, on our coast.

6. Derbesia Sol.

Filaments unseptate, or with occasional partitions, multinucleate, simple or branched, with no differentiation of axis and branches; chromatophores numerous, discoid, with or without pyrenoids; non-sexual reproduction by large, multiciliate, stephanokont zoospores, each with a single nucleus, formed in lateral globose to pyriform zoosporangia; sexual reproduction unknown.

Solier, Sur deux alg. zoosp. form. le nouv. genre *Derbesia*, 1846, p. 453 (cf. also Bot. Zeit., vol. 4, 1846, p. 497), Mém. sur deux algues, 1847, p. 158.

Little remains to be said of the genus *Derbesia* after the description of the family, since there are only two genera included in Derbesiaceae. The genus *Derbesia* was founded on *D. marina* and *D. Lamourouxii*, of which the former is given first, and may properly be considered as the type. *D. marina* Solier, however, is judged not to be identical with *Vaucheria marina* Lyngbye and is now known as *D. tenuissima* (De Not.) Crouan. The genus at present consists of eight to ten species widely distributed chiefly in tropical and subtropical waters. It differs from *Bryobesia* in having the sporangia (?) lateral. In *Bryobesia* after the terminal sporangium is emptied it is forced to one side by the continued growth of the filament beneath.

Unfortunately we have had no opportunity of studying any of our Pacific Coast species of *Derbesia* in the living condition and must draw upon the publications of others for all details.

KEY TO THE SPECIES

1. Filaments 50–70μ in diameter .... 1. **D. marina** (p. 165)
1. Filaments 100–600μ in diameter .... 2. **D. Lamourouxii** (p. 165)

1. **Derbesia marina** (Lyngb.) Kjellm.

Plate 15, fig. 3

Filaments arising from a creeping base, bright green, $50–70\mu$ diam., simple or usually with a few lateral branches similar to the axes, continuous but with a short segment separated by partitions near the base of a branch or occasionally in the axils just above a branch, little smaller than the branch itself and about as long as broad; sporangia occupying the place of branches, ovoid to subspherical, $150–250\mu$ long, $90–200\mu$ broad; pedicel varying from $30–70\mu$ in length, $30–35\mu$ in diameter, about as long as broad; spores 20 or more in a sporangium.

Alaska to southern California.

Kjellman, Alg. Arctic Sea, 1883, p. 316 (not of Solier, fide J. G. Agardh, Till. Alg. Syst., 5th part, VIII, 1886, p. 34); Saunders, Alg. Harriman Exp., 1901, p. 415; Setchell and Gardner, Alg. N.W. Amer., 1903, p. 230; Collins, Green Alg. N.A., 1909, p. 407. *Derbesia tenuissima* Collins, Holden and Setchell, Phyc. Bor.-Amer. (Exsicc.), no. 574. *Vaucheria marina* Lyngbye, Hydr. Dan., 1819, p. 79, pl. 22 A.

We suspect that the *Derbesia vaucheriaeformis* of Saunders (1899, p. 3, pl. 350, f. 4 a-d) from "Point Lobos" (really Point Carmel) near Monterey, California, is to be placed rather under *D. marina* as understood by J. G. Agardh. Saunders describes his species as possessing filaments from 30μ to 40μ broad and elliptical, obovate or pyriform sporangia 140μ to 200μ long and 50μ to 80μ wide. The sporangia are distinctly pedicellate but the diameter of the pedicel is not given. Judging from his figure 4a, the pedicel is one half the diameter of the filament, or 15μ to 20μ broad. The zoospores, also judging from the same figure, number decidedly more than twenty. The branching is represented as dichotomous. Saunders, also, in the Algae of the Harriman Expedition to Alaska (1901, p. 415) refers a plant from Yakutat Bay to *D. vaucheriaeformis,* but did not observe mature zoosporangia.

2. **Derbesia Lamourouxii** (J. Ag.) Sol.

Filaments arising from a creeping base, a few centimeters to 2 dm. high, $100–600\mu$ diam., dark green, rather stiff, sometimes simple, sometimes with more or less numerous irregular branches; sporangia spherical, $300–550\mu$ diam., sessile or on short and slender pedicels.

Southern California.

Solier, Mém. sur deux Algues, 1847, p. 162, pl. 9, f. 18–30; Collins, Green Alg. N. A., 1909, p. 407. *Bryopsis Balbisiana* var. *Lamourouxii* J. Agardh, Alg. Med., 1842, p. 18.

This species is represented in the Herbarium of the University of California by two sterile specimens collected by Mrs. E. A. Lawrence, five miles south of the boundary between southern and Lower California. It is decidedly coarser than any other of the described species of the genus.

FAMILY 6. CODIACEAE (TREVIS.) ZANARD.

Thallus dark green, spongy, subspherical, applanate or erect, cylindrical, flattened, or jointed, simple or dichotomously branched, at times incrusted with lime, composed of intertwined branching filaments, the peripheral branchlets forming a palisade or pavement-like external layer; septa (diaphragms) frequent but in connection with formation of reproductive organs or in older filaments; chloroplasts parietal, small, very numerous, especially at the apices of the branches, destitute of pyrenoids; multiplication through fragmentation and by zoospores and anisoplanogametes; zoosporangia and gametangia differentiated and variously situated.

Zanardini, Sagg. di class. nat. d. Ficee., 1843 (table opposite p. 17). *Codieae* Trevisan, Prosp. fl. Eugan., 1842, p. 50, Flora, 1843, p. 465 (in part).

The family, originally separated as Spongodiées by Lamouroux (1813, p. 280 or p. 71 of repr.), has long been recognized as distinct among the Chlorophyceae. It contains the most highly differentiated of the genera of the marine Green Algae, both as to complexity of thallus and as to differentiation of the reproductive cells, its only competitors being the Dasycladaceae and the Vaucheriaceae. The thallus is made up of interwoven coenocytic filaments, the peripheral branchlets of which are distinctly and variously differentiated and combined into a distinct external layer. The sporangia and gametangia are formed from modified lateral branchlets of the coenocytic filaments and the gametes are distinctly unlike in some species at least. Many of the species are heavily incrusted with lime and are important agents in the building up of coral reefs. The majority of the species are strictly tropical but some of the species of *Codium* are to be found in subtropical, temperate, and perhaps even in frigid waters.

KEY TO THE GENERA

1. Thallus without joints or calcification..7. **Codium** (p. 167)
1. Thallus with distinct joints and more or less calcified............8. **Halimeda** (p. 176)

7. **Codium** Stackh.

Thallus spongy, not incrusted with lime, applanate, subspherical or cylindrical, simple or dichotomously branched, attached, dark green; medullary filaments vertically intertwined, giving rise to horizontal branchlets whose tips, swollen into "utricles," form a continuous external palisade layer; multiplication by fragmentation of the thallus; sexual reproduction through 2-ciliated anisogametes produced in gametangia situated laterally on the utricles; dioecious or occasionally monoecious.

Stackhouse, Nereis Brit., 1797, p. xvi. *Lamarckia* Olivi, *in* Olivi, Zool. Adriat., 1792, p. 258, and *in* Usteri, Ann., part 7, 1794, p. 76. *Spongodium* Lamouroux, Essai, 1813, (p. 72 repr.).

The designation of this genus presents certain difficulties. The earliest name proposed seems undoubtedly to be *Lamarckia* of Olivi (1792, p. 258 and 1794, p. 76). There are, however, several other genera dedicated to Lamarck and the generic names have been spelled in various ways. The first of these was proposed by Medicus in 1789 (p. 28), but is now regarded as a synonym of the Malvaceous genus *Sida*. *Lamarkia* of Mönch, proposed in 1794 (p. 201) is still recognized as a genus of grasses, and has been adopted by the International Botanical Congress at Vienna as a nomen conservandum (cf. Briquet, 1906, p. 73, and 1912, p. 79). *Codium* was proposed by Stackhouse in 1797 in the first edition of the Nereis Britannica (2d fascicle, p. xvi), but in the second edition (1816, p. xii) evidently abandoned in favor of "*Lemarkea.*" There is an earlier generic name, *Codia* (Forster and Forster, 1776, p. 59), still used for a genus of Saxifragaceae, and *Codiaeum* of Rumphius (1743, p. 65) is still current among the Euphorbiaceae. Otto Kuntze (1891, p. 900) argues for "*Lamarckia*" as the proper designation, but *Codium*, properly diagnosed (for the period), has been in almost universal use for nearly, if not quite, a century, and has the right of way now that the status of the name of *Lamarkia* has been settled as indicated above.

The genus *Codium* contains somewhat over twenty-five described species agreeing closely in microscopic structure, but differing very decidedly in habit. Some are flat expansions, some are expanded but cushion-shaped, some are spherical and hollow, while some are either cylindrical or flattened but erect and branching. J. G. Agardh (1886, p. 35 *et seq.*) has subdivided the genus according to these differences. After habit, good characters for distinction of the species have been sought for in the varying size, shape, proportions, and modification of the tips of the utricles. Further and more careful study in this direction, not only of plants of different species, but also of plants of varying ages of the same species is very much to be desired to determine the limits of these variabilities.

The reproduction of *Codium* is known in detail only for the European *C. tomentosum* (cf. Oltmanns, 1904, p. 301, and West, 1916, p. 241). In this species the spore reproduction is exclusively sexual. There are two kinds of gametangia giving rise respectively to larger motile 2-ciliated female gametes, and smaller but similar male gametes. Conjugation has been observed and a thick walled zygote is formed which germinates later. Nothing has been undertaken, thus far, towards the study of the reproduction of our species. Gametangia (?) have been seen, in most species credited to our coast, but further stages have not been observed.

KEY TO THE SPECIES

1. Thallus prostrate... 2
1. Thallus erect.. 3
    2. Thallus applanate, flat..1. **C. Setchellii** (p. 168)
    2. Thallus cushion-shaped, rounded.................................2. **C. Ritteri** (p. 169)
3. Utricles more or less mucronate...................................3. **C. fragile** (p. 171)
3. Utricles never mucronate... 4
    4. Thallus cylindrical or flattened only below the axils...................................... 5
    4. Thallus cylindrical only at base, decidedly flattened above 6. **C. latum** (p. 175)
5. Utricles 400μ or more in maximum diameter .... 4. **C. decorticatum** (p. 172)
5. Utricles 250μ or less in maximum diameter................. 5. **C. tomentosum** (p. 174)

1. Codium Setchellii Gardner

Plate 30, and plate 9, figs. 10, 11

Thallus forming dense, compact, spongy, irregular cushions, 6–10 mm. up to 15 mm. thick, adhering firmly to rocks; color dark glossy green; medullary filaments 12–30μ diam.; utricles variable in shape, clavate, cylindrical, or sometimes constricted below the apex, truncate or slightly rounded above, 65–75μ wide, walls thin throughout when young, the outer ends 6–16μ thick and lamellose when older; gametangia cylindrical or slightly fusiform, 300–330μ long, 45–55μ diam., growing singly on the utricles; trichomes wanting.

Growing on rocks in the lower littoral belt. Central California (possibly extending to southern California) and northward to Sitka, Alaska.

Gardner, New Pac. Coast Mar. Alg. IV, 1919, p. 489, pl. 42, f. 10, 11. *Codium adhaerens* Anderson, List of Calif. Mar. Alg., 1891, p. 217; Howe, A month on the shores of Monterey Bay, 1893, p. 63; McClatchie, Seedless Plants, 1897, p. 351; Saunders, Four Siphon. Alg., 1899, p. 2, pl. 350, f. 3 a, b, c, Alg. Harriman Exp. 1901, p. 416; Setchell and Gardner, Alg. N. W. Amer., 1903, p. 231; Collins, Green Alg. N. A., 1909, p. 387 (not of Agardh). *Codium dimorphum* Hurd, Pug. Sound Mar. Stat. Publ., vol. 1, 1916, p. 211–217, pl. 38, f. 1–13; Collins, Green Alg. N. A., Supl. 2, June, 1918, p. 88 (not of Svedelius).

Codium Setchellii represents the *adhaerens* group of J. Agardh (1886, p. 37) on our coast. For many years all the collections of this group from the Pacific Coast of North America passed under the name of *C. adhaerens* (Cabr.) Ag. Many different species, however, have been and still are referred to *C. adhaerens,* and much careful study and comparison will be necessary before they can be satisfactorily separated.

Miss Hurd (*loc. cit.*) was the first to throw doubt on the relationship of our plant with the *C. adhaerens* of the European coast. She, however, concluded from her studies that our plants, particularly from the region of the San Juan Group of Islands, Washington, are identical with *C. dimorphum* Svedelius (plate 9, figs. 7, 8) from West Patagonia. A careful study and examination of authentic material show sufficiently constant differences between that species and our plants to seem to make it necessary to consider ours distinct from that species and from the European *C. adhaerens* as well. It has consequently been described as new by Gardner (*loc. cit.*), its type locality being Monterey, California, since from that general locality only have fruiting specimens been collected. It is highly desirable that material be studied at different seasons throughout its range and fruiting material found with a view of determining whether we have one or more species on our coast.

2. **Codium Ritteri** S. and G.

Plate 16, fig. 5

Thallus spongy, globose when young, becoming flattened, expanded and variously lobed when older, 1.5–2.5 cm. thick, attached by a broad base; the center and lower portion consisting of tortuous, loosely interwoven, rhizoidal filaments, 50–65μ diam.; utricles clavate, usually branching, rarely swollen in the middle and fusiform, mostly truncate,

150–250μ, up to 400μ diam.; 2–4 mm. long, end wall slightly thickened and with a small depression in the center; gametangia unknown.

Growing on rocks in the lower littoral and upper sublittoral belts. Extending from Kadiak Island, Alaska, to the west coast of Vancouver Island, British Columbia.

Setchell and Gardner, Alg. N.W. Amer., 1903, p. 231, pl. 17, f. 8–11. *Codium adhaerens* Tilden, Amer. Alg. (Exsicc.), no. 370 (not of Agardh).

The type specimen is a small plant about three centimeters in diameter, or possibly it may be a small lobe of an old thallus that had become loosened at the base and by wave action had been torn from the remainder of the thallus. Since the publication of the species, excellent material, several inches across, has come from Alaska to the Herbarium of the University of California through the courtesy of T. C. Frye and G. B. Rigg. These show that the species is considerably expanded and variously lobed, though not adhering so firmly and closely to the rock as do other incrusting forms of this genus. The utricles in the upper portion away from the margin are relatively much longer and narrower than those figured for the type (Setchell and Gardner, *loc. cit.*).

New utricles arise from the older one by lateral branching, as many as four and five appearing at a time. The point of origin remains constricted, and a plug cuts off the new utricular protoplast from the old one. In due course these branches may drop off, the old utricle continuing to grow, and later giving rise to others. From their bases the new utricles give rise to several rhizoidal filaments. The species probably fruits in the winter, since all the material thus far collected is sterile, and the collections have all been made in the summer.

Setchell and Gardner (*loc. cit.*, p. 232) suggested that possibly the plant distributed by K. Okamura (Algae Japonicae Exsiccatae, no. 49) under *C. mamillosum* belongs under *C. Ritteri*. Okamura (1915, p. 152) dissents from this suggestion. Since having had opportunity for further study of *C. Ritteri*, reëxamination of the Japanese plant leads us to conclude with Okamura that his plant is not the same as *C. Ritteri*, but we do not agree with him that it is the same as *C. mamillosum* of Agardh.

We have not seen the specimen of *Codium mamillosum* of Coville and Rose (1898, p. 353) from Preobrazhenskoye, Copper Island, in the Commander Group, Siberia, but, on account of the low temperature of the water at that point, we suspect that their plants may belong rather to *C. Ritteri* than to *C. mamillosum* which is a subtropical species.

3. **Codium fragile** (Suring.) Hariot
Plates 28, 29

Fronds one to several from a broad spongy disk, cylindrical, profusely and dichotomously branched, 25–40 cm. high, 2–10 mm. in diameter, glossy, dark green, finely rugose on the surface or, at times, densely tomentose with long hyaline hairs; utricles cylindric clavate, 150–350μ (occasionally 630μ) in maximum diameter, and 5–10 times as long as broad, provided (at least when young) with a more or less distinct mucro; gametangia ($\female$?) fusiform, 1–3 to each utricle, 250–450μ long and 75–150μ in diameter.

Growing on exposed rocks and in small pools in the littoral belt. Alaska to Mexico.

Hariot, Algues du Cap Horn, 1889, p. 32; De-Toni, Phyc. Japon. Novae, 1895, p. 64. *Acanthocodium fragile* Suringar, Algarum Japonicaeum, Index praec. 1867, *ibid.*, Hedwigia, vol. 7, 1868, p. 55; Algae Japonicae, 1870, p. 23, pl. 8; *ibid.*, Hedwigia, vol. 9, 1870, p. 133. *Codium mucronatum* J. Agardh, Till. Alg. Syst., 5th part, 1886, pp. 43, 44; Hurd, Pug. Sound Mar. Stat. Publ., 1916*a*, vol. 1, pp. 109–135, pl. 19–24; Setchell and Gardner, Alg. N.W. Amer., 1903, p. 232; Saunders, Alg. Harriman Exp., 1901, p. 416. Four Siphon. Alg., 1899, p. 1, pl. 350, f. 1 a, b, c, d; Collins, Green Alg. N. A., 1909, p. 389; Collins, Holden and Setchell, Phyc. Bor.-Amer. (Exsicc.), no. 229. *Codium tomentosum* Tilden, Amer. Alg. (Exsicc.), no. 281 (not of Stackhouse).

The mucronate tip of the utricle of this plant is a prominent specific character, but this character is subject to extreme variation. J. Agardh (1886, pp. 43, 44) separated the species into three varieties based chiefly upon the character of the mucro. We have studied and compared plants from a considerable number of different localities ranging from Alaska to Mexico, and have come to the conclusion that the species cannot be split into varieties based upon that character. Single plants may be found producing mucronate tips covering the entire range of shapes, as regards thickening of the walls and acuteness of the tip, that have been used to designate the varieties. Miss Hurd (1916*a*, p. 109) made a critical study of the species as occurring in the vicinity of the Puget Sound Marine Station, and came to the conclusion that they exhibited all the possible variations assigned to the varieties *novae-zelandiae* and *californicum*, both of which have been accredited to our coast. Cotton (1912, p. 114–119, pl. 7, 8, f. 3–5) has discussed the species as found on the coasts of Ireland and of Scotland with remarks on the varieties described by himself and J. G. Agardh.

As usually found, *Codium fragile* gives no impression of being tomentose, but occasionally plants are found which are covered with a thick coating of hairs. Miss Hurd (1916a, p. 114–116) has discussed the hairs on the utricles of this species in some detail, but did not find the extreme tomentose condition found by one of us (Gardner) at Redondo, California where the hairs were 2 mm. long, and so densely covering the whole plant as to make it seem as if parasitized. It seems that the conditions causing such extreme growth of hairs have not as yet been definitely ascertained (cf. Hurd, *loc. cit.*, and Oltmanns, 1905, p. 239). There is still a fertile field for observation and experimentation in this subject.

Codium fragile seems to be a widespread species. J. G. Agardh described it as occurring on the west coast of North America, in New Zealand, in Australia and in Tasmania. It has been found also in the Cape of Good Hope region and in the region of the Straits of Magellan (Hariot, 1889, p. 33, and Svedelius, 1900, p. 299) and on the coasts of Scotland and of Ireland (cf. Cotton, 1912, p. 115). This is certainly a wide distribution in widely separated waters, but at least the waters have approximately the same temperatures for certain portions of the year. It is interesting to compare in this connection *Codium divaricatum* f. *hybrida* Okamura (1915, p. 157, pl. 135, f. 17), which is suggested as being a hybrid between *C. divaricatum* Holmes and *C. fragile* (Suring.) Hariot.

4. Codium decorticatum (Woodw.) Howe

Thallus sparingly branched, dichotomous, up to 5 dm. or more long, usually decidedly flattened under the dichotomies; peripheral utricles obovoid to broadly clavate, thin walled throughout, obtuse, 135–520μ maximum diameter, 500–700μ long.

La Paz, Lower California.

Howe, Phyc. Studies V, 1911, p. 494. *Ulva decorticata* Woodward, Observations upon the generic characters of *Ulva*, 1797, p. 55.

Howe (*loc. cit.*) refers a plant from La Paz, Lower California, to this species, apparently with some confidence, and at the same time he selects *Codium decorticatum* as the correct name instead of *C. elongatum*. We have not seen Howe's plant, but certain questions arise as to the exact nature of *C. decorticatum* and of *C. elongatum*. The type specimen of *Ulva decorticata* Woodward (*loc. cit.*) is unknown and the type region is given as the Mediterranean Sea with the statement that the exact locality was unknown to the author. The type locality for *C. elongatum* Ag. (1822, p. 454) is Cadiz, Spain. The species is related to *C. tomentosum*, but differs in being more elongated, with fewer and longer branches, and in being more or less distinctly dilated and flattened just below the axils (or some of them). Such a plant is figured by Kuetzing (1856, pl. 96 b), but his plant came from Rio Janeiro, Brazil, a tropical locality. Bornet (1892, pp. 216, 217) has given his experience with *C. elongatum* and his attempts to ascertain the characters upon which separation could be made from *C. tomentosum*. He considered two sets of characters, viz., the extent of infra-axillary dilation and the magnitude of the utricles. Bornet decided to separate the species according to the presence or absence of dilations, and to subdivide the species with dilations (*C. elongatum*) into a variety with large utricles and one with small utricles, since this fulfilled the idea of C. Agardh and also corresponded with the geographical distribution. *Codium elongatum* with small utricles extends north along the Atlantic coast of France and to England, while *C. elongatum* with large utricles does not extend north of Cadiz. In this connection it is interesting to note again that Cadiz is the type locality of *C. elongatum*, and to note also that the type specimen of *C. elongatum* is, according to Howe (1911, p. 495), provided with slender utricles. The tropical forms referred to *C. elongatum* (or *C. decorticatum?*) have stout utricles as Kuetzing has described and figured (*loc. cit.*). We suspect that there may be two overlapping species of somewhat similar habit, but differing in utricles and in geographical distribution, represented under *C. decorticatum*. It may be that *C. decorticatum* (Wood.) Howe, being a Mediterranean species (subtropical), may finally be separated from *C. elongatum* Ag., the more northern (temperate) species which reaches its southern limit near Cadiz and Tangiers, and there intermingles with the large utricled form (*C. decorticatum?*) as Bornet has found it. Howe's plant from La Paz is described as having large utricles (up to 520μ diam.) and must therefore be arranged with the larger form of Bornet.

5. Codium tomentosum (Huds.) Stackh.

Thallus rather slender, much branched, 22–37 cm. high when growing in pools, 48–60 cm. high when growing in deep water, 3–4 mm.
thick, cylindrical, often slightly flattened at the axis, dichotomous,
surface often very tomentose, becoming smooth with age, color dark
green; utricles cylindrical, small, 500–650μ long, 120–170μ (rarely to
220μ) wide, apex usually distinctly thickened, blunt; smaller utricles
sometimes pointed, but never mucronate; gametangia (♀) small, 200–
250μ long, 40–70μ wide; gametes 20–22μ long, 10–12μ wide.

La Paz, Lower California.

Stackhouse, Ner. Brit. (fasc. 3), 1801, p. xxiv; pl. 7; Howe,
Phyc. Studies V, 1911, p. 493; Collins, Green Alg. N. A., 1909, p. 388;
Harvey, Phyc. Brit., 1846, pl. 93; Vickers, Phyc. Barb., 1908, p. 22,
pl. 26. *Fucus tomentosus* Hudson, Flora Anglica, 1778, p. 584.

In attempting to arrange the erect branched species of our coast
with non-mucronate utricles, we are confronted with a problem of
which the solution seems impossible at present. In the first place,
the material available to us is slight; in the second place, the reference
of the similar species of Europe and other parts of the world is not
at all satisfactory; and in the third place it is impossible at present
to examine the types of the hitherto described species of this group.
When we add to this a lack of knowledge as to the possible variation
in habit and size of utricle of the species of *Codium*, it seems sufficient
to prevent us from presenting any but a tentative, and by no means
satisfactory, arrangement. We have decided to refer our plants of
this group under two species, viz., *C. tomentosum* (Huds.) Stackh.
and *C. decorticatum* (Woodw.) Howe.

Codium tomentosum was originally described from Exmouth in
Devon, on the south coast of England. The type specimen will presumably be found in the Buddle Herbarium in the British Museum,
but no account has been published as to its exact nature. Cotton
(1912, p. 114) has published an exact description of the Clare Island
plant which, presumably, is true *C. tomentosum*, and we have adopted
this in our diagnosis. We may assume that the typical form is a
slender, much branched plant, of varying length, cylindrical, or often
slightly flattened just below the axils and with slender utricles, "120–
170μ (rarely to 220μ) wide," with apex distinctly thickened and blunt
or at times pointed, but never mucronate. We have never seen a
plant from our coast answering to this description. Howe (*loc. cit.*)
has referred here a plant from La Paz, Lower California. He also

seems to be inclined to refer here no. 628 of the Phyc. Bor.-Amer. from La Jolla, California, which was distributed under the name of *Codium Lindenbergii* (cf. plate 31). We have examined the specimens of no. 628 in our copies of the distribution. Altogether there are three specimens available and all are much branched, rather broad (up to 1.5 cm.) plants which seem distinctly flattened. The utricles vary more than we have found to be the case in European *C. tomentosum*. In fact there is almost a "dimorphism" and the occasional larger type of utricle reaches a diameter of 150–330μ while the diameters of the smaller, somewhat differently shaped, utricles range from 45–80μ or more. The walls of both sorts of utricles may be thickened at the top, even to 28μ. The gametangia (?) are broadly fusiform and measure close to 200μ long and 100μ wide. More information is needed concerning these plants and it is hoped that further collections may be made. They do not seem to belong to *C. Lindenbergii* since they, although flattened, differ decidedly in details of branching and in breadth. The utricles of *C. Lindenbergii*, as described and figured by Kuetzing, are probably at least 250μ in maximum diameter, and in specimens distributed by Tyson (no. 55) occur up to at least 380μ in diameter, while the gametangia (?) are 200μ to 228μ long and 76μ to 95μ wide. It should be mentioned, however, that there are three flattened species described from Cape Colony, viz., *C. Lindenbergii* Kuetz., *C. damaecorne* (Bory) Kuetz., and *C. platylobium* Aresch., usually combined (cf. J. G. Agardh, 1886, p. 46) under the name of the first species, but possibly without good reason.

Bornet (1892, pp. 216, 217) discusses at length a similar problem connected with Schousboe's plants from Tangiers and refers *C. Lindenbergii* as a form with smaller utricles under *C. elongatum* Ag. More has been said of this under *C. decorticatum*.

In conclusion, we may say that we are not satisfied in referring the plant of southern California (as represented by no. 628, Phyc. Bor.-Amer.) to *C. tomentosum* and feel that it is probably an undescribed species. More experience with the living plant is needed, however, satisfactorily to determine its exact status.

6. **Codium latum** Sur.

Plate 15, fig. 6

Thallus arising from a small, spongy disk, 15–25 cm. high, or more, the lower stipitate portion cylindrical, 3–5 mm. diam., 2–3 cm. long, more or less branched and abruptly flattened into broad, di-trichotomously branched lobes rounded at the apices; lobes 2–3 mm. thick

and up to 5 cm. or more wide; utricles 500–600μ long, 50–110μ, up
to 160μ, diam., nearly cylindrical when young, with a constriction just
below the rounded apex, at maturity, with a decided shoulder below
the constriction bearing a whorl of 4–6 hairs, membrane usually thickened at the apices, up to 25μ thick; gametangia (?) borne below the
middle of the utricles, fusiform, 220μ long, 60–75μ diam., membrane
thin.

Guadalupe Island, Mexico.

Suringar, Algae Japonicae, 1870, p. 22, pl. 7; Okamura, Icon.
Japan. Algae, 1915, vol. 3, no. 9, p. 158, pl. 142.

The inclusion of *Codium latum* Sur., a Japanese species, in our
account is based upon several specimens in the Daniel Cady Eaton
Herbarium of Yale University. These specimens were collected on
the shores of Guadalupe Island by E. Palmer in 1875. The specimens
are nearly cylindrical and dichotomous or dichotomo-fastigiate below,
but soon expand into long, broad, flattened lobes or branches. The
particular character which makes the reference of these specimens to
C. latum seem plausible, is the existence of a distinct whorl or verticil
of hairs (shown in older specimens by projecting scars) a little below
the broad apex of each utricle. These are distinctly represented in the
illustrations of the species by Okamura (1915, pl. 142, f. 4, 6). The
specimens resemble those of the *C. Lindenbergii* complex, but differ
decidedly in the shape and size of the utricles as well as the arrangement of hairs upon them. The Guadalupe plants do not approximate
the extremes of either length or breadth given by Okamura (*loc. cit.*)
for his Japanese specimens but are very similar to the dimensions
given by Suringar (*loc. cit.*).

8. **Halimeda** Lamour.

Fronds jointed, freely branching from near the base, attached by
a dense mass of rhizoidal filaments usually strongly calcified except
at the nodes; segments from slightly to very much flattened and
expanded, flattened cylindrical, cuneate, orbicular or reniform, entire
or variously lobed; medullary tissue a strand of longitudinal, slender,
branched, unseptate filaments, expanding in the segments by lateral
branchlets whose terminal cells (utricles) cohere more or less tightly
to form a continuous layer, but unchanged, although often anastomosing, at the nodes and at the apex; reproduction by globose or ovoid
sporangia (?) borne on slender filaments projecting beyond the sur-

face of segments and producing 2-ciliated zoospores (?) whose further development has not been followed.

Lamouroux, Class. Polypes, 1812, p. 186.

The genus *Halimeda* is very well marked on account of its calcified, jointed structure, most of the species are strongly calcified, but in the single species thus far detected on our coast the calcification is slight. The species are strictly tropical, occurring in abundance on coral reefs and assisting materially in their formation. Otto Kuntze (1891, p. 908) has urged the substitution of the earlier name *Opuntiodes* of Ludwig (1737, pl. 138), but since *Halimeda* has been in practically undisputed use for over a century, it seems best to retain it as a *"nomen conservandum"* if necessary.

Halimeda discoidea Dec'ne
Plate 13, fig. 3

Fronds branched in one plane, up to 15 cm. high, nearly orbicular in outline, very slightly calcified, color bright green, fading on drying; segments mostly quadrangular-oblong or cuneate-obovoid, the longer axis longitudinal rather than transverse, thin, smooth; central filaments fused in twos, rarely threes, at the nodes; utricles of subcortical layer 68–175μ diam., larger than the interlocked, often fused, peripheral utricles.

La Paz, Lower California.

Decaisne, Mém. sur Corall., 1842*a*, p. 102; Howe, Phyc. Studies III, 1907, pp. 495–500, pl. 25, f. 11–20, pl. 26, Phyc. Studies V, 1911, p. 492.

Only a single species of *Halimeda* has as yet been credited to our coast and from a single locality, viz., La Paz in California Baja (or Lower California), Mexico. From this locality it is reported by Howe (*loc. cit.*), and we have also specimens collected at the same locality by Dr. and Mrs. Marchant. It is to be distinguished fairly readily from other flat, jointed species by its slight calcification and by the large rounded utricles of the subcortical layer. Other species are to be looked for along the tropical portion of our coast.

FAMILY 7. VAUCHERIACEAE DUMORT.

Filaments simple or usually more or less dichotomously branched, cylindrical throughout or with frequent constrictions, without septa, often gregarious into expanded tufted or feltlike masses, attached at first by colorless rhizoidal branches; chromatophores numerous, small, lenticular, destitute of pyrenoids; akinetes (?) thick-walled formed within the continuity of the filaments; aplanospores large, ellipsoidal, formed at the ends of the branches; zoospores large, formed at the ends of the branches, provided with many cilia arranged in pairs with a nucleus immediately beneath each pair; sexual reproduction anisogamous; antheridia tubular, usually curved, emitting many very small uninucleated 2-ciliated male gametes; oogonia swollen, globular to ovoid, sessile or pedicellate, single or several together, producing a single large uninucleate non-motile female gamete amply provided with chromatophores, fertilized in position; monoecious or dioecious.

Dumortier, Comm. bot., 1822, p. 71, Analyse fam. pl. 1829, p. 77. *Vaucherideae* Gray, Arr. Brit. Pl., vol. 1, 1821, p. 288.

The family of the Vaucheriaceae, while simple in its coenocytic structure and form of filament, is probably the most complex in its reproduction among the Chlorophyceae. The zoospore is distinctly coenocytic, being large and covered with cilia which, however, are arranged in distinct pairs, each pair associated with its own nucleus. The male gametes, on the other hand, and the female gametes, so far as examined, possess at maturation only a single nucleus. They are very different in size and, while the female gamete or egg is non-motile, the sperms or male gametes are motile, with two equal cilia widely separated and pointing in opposite directions.

There are two genera usually included in the family, viz., *Vaucheria* with continuous and unconstricted filaments and *Dicho-tomosiphon* with more or less interrupted and constricted filaments.

9. Vaucheria DC.

Filaments continuous, without constrictions; reproduction as indicated for the family.

De Candolle, Extrait d'un rapport sur les Conferes 1801.

The genus *Vaucheria* includes both fresh-water and brackish-water species. A few are truly marine. There are both marine and brackish water species about San Francisco Bay, but careful search has as yet failed to reveal any trace of sexual reproduction in these. Without a knowledge of the details of this process, it is impossible to be certain of their identification. Specimens of this genus are better preserved for future study in formalin solution or in alcohol, since dried specimens are difficult to restore to normal appearance on moistening.

Otto Kuntze (1891, p. 926) raises the query as to whether *Vaucheria* DC. or *Ectosperma* Vauch. is the older name. We are unable to settle this question.

ORDER 3.   SIPHONOCLADIALES (BLACKMAN AND TANSLEY)
OLTMANNS

Thallus usually of abundantly branched filaments, or of slightly branched sacks, septate, divisions multinucleate; chromatophores single and reticulate or numerous and lenticular; sexual reproduction by isoplanogametes.

Oltmanns, Morph. und Biol. der Algen, vol. 1, 1904, p. 134. *Siphonocladeae* Blackman and Tansley, Class. Green Algae, 1902, p. 119.

The order Siphonocladiales consists of plants with septate coenocytes, thus differing from the Siphonales which have unseptate coenocytes and from the Ulotrichales whose filaments are made up of cells in the restricted sense. Siphonocladiales are large, richly branched, filamentous species, although a few are unbranched or branched only slightly, while some species are sack-shaped with few and short segments separated from the main portion of the plant. The larger number and more complex families of this order are tropical and marine, but some families are well represented in fresh water also, and in extra-tropical as well as in tropical waters. Almost nothing is known of the tropical forms of our coast.

The recognition of the group of septate coenocytes among the Chlorophyceae as separate from the group of unseptate coenocytes, is due to Schmitz (1878, 1879, p. 273), who designated it as family Siphonocladiaceae. The placing of this group as a subseries (?) and as being made up of separate families is due to Blackman and Tansley as quoted above, although the content was not exactly coincident with that now generally assigned to the order. Although the idea of a separate group originated with Schmitz, Blackman and Tansley were the first to view it as practically a suborder of Siphonales and as made up of families.

FAMILY 8.   CLADOPHORACEAE (HASS.) DE-TONI (AMPL.)

Fronds of simple or branching monosiphonous filaments, free or more or less united laterally; septa frequent, enclosing segments with few to many nuclei; chromatophores broad, reticulate or polygonal-lenticular, but arranged in a network and at times connected by slender strands; multiplication by fragmentation and by akinetes; reproduction by 4-ciliated (or possibly 2-ciliated) zoospores and by 2-ciliated isogametes, formed in segments slightly, if at all, differentiated.

180   *University of California Publications in Botany*   [Vol. 8

De-Toni, Syll. Alg., vol. 1, part 1, 1889, p. 264. *Cladophoreae* Hassall, Brit. F. W. Algae, vol. 1, 1845, p. 213.

The family of the Cladophoraceae is understood in various ways, but it has seemed best to understand it in the sense used by Oltmanns (1904, p. 255) and by Collins (1909, p. 321). It includes all the forms of the strictly filamentous Siphonocladiales which are either simple or, if branched, have branches which are septate at their bases, usually with no distinct main axis in the branched forms, and with all axes of indefinite growth. Thus we include *Microdictyon, Boodlea* and *Anadyomene*. Possibly *Struvea* also should be included in this family but it has a very distinct main axis and lateral axes of definite growth. In these respects *Struvea* resembles the Dasycladaceae, but it differs from the members of this family in its lack of calcification and of specialized zoosporangia (or gametangia) while it differs from the members of the Siphonocladiaceae (in the narrower sense) by having septa at the the bases of the branches. In *Cladophoropsis,* there occur more or less basal septa, but more frequently they are absent. We have followed West (1916, p. 305) in referring *Gomontia* to the Ulotrichales.

KEY TO THE GENERA

1. Filaments simple.. 2
1. Filaments branched.. 6
   2. Filaments usually stiff or rigid.. 3
   2. Filaments flaccid.. 5
3. Filaments large, over 100µ diam.......................12. **Chaetomorpha** (p. 198)
3. Filaments smaller, under 100µ diam... 4
   4. Filaments attached............................12. **Chaetomorpha** (p. 198)
   4. Filaments unattached, prostrate..........10. **Rhizoclonium** (p. 180)
5. Filaments attached..11. **Hormiscia** (p. 187)
5. Filaments unattached, prostrate ....10. **Rhizoclonium** (p. 180)
   6. Branches free.. 7
   6. Branches anastomosing, forming a network.......15. **Microdictyon** (p. 231)
7. Filaments not held together by special rhizoidal or hooked branchlets..................
       ....13. **Cladophora** (p. 207)
7. Filaments held together by special rhizoidal branchlets or by hooked branchlets
   or by both..14. **Spongomorpha** (p. 220)

10. **Rhizoclonium** Kuetz.

Filaments usually prostrate, or slightly ascending, of a single series of segments, unbranched, or occasionally slightly branched, with few to many rhizoidal branchlets composed of one to few segments; segments with one to several nuclei (rarely one) and a single reticulate, parietal chromatophore with numerous, more or less regularly spaced,

pyrenoids; multiplication by fragmentation and by akinetes; reproduction by 2-ciliated zoospores with stigmata; gametes unknown.

Kuetzing, Phyc. Gen., 1843, p. 261, Ueber syst. Eintheil. der Algen,
1843*a*, p. 75 (nomen nudum).

The genus *Rhizoclonium,* as founded by Kuetzing but without
special indication of type species, is one of the simplest of the Cladophoraceae. It consists ordinarily of species with simple, unbranched
filaments whose segments are provided with comparatively few nuclei.
The rhizoidal branches may, or may not, be readily found. In the
absence of these it is sometimes difficult to be certain of the genus.
The species may usually be distinguished from *Chaetomorpha* by their
characteristically cylindrical, never swollen, segments with a smaller
number of nuclei in each, as well as by their different texture and
tendency to a horizontal habit. The branched species resemble *Cladophora,* but the branches push aside the main axis and continue the
direction of the main filament. The species of *Rhizoclonium* inhabit
both the fresh and the salt waters.

The separation of the species of *Rhizoclonium* one from another
presents certain difficulties which are increased by the fact that very
little seems to have been done in the culture of species of this genus.
The following characters have been considered: (1) color, (2) texture,
(3) straightness or crispate character of the filaments, (4) diameters
of the segments, (5) proportions of length to breadth of the segments,
(6) varying thickness of the walls of the segments, (7) varying number of nuclei in the segments, (8) presence or absence of long (true)
branches, and (9) presence or absence of rhizoids. Of these characters, it seems from our present imperfect knowledge that 1, 2, 3, 8
and 9 are the more dependable characters, and that 4, within certain
limits, is very helpful. Characters of 5, 6, and 7 will vary within wide
limits in the same plants, according to whether they are actively dividing or passing into a quiescent condition, as one of us (Gardner) has
experienced in growing *Rhizoclonium lubricum* in the laboratory, and
as Brand (1908, p. 66) has stated as the results of his cultures.

The principal account of *Rhizoclonium* of more recent years is that
of Stockmayer (1890), but his arrangement of the marine species is
not satisfactory to most students. The general disposition of the
marine species made by Rosenvinge (1893, p. 911 *et seq.* and 1894,
p. 126–129) seems based on more certain characters and has been
generally followed. F. Brand (1908, p. 45 *et seq.*) has made some
important studies through cultures, testing certain of the characters

mentioned above as to their constancy or variability. After considering all the data possible it has seemed best to take a somewhat narrower view of specific limits than has been prevalent and divide our
west coast plants among five seemingly distinct species.

KEY TO THE SPECIES

1. Filaments flaccid or lubricous, straight or flexuous, light or yellowish green........ 2
1. Filaments rigid, contorted, dark green............................5. **R. tortuosum**.. (p. 185)
    2. Layer fleecy, filaments flexuous, 10–35μ...................................... 3
    2. Layer lubricous, filaments straight, 25–50μ ....4. **R. lubricum** (p. 185)
3. Rhizoidal branches frequent, often 2–3 septate..................1. **R. riparium** (p. 182)
3. Rhizoidal branches scarce, when present non-septate............................ 4
    4. Filaments 10–14μ..3. **R. Kerneri** (p. 185)
    4. Filaments 20–30μ............. ....2. **R. implexum** (p. 183)

1. Rhizoclonium riparium (Roth) Harv.

Filaments pale green, expanded on the substratum, flexuous, intertwined into a fleece; segments 20–25μ diam., rarely slightly greater
or smaller, usually once or twice as long as broad; branches and
rhizoids frequent, often 2–3-septate.

In skein-like masses on cliffs or hard clay banks, often among other
algae in the littoral belt. Alaska to central California.

Harvey, Phyc. Brit., vol. 2, 1849, pl. 238 (binomial attributed to
Kuetzing); Collins, Green Alg. N. A., 1909, p. 327 (in part). *Conferva riparia* Roth, Cat. Bot., vol. 3, 1806, p. 216; Dillwyn, Brit. Conf.,
1809, p. 69, pl. E. *Rhizoclonium riparium* var. *polyrhizum* Rosenvinge, Groenl. Havalg., 1893, p. 913, f. 32; Collins, Green Alg. N. A.,
1909, p. 328; Collins, Holden and Setchell, Phyc. Bor.-Amer. (Exsicc.),
no. 2238.

There is no question in our minds that the situation as to species
of *Rhizoclonium* on the Pacific Coast of North America is very similar
to, if not perhaps identical with that on the Atlantic Coast of the same
continent and on the Atlantic shores of Europe. There are at least
two sets, or groups, of plants differing somewhat in dimension and
in habit. One of these sets of forms is usually referred to *Rhizoclonium
riparium* and the other to *R. tortuosum*. This would be reasonably
satisfactory, if no previous conceptions existed to be considered. The
two species, however, as far as names are concerned, date back to the
early portion of the nineteenth century, viz., to the publications of
Roth (1806) and Dillwyn (1809) respectively. Unfortunately no
type specimens are available to us and we can simply follow the

custom, with such information as we may glean from the literature. The *Conferva riparia* Roth was known to Dillwyn (1809, p. 69, under no. 111) by an authentic specimen in Turner's herbarium. Dillwyn says that the English plants referred by him to Roth's species and figured under the name were treated so on authority of an authentic specimen. Dillwyn figures (*loc. cit.*, pl. E) a plant which has shortly flexuous, but not contorted, filaments and frequent rhizoidal branchlets of several segments each. Harvey (1849, pl. 238) figures a similar plant which he states is certainly that of Dillwyn, since it is drawn from a specimen belonging to Miss Hutchins (from Bantry Bay) referred to by Dillwyn. The type of *Rhizoclonium riparium* (Roth) Kuetz., then, may be properly inferred to be the variety with frequent and complex rhizoids which Rosenvinge (1893, p. 913) named var. *polyrhizum*. This is certainly a *Rhizoclonium,* being provided with the rhizoidal branches characteristic of the genus. In general the various authors agree that the diameters of the segments vary from 20μ to 35μ, but most commonly are from 20μ to 25μ.

It seems best to us to keep *Rhizoclonium riparium* as thus characterized distinct and separate from other varieties (than *polyrhizum*) usually referred to it. The var. *implexum* (Dillw.) Rosenvinge, while very similar to the type of *R. riparium* in both texture and diameter of its filaments, is to be distinguished by the scarcity and structure of the rhizoidal branchlets and is treated here as a distinct species. The *Rhizoclonium riparium* var. *validum* Foslie (1890, pp. 138, 139) is decidedly coarser than the type and is probably to be separated from *R. riparium* in the more restricted sense, although probably closely related to it.

2. **Rhizoclonium implexum** (Dillw.) Kuetz.

Filaments simple, $20–30\mu$ (rarely 40μ) in diameter, yellowish or light green, forming a horizontal fleecy layer; segments 1.5–2.5 times as long as broad; rhizoidal branches few or wanting, when present short, non-septate and usually continuous with the segments from which they arise.

Forming fleecy masses on mud or on various objects in the littoral belt. Alaska to central California.

Kuetzing, Phyc. Germ., 1845, p. 206 (at least as to plant of Dillwyn); Batters, Alg. Clyde Sea Area, 1891, p. 230, repr., p. 8. *Conferva implexa* Dillwyn, Brit. Conf., 1805, p. 46, pl. B; Harvey, Phyc.

Brit., vol. 1, 1846, pl. 54 B. *Rhizoclonium riparium* var. *implexum* Rosenvinge, Groenl. Havalg., 1893, p. 915; Saunders, Alg. Harriman Exp., 1901, p. 414; Setchell and Gardner, Alg. N.W. Amer., 1903, p. 222; Collins, Holden and Setchell, Phyc. Bor.-Amer. (Exsicc.), no. 976; Collins, Green Alg. N. A., 1909, p. 328. *Rhizoclonium riparium* Tilden, Amer. Alg. (Exsicc.), no. 379 (not of Roth or Kuetzing).

A plant very similar to *Rhizoclonium riparium,* but destitute of, or provided with very few and simple, rhizoids, is found on the shores of the Pacific Coast of North America. This seems to be the same as the plants from both the European and North American Coasts which have passed under the name of *Rhizoclonium riparium* var. *implexum.* The filaments are nearly the same in diameter as those of the preceding, but possibly average slightly smaller. The rhizoidal branches are often entirely wanting and in no case are really abundant. When present they alway lack septa and generally are not cut off from the segment from which they arise. The growth is generally more entangled and fleecelike than that of the preceding species. It seems best to us to keep this form, which seemingly has a wider distribution along the Pacific Coast than *R. riparium,* separate. In adopting the name of *R. implexum,* we are guided by the descriptions of others. The type specimen of Dillwyn is unknown to us, but this seems to be the plant of Harvey and possibly also of Kuetzing and Rosenvinge. Harvey evidently founded his description on Miss Hutchins's specimen from Bantry Bay which is a topotype, possibly even a cotype of Dillwyn's species. Harvey and Dillwyn give no measurements of *Conferva implexa,* but Harvey states that the filaments are about two-thirds of the thickness of those of *C. tortuosa.* The latter is probably 40μ to 70μ in diameter. Kuetzing assigns to his *Rhizoclonium implexum* a diameter of $\frac{1}{200}'''$–$\frac{1}{150}'''$ or about 11μ to 12μ, which is very much more slender than the *Rhizoclonium riparium* var. *implexum* of Rosenvinge which is described as being 20μ to 30μ (or 40μ) in diameter.

Our plants agree well with no. 142 and even with no. 190 of Wyatt's Algae Danmonienses, issued under the names of *Conferva implexa* and *C. tortuosa* respectively. No. 142 shows no rhizoids while no. 190 shows frequent unseptate rhizoids. The segments in no. 142 vary from 35μ to 45μ, while those of no. 190 vary from 22μ to 27μ. They also agree in general with no. 624 of Wittrock and Nordstedt's "Algae aquae dulcis exsiccatae," distributed under the name of *Rhizoclonium riparium* f. *valida* Foslie.

3. **Rhizoclonium Kerneri** Stock.

Filaments pale yellowish green, segments 10–14μ diam., 3–7 diameters long, free from rhizoids or branches.

Growing in loose masses in tide-pools. Victoria, Vancouver Island, British Columbia.

Stockmayer, Ueber die Algengat. *Rhizoclonium,* 1890, p. 582; Collins, Mar. Alg. Vancouver Island, 1913, p. 103.

We have not seen any specimens of this species and are including it upon the authority of Collins (*loc. cit.*). It is decidedly more slender than the other four species of *Rhizoclonium* thus far detected on our coast.

4. **Rhizoclonium lubricum** S. and G.

Plate 9, figs. 5*a, b*

Filaments flaccid, lubricous, straight, cylindrical throughout, 3–4.5 dm. long, pale green; segments 35–50μ, mostly 40μ diam., resting segments 4–6 diam. long, after division segments 1–2 diam. long; chromatophore a coarse, parietal network; pyrenoids small, numerous, 40–50 in resting segments; wall 2μ thick, homogeneous; rhizoids short, mere prolongations of cells, non-septate, rare; zoospores and gametes unknown.

Growing attached in mud or floating on mud flats between tides. Roche Harbor, Washington, and Berkeley and Alameda on the shores of San Francisco Bay, California.

Setchell and Gardner, *in* Gardner, New Pac. Coast Mar. Alg. IV, 1919, p. 492, pl. 42, f. 5; Collins, Holden and Setchell, Phyc. Bor.-Amer. (Exsicc.), no. 2289.

This form closely resembles *R. riparium* f. *validum* Foslie, but is practically free from rhizoids, has thinner walls and larger and longer segments. From *R. implexum* it differs in having broader, straighter filaments of very different consistency, as well as, usually, longer segments. Unlike other species of *Rhizoclonium* it is very lubricous, in mass, having the consistency of a *Spirogyra*.

5. **Rhizoclonium tortuosum** (Dillw.) Kuetz.

Filaments rigid, crispate and contorted, dark green, 40–70μ diam., forming woolly skeinlike or ropelike horizontal masses; segments 1–2, up to 6 times as long as broad, wall thick, indistinctly lamellose; rhizoids short, few or, more usually, none.

On various algae in the middle and upper littoral belts. Alaska to California.

Kuetzing, Phyc. Germ., 1845, p. 205 (at least as to the plant of Dillwyn) ; Farlow, New Eng. Alg., 1881, p. 49 ; Setchell and Gardner, Alg. N.W. Amer., 1903, p. 223 ; Collins, Green Alg. N. A., 1909, p. 328. *Conferva tortuosa* Dillwyn, Brit. Conf., 1805, p. 46, pl. 46 ; Harvey, Phyc. Brit., vol. 1, 1846, pl. 54 A. *Chaetomorpha tortuosa* Kuetzing, Spec. Alg., 1849, p. 376 ; Harvey, Ner. Bor.-Amer., part 3, 1858, p. 88, pl. 46 B.

Kuetzing has described two plants, both founded, in the final analysis, upon *Conferva tortuosa* Dillwyn. One of these plants he bases directly upon Dillwyn's species and refers it to *Rhizoclonium* (1845, p. 205 ; 1849, p. 384). It is credited by Kuetzing to the North Sea. Kuetzing gives $\frac{1}{70}'''-\frac{1}{60}'''$ as the diameter of his *Rhizoclonium tortuosum* which approximates 32μ to 35μ. The *Conferva tortuosa* J. Ag. (1842, p. 12) which is the *C. tortuosa* C. Ag. (1824, p. 98) and which, in turn, is founded, as to name at least, on *C. tortuosa* Dillw., is referred by Kuetzing (1849, p. 376) to the genus *Chaetomorpha* as *Chaetomorpha tortuosa* and the diameter of the filaments is given as $\frac{1}{45}'''-\frac{1}{40}'''$, or approximately 46μ to 56μ. It is restricted by Kuetzing to the Mediterranean and Adriatic Seas.

Dr. Anna Weber-van Bosse has kindly allowed us to examine the specimens in Herbarium Kuetzing under the names of *Rhizoclonium tortuosum* and *Chaetomorpha tortuosa*. There are three of the former, from the Faeröes, England and Cherbourg respectively. They all seem to be *R. tortuosum* in the sense in which we use the name : The specimen of *Chaetomorpha tortuosa* from the Kuetzing collection is from Nice and while it resembles fairly closely the Kuetzing specimens of *R. tortuosum*, it has much thicker walls and possibly may be found to belong to another species.

We feel convinced of the likelihood of *Conferva tortuosa* Dillw. being the coarse, crispate plant which has been described by Harvey (1846, pl. 54 A) as *Conferva tortuosa* and (1858, p. 88) as *Chaetomorpha tortuosa* and finally by Farlow (1881, p. 49) as *Rhizoclonium tortuosum*.

As we understand this species, which is not common on our coast, it forms woolly skeinlike or loose ropelike masses of a dark green color on other algae in the littoral belt of exposed coasts. The filaments are rigid, harsh to the touch and crispate or contorted. In diameter its filaments exceed those of both *R. riparium* and *R. implexum* by one-third or more. It varies in the length-breadth proportions of the segments, and plants with longer segments have been named forma *longiarticulatum* by Collins (Phyc. Bor.-Amer. (Exsicc.), no. 1735).

11. **Hormiscia** Fries

Filaments simple, attached at the base by rhizoids developing from a few of the basal segments, either intramatrical, extramatrical or both; segments multinucleate, all above the few basal ones similar and capable of division and producing zoospores or gametes; chromatophore covering the segment wall, entire or more or less coarsely reticulate, with few to many pyrenoids; multiplication by akinetes formed by breaking up of the filaments into individual segments with thick walls, either producing new filaments or zoospores; reproduction by macrozoospores, by microzoospores and by gametes, produced many in a segment; macrozoospores obovoid, extending posteriorly into a long "tail," provided anteriorly with 4 cilia; microzoospores smaller with less obvious tail and 4 cilia; sexual reproduction by 2-ciliated (possibly in some species by 4-ciliated) iso- or hetero-gametes.

Fries, Flora Scanica, 1835, p. 327. *Urospora* Areschoug, Observ. Phyc., 1866.

The genus *Hormiscia,* as constituted by Fries, comprised two species, viz., *H. penicilliformis* and *H. Wormskioldii.* The identity of the second species is clear enough, but that of the first rests somewhat in doubt. It seems likely, however, that the *Conferva penicilliformis* Roth is, in part at least, made up of the species more recently assigned under that name as well as under that of *Urospora mirabilis* Aresch. In such case, it seems best to follow Hazen (1902, pp. 146, 147) and adopt the name *Hormiscia* Fries rather than *Urospora* Aresch.

The genus *Hormiscia,* as at present understood, includes about fifteen described species inhabiting the cooler waters of the Northern Hemisphere. The filaments are characteristically simple but occasionally may be branched. The branches are the result of injuries or, at least, of some disturbance of the normal course of development. In our own experience they are rare. Usually the branches are short and rhizoidal, sometimes occurring in pairs, but at times they are longer, with evidence in both cases of some disturbance of normal growth at the point of origin.

Hormiscia is closely related to both *Chaetomorpha* and *Rhizo-clonium*. From both of these it differs particularly in the pointed posterior ends, or "tails," of the 4-ciliated macrozoospores. The determination of this character is most satisfactorily made from living specimens. It is possible, however, to be certain of it in dried speci-mens of the mature fruiting stages if those are properly soaked out and stained. In such specimens, it is often possible to ascertain the shape of the zoospores with the conspicuous "tail" and in some cases even to determine the number of cilia present if specimens have been preserved in formalin solution. From *Rhizoclonium, Hormiscia* differs in habit, in being attached at one end and erect, resembling *Chaeto-morpha* in this respect. From *Chaetomorpha, Hormiscia* differs in texture, being more flaccid or more lubricous.

The species of *Hormiscia* resemble one another closely and are to be separated by several minor characters. Those usually employed are as follows: (1) chromatophores, (2) mode of attachment, (3) diameters of the segments, and (4) shape and proportions of the seg-ments. Besides there are some differences in height and texture to be considered.

The chromatophores of the species of *Hormiscia* are annular, parietal, and single in the segment. They contain a larger or smaller number of pyrenoids, more or less regularly placed. Some chromato-phores seem practically imperforate, some are finely reticulate with few or many small openings, while others are coarsely reticulate with few or many large openings. Some species have thin, membranaceous chromatophores, while others have thick, solid chromatophores. The characters of the chromatophore also vary somewhat according to the metabolic or reproductive condition of the segment. The study of the chromatophore of specimens which have been dried presents cer-tain difficulties, but usually they can be swollen up sufficiently and stained (we use acid Fuchsin) to show their nature. Ordinarily it is a fairly certain matter to determine whether a chromatophore is coarsely reticulate or not. With the more finely perforate chromato-phores, the difficulties are sometimes considerable. When dried speci-mens are mounted fresh from the water before any collapse of the protoplast has taken place, the chromatophore will swell out with water and may plainly be seen. When segments in the filament approach the reproductive condition, the chromatophore becomes less and less perforate and also thicker, finally breaking up into areolae (as seen from the surface) which precede the formation of either zoospores or gametes as the case may be. There seems also to be a very considerable multiplication of pyrenoids at about the same time. In the following account we have described the chromatophores as well as the material allows, pending investigation of living plants.

The filaments of *Hormiscia* are attached at the base, at least at first. They are usually attached to rock or wood, but may be, in some species at least, attached to other algae. Kjellman (1897, p. 8 *et seq.*) was the first to call attention to the variety in the method of attachment of the filaments of *Hormiscia*, contrasting those of *H. penicilliformis* with those of *H. incrassata*. The differences depend upon the behavior of the rhizoidal outgrowths from the basal segments. In some species only a few of the lower segments emit rhizoids, while in others twenty or more of the basal segments may produce them. In some species the rhizoids do not emerge from the outermost walls of the filament but descend along the segments below them, still enclosed within the walls, until the very base or near it, when they may or may not emerge and become free. We have designated such rhizoids as *intramatrical*. In some species, however, the rhizoids do not descend for any considerable distance within the outer wall of the filament, but emerge at once and either extend at any angle to the filament or descend closely applied to it. Such rhizoids are designated by as as *extramatrical*. (For the terms intramatrical and extramatrical see Jónsson, 1903, p. 360.)

In diameter the filaments vary considerably, both in regard to the different species and in regard to the segments in different portions of the same filament as well as in filaments of different ages. The filaments taper from the base upwards and then often narrow again more or less towards the apex. The tapering upwards is slight in some species but more considerable in others. In most species it is gradual but in other species it is abrupt. The diameters of the reproductive segments are usually the most distinctive.

The shape and proportions of the segments, especially of the fertile segments, vary considerably among the species, but are sufficiently constant to furnish valuable characters for diagnoses. The thickness and structure of the wall, particularly of the reproductive segments, offers additional points of difference between species. The consistency of the plants, as observed in mass, varying from lubricous to fleecy, is worthy of attention, whether judged from observation of the living specimens or from the extent to which they adhere to paper in drying.

In attempting to arrange the plants of our coast, it has been necessary to rely largely upon dried specimens or those preserved in formalin. The resulting account which follows must, therefore be regarded as tentative, but as put forth to indicate possibilities to be tested through study of living material. The tendency of two or more species to grow intermingled has often caused confusion in the past, and may have passed unsuspected in our own examinations.

As to reproductive processes, we have practically nothing new to offer from our own observations. Each species, however, should be studied in cultures to determine its correspondence with, or deviation from, the published statements for such species as have been investigated. The size and shape of the zoospores and of the gametes should be noted as well as the number of cilia present. This is especially necessary in the case of the gametes as will be noted below under *Hormiscia tetraciliata* Frye and Zeller.

In considering the specimens studied, as well as the species described, it seems possible to separate the genus *Hormiscia* into three fairly readily distinguishable sections, as follows:

Section 1. *Penicilliformes.* Segments shorter, or at most, very little longer than broad, fertile cells only slightly swollen; holdfast consisting of extramatrical, generally divaricate rhizoids, or of intramatrical rhizoids; chromatophore varying in the different species from nearly imperforate to coarsely reticulate. To this section we are inclined to refer the following described species: *H. penicilliformis* (Roth) Fries, *H. incrassata* (Kjellm.) Collins, *H. collabens* (Ag.) Rab., *H. Hartzii* (Rosenv.) Collins, *H. crassa* (Kjellm.) Collins and *H. tetraciliata* Frye and Zeller, as well as *Urospora bangioides* (Harv.) Holmes and Batt., *U. claviculata* Kjellm. and *U. acrogona* Kjellm.

Section 2. *Grandiformes.* Segments elongated, especially the fertile segments, usually twice or more as long as broad in fertile condition and not swollen; holdfast of intramatrical rhizoids (at times free at the very base) and arising from a considerable number of the lower segments; chromatophores, so far as known, coarsely reticulate. To this section we are inclined to refer *Hormiscia grandis* (Kylin) S. and G. and *Urospora elongata* (Rosenv.) Hagem.

Section 3. *Wormskioldiiformes.* Segments shorter or only slightly longer than thick, the fertile swollen to spherical or to more or less broadly or even ventricosely ellipsoidal; holdfast of intramatrical rhizoids arising from even as many as 25–30 of the lower segments and, at times, free at the very base; chromatophore from finely perforate to coarsely reticulate; coarse plants with fertile segments from

0.5 mm. to 3 mm. in diameter. To this section we are inclined to refer *H. Wormskioldii* (Mert.) Fries, *H. sphaerulifera* S. and G., and *H. vancouveriana* (Tilden) S. and G.

1. **Hormiscia penicilliformis** (Roth) Fries

Plate 9, fig. 4

Filaments dark green, attached by extramatrical rhizoids from a few of the lower segments, 30–60μ up to 90μ diam., 0.3–2.5 times as long as the diameter; vegetative segments mostly cylindrical, fertile more or less swollen to barrel-shaped; chromatophore usually dense, a continuous parietal band, or at times somewhat fenestrate, pyrenoids relatively few and large.

Growing on rocks and timbers exposed to the surf. Alaska (Port Clarence, etc.) to central California.

Fries, Flora Scanica, 1835, p. 327; Collins, Green Alg. N. A., 1909, p. 368. *Conferva penicilliformis* Roth, Cat. Bot., III, 1806, p. 272. *Urospora penicilliformis* Setchell and Gardner, Alg. N.W. Amer., 1903, p. 220 (in part). *Urospora incrassata* Setchell and Gardner, Alg. N.W. Amer., 1903, p. 221 and of Collins, Holden and Setchell, Phyc. Bor.-Amer. (Exsicc.), no. 1125 (not of Kjellman). *Hormiscia incrassata* Collins, Green Alg. N. A., 1909, p. 369 (excl. syn. of Kjellman).

The exact nature of the *Conferva penicilliformis* Roth is a subject of doubt, and no one has published any statement either as to the existence of a type specimen or as to its nature. It seems best to retain the name for the present, however, and to assume its identity with *Conferva isogona* of the English Botany (1808, pl. 1930) whose type (cf. Batters, 1894, p. 116) is in existence and has been recently examined. There are several conflicting statements concerning the details of structure of *H. penicilliformis* and it seems probable that more than one species may be included under the name. We are inclined at present to include under this name all the specimens from the western coast of North America accessible to us which are under 100μ in the maximum diameter of the fertile segments. These specimens all have segments which are nearly isodiametric and with the fertile segments slightly swollen and barrel-shaped. Rosenvinge (1893, p. 918, 1894, p. 30) and Jónsson (1903, p. 360) include forms with elongated segments, but it seems probable that such forms may be referred rather to members of the section *Grandiformes*. *H. grandis* (Kylin) S. and G. sometimes occurs intermixed with *H. penicilliformis* (cf. Kylin, 1907, p. 20). The base is attached by extramatrical and divaricate rhizoids which arise from a few of the basal segments. Concerning the latter statement it is proper to call attention to the conflicting statement and figure of Börgesen (1902, pp. 500, 501, f. 100) and Jónsson (1903, p. 360). These may possibly be explained as resulting from a confusion or admixture of species. The chromatophore in our specimens shows very slight perforation at any time, being nearly imperforate. Kjellman (1897, p. 12) was among the first to call attention to the chromatophore as varying among the species. He states that the chromatophore of what he regards as true *"Urospora penicilliformis"* differs from that of *"U. incrassata"* and *"U. Wormskioldii"* in having only small openings. The general status of the nature of the chromatophore of *Hormiscia penicilliformis* has been discussed by Jónsson (1903, p. 361) who passed in review the different ideas and states that "especially in the elongated cells, the chromatophore is distinctly reticular, often with great meshes." This brings the type of chromatophore in this species near to that of *H. Wormskioldii*. We feel that it is likely that there has been a confusion of species and have restricted our idea of the chromatophore to one at most perforated with small angular openings. We have found none with large or elongated openings such as Jónsson mentions and such as Hagem (1908, p. 294, pl. 1, f. 2) describes and figures.

The Pacific Coast specimens vary from about 45μ to 90μ in maximum diameter of fertile segments but otherwise seem to agree and to be referable to *H. penicilliformis.* Some of the specimens included here by us have been referred to *Hormiscia incrassata* (Kjellm.) Collins, but that is a somewhat larger species with a coarsely reticulate chromatophore. Some of the specimens referred by us previously (1903) to *Urospora penicilliformis* seem to us now to be placed more satisfactorily under *Hormiscia grandis* (Kylin) S. and G.

2. **Hormiscia doliifera** S. and G.

Filaments 3–4 cm. long, nearly cylindrical throughout when young, tapering only at the base, attached by extramatrical rhizoids from a few of the lower segments; color dark green; fertile segments 80–130μ, up to 184μ diam., 0.75–1.25 times as long, doliiform, with thin, 5–7μ thick, hyaline, homogeneous walls; chromatophore a thin, fenestrate, parietal band, with numerous small pyrenoids.

Growing on rocks in the upper littoral belt. San Francisco, California.

Setchell and Gardner, Phyc. Cont. I. 1920, p. 279.

Hormiscia doliifera resembles most closely *Urospora Hartzii* Rosenvinge (1893, p. 922) and *U. incrassata* Kjellman (1897, p. 7). From each of these species it differs in having filaments of larger diameter, in having more uniformly swollen, sometimes almost spherical, fertile segments, and in having the segments more nearly uniform in length, averaging a little less than quadrate. From *U. incrassata* it differs also in its strictly extramatrical rhizoids. It approaches also the little known *U. crassa* Rosenv., but its segments seem never so short as represented for that species. It is much too slender for *Hormiscia collabens* (i.e., up to 450μ) as indicated by Batters (1894, p. 114). The filaments are decidedly larger than any dimensions given for *H. penicilliformis* (Roth) Fries and the chromatophore is thinner, usually more coarsely reticulate, and with many more and much smaller pyrenoids.

3. **Hormiscia tetraciliata** Frye and Zeller

Filaments flaccid, slightly clavate, 25μ at the base, up to 220μ at the apex, 5–8 cm. long, attached by numerous intramatrical rhizoids, 1–2 from each of the lower 8–15 segments, passing out of the lower disintegrating sheath, or segment wall; segments 0.5–2 times as long as the diameter, cylindrical below, becoming decidedly barrel-shaped above; walls up to 15µ thick, homogeneous, hyaline; chromatophore a thin, parietal, finely reticulate band, with numerous small pyrenoids.

Growing on stones, shells, and other algae. San Juan County, Washington.

Frye and Zeller, *Hormiscia tetraciliata*, sp. nov., 1915, pp. 9–13, pl. 2. "*Hormiscia Wormskjoldii*" Collins, Green Alg. N. A., Suppl. II, 1918, p. 86 (in part); Collins, Holden and Setchell, Phyc. Bor.-Amer. (Exsicc.), no. 2237.

Hormiscia tetraciliata Frye and Zeller presents a number of interesting and puzzling characters as described and figured. It seems to be a species of the *Penicilliformes* section, in that the fertile cells are short and only slightly swollen. It differs from all others of the species of that section except *H. collabens* (Harv.) Collins in the greater maximum diameter attained by the fertile cells. Frye and Zeller give this diameter as 220µ for *H. tetraciliata* while Batters (1894, p. 114) states that *H. collabens* reaches a maximum diameter of 450µ (verified by an examination of the type specimen, *loc. cit.*, p. 115). The *Urospora bangioides* (Harv.) Holmes and Batters (cf. Batters, 1894, pp. 115, 116) seems too slender (up to 150µ, even perhaps to 180µ in maximum diameter), but the plant distributed under the name of *Urospora collabens* by Collins (Phyc. Bor.-Amer. (Exsicc.), no. 970) shows sterile segments up to 220µ in diameter and agrees closely with the description of *H. tetraciliata*. We do not find any fertile segments, however, in the Collins specimen, and the maximum diameter of the fertile segments may be much greater.

The plant distributed under no. 970 of the Phycotheca Boreali-Americana is supposed to be *H. tetraciliata* since it was collected under the supervision of Professor Frye and received his sanction. It seems to have its fertile segments rather more swollen and elongated than is indicated by the drawings of Frye and Zeller (*loc. cit.*, pl. 2, f. 6 and 17) or as provided for in the description. The distributed specimens seem rather to belong to the section *Wormskioldiiformes*. In this connection, it may also be mentioned that a later gathering, by Professor Frye, from the same locality has yielded no *H. tetraciliata*, but a mixture of what we refer below to as *H. grandis* and *H. Wormskioldii*.

The unusual character of *Hormiscia tetraciliata* is shown by the gametes. These, according to Frye and Zeller (*loc. cit.*, pp. 10, 11) have four cilia each and, judging from their figure, fuse in pairs (*loc. cit.*, f. 20) in an unusual way, i.e., not at the tips, but at the posterior extremities. It seems possible that there may be some error of interpretation about this and Collins (1918, p. 86) suggests, with good reason, as it seems to us, that the supposed gametes may be microzoospores and that "figure 20, 'gametes fusing,' would seem rather to represent two imperfectly separated microzoospores." The matter of microzoospores and gametes especially needs further investigation in this as well as in other species of *Hormiscia*.

The holdfast of *H. tetraciliata* is, as represented by Frye and Zeller (*loc. cit.*, p. 9, pl. 2, f. 3), composed of intramatrical rhizoids arising from a considerable number of cells and simple or branched, descending within the outer wall, or sheath, of the filament, but emerging and spreading to some extent at the very base and becoming trumpet-shaped. This combination of intra- and extramatrical rhizoids seems like a combination of the characters of the species of the *Penicilliformes* section and those of the *Wormskioldiiformes* section.

4. **Hormiscia grandis** (Kylin) S. and G.

Plate 9, fig. 3

Filaments attached by intramatrical rhizoids from 8–14 segments above the base, flaccid, 8–10 cm. long; segments cylindrical, constricted at the joints, 45–70μ thick at the base, usually 2 times as long as broad, 125–175μ, up to 200μ thick at the upper end, 1–3.5 times as long as broad, walls 12–18μ thick, lamellate, fertile segments slightly swollen; chromatophore a thin, parietal, reticulate band, with numerous small pyrenoids.

Growing on rocks in the upper littoral belt. West shore of Amaknak Island, Bay of Unalaska, Alaska, to Puget Sound, Washington.

Setchell and Gardner, *in* Gardner, New Pac. Coast Mar. Alg. IV, 1919, p. 494. *Urospora penicilliformis* Setchell and Gardner, Alg. N.W. Amer., 1903, p. 220 (in part). *Urospora Wormskioldii* Setchell and Gardner, Alg. N.W. Amer., 1903, p. 221 (in part). *Urospora grandis* Kylin, Studien ueber Algenflora, etc., 1907, p. 18, f. 3.

In reëxamining the specimens previously referred by us (1903, p. 220) to *Urospora penicilliformis* and *U. Wormskioldii*, we found several which differed in having the fertile segments long and not swollen. They also have thick, lamellate walls. These characters agree so well with those given for *Urospora grandis* that we feel reasonably safe in referring them to that species and transferring it to the genus *Hormiscia*. The type locality of the species is Christineberg on the west coast of Sweden, where it occurs mixed with *Hormiscia penicilliformis* and *Ulothrix flacca*. On the West Coast of North America it occurs mixed with the same species and also with *Hormiscia Wormskioldii*. It is to be compared with *Urospora elongata* (Rosenv.) Hagem but is of much greater diameter and with thicker walls in the fertile segments. The chromatophore is very coarsely reticulate.

5. Hormiscia sphaerulifera S. and G.

Plate 9, fig. 2

Filaments very lubricous, bright green, protoplast becoming dark when dry, 4–6 cm. long. up to 700μ diam., tapering abruptly at the base, of nearly uniform diameter above, attached by intramatrical rhizoids; segments 0.5–2 times as long as the diameter, cylindrical when young, becoming almost spherical at maturity; chromatophore a thin, unbroken band, pyrenoids small, numerous.

Growing on boulders in the extreme lower littoral belt. West coast of Whidbey Island, Washington.

Setchell and Gardner, *in* Gardner, New Pac. Coast Mar. Alg. IV, 1919, p. 493, pl. 42, f. 2. *Urospora Wormskioldii* Setchell and Gardner, Alg. N.W. Amer., 1903, p. 221 (in part); Collins, Holden and Setchell, Phyc. Bor.-Amer. (Exsicc.), no. 915 (not *Conferva Wormskioldii* Mert.).

In a careful study of the west coast specimens previously referred to *Hormiscia* (or *Urospora*) *Wormskioldii*, considerable differences were found which have led to their segregation among three species. The present species is the most slender of the group, reaching a maximum diameter, even in the fertile segments, of 700μ. The chromatophore seems to consist of a thin, unbroken, annular band. The fertile segments are swollen into a spherical shape and the upper portions of the filament are in the form of a necklace. The filaments are very lubricous and adhere tightly to paper.

6. Hormiscia Wormskioldii (Mert.) Fries

Filaments attached by intramatrical rhizoids arising from 12 to 15 of the basal segments, dark green, 10–20 cm. high, distinctly clavate; segments cylindrical below or when young, 30–60μ in diameter, increasing above in fertile segments to 500μ or even to 1 mm. in maximum diameter; fertile segments swollen to spherico-ellipsoidal, long ellipsoidal, or even ventricose-ellipsoidal; chromatophore a coarsely reticulate, annular band.

On rocks, etc., in the lower littoral belt. Friday Harbor, Washington (*Frye*), to Gualala, California (*Brandt*).

Fries, Flora Scanica, 1835, p. 328 ("*Wormskjoldii*). *Conferva Wormskioldii* Mertens, *in* Hornemann, Flora Danica, vol. 9, fasc. 26, 1816, p. 6, pl. 1547 (not *H. Wormskioldii* Setchell and Gardner, Alg. N.W. Amer., 1903, p. 221, and not of Collins, Green Alg. N. A., 1909, p. 368 as to Pacific Coast references and localities). *Urospora Wormskioldii* Rosenvinge, Groenl. Havalg., 1893, p. 920, f. 36.

It has been the practice to refer all the coarser species of *Hormiscia,* especially those with much swollen fertile segments, to *Hormiscia Wormskioldii.* In examining the specimens from the western coast of North America and comparing them with specimens from the type locality (Gothaab or Godthaab, on the southwestern coast of Greenland) collected in 1831 by Vahl, we find sufficient differences to make it seem best to distinguish three separate species. None of the specimens from the West Coast previously referred by us, or others, to *Hormiscia Wormskioldii* seems to us, at present, to be referable to it. We have found specimens seemingly very close to it, however, in two recent collections, as indicated above.

Hormiscia Wormskioldii is to be distinguished from the other two species we have referred to the section *Wormskioldiiformes* by its coarsely reticulate chromatophore and its intermediate size (up to 1 mm. in maximum diameter of the fertile segments). The fertile segments are usually long ellipsoidal.

7. **Hormiscia vancouveriana** (Tilden) S. and G.

Filaments attached by intramatrical rhizoids arising from 14–20 basal segments, soft, gelatinous, dark green, 10–15 cm. high, distinctly clavate; segments cylindrical when young, 75–130μ diam., quadrate, or up to 3 times as long as broad, soon becoming decidedly moniliform, the upper segments almost spherical, up to 3 mm. in diameter; chromatophore thin, very slightly if at all perforate.

Growing on stones and shells, in the lower littoral belt. On a small island east of Oak Bay, Vancouver Island, British Columbia, July.

Setchell and Gardner, *in* Gardner, New Pac. Coast Mar. Alg. IV, 1919, p. 494. *Urospora Wormskioldii* Setchell and Gardner, Alg. N.W.

Amer., 1903, p. 221 (in part). *"Hormiscia Wormskjoldii"* Collins, Green Alg. N. A., 1909, p. 368 (in part). *"Urospora Wormskjoldii* f. *vancouveriana"* Tilden, Amer. Alg. (Exsicc.), no. 381.

The plants distributed in Tilden's American Algae under no. 381 are by far the most robust of any of the genus *Hormiscia* yet seen. In several characters they resemble *H. Wormskioldii* closely, but differ decidedly in diameter and in the character of the chromatophore. The maximum diameter of the fertile segment is 3 mm. or even more. The chromatophore seems to be a close, very little, if at all, perforate ring. From *H. sphaerulifera, H. vancouveriana* differs both in its greater thickness and in the proportionally longer fertile cells.

Hormiscia vancouveriana is known only from the type locality and, as far as we are aware, from a single collecting. It is to be hoped that it may be rediscovered and studied in the living condition.

12. **Chaetomorpha** Kuetz.

Filaments composed of a single unbranched series of multinucleate segments, all except, usually, a few long basal segments capable of division and reproduction, attached by more or less branched, shorter or longer rhizoids, often coalescing, always attached at first, later sometimes loosening and continuing in a free state; membrane of segments thick, firm, usually distinctly lamellate in age; chromatophore a parietal band, more or less perforate (or in age broken into small disks), with numerous pyrenoids; reproduction by 4-ciliated zoospores produced in little changed segments, and by 2-ciliated isogametes; thick-walled akinetes formed from single segments; mostly marine species.

Kuetzing, Phyc. Germ., 1845, p. 203.

The genus *Chaetomorpha* was founded by Kuetzing in 1845 (May), as indicated above, with fifteen species referred to it, all of which still seem to be closely generically related. It preceded by only two months (cf. Harvey, 1858, p. 84) the publication of Hassall's *Aplonema* (1845, p. 213, July). It seems to be generally recognized as a genus and to be retained in spite of the fact that it approaches *Rhizoclonium* on the one side and *Hormiscia* on the other. From *Rhizoclonium* species in typical form it is readily to be distinguished by the lack of branches of any kind, rhizoidal or otherwise. Some species of *Rhizoclonium*, however, lack all branches and consequently resemble closely the unattached and expanded species or forms of *Chaetomorpha*.

The only differences in such cases are the usually more regularly cylindrical segments of the species of *Rhizoclonium* and the fewer nuclei in each segment. Even these fail in some cases. From *Hormiscia*, *Chaetomorpha* differs essentially in the characters of the zoospores as indicated under the description of that genus.

The genus *Chaetomorpha* is credited with about fifty species, but some of them will probably be found to be invalid. While we are able to enumerate six species as having been found within our territory, we feel certain that careful search will reveal additional species and varieties.

Chaetomorpha consists of attached, erect species and of horizontal, entangled species unattached at maturity. It is possible that some, at least, of the latter are simply states of the former. Such a relationship is supposed to exist between *Chaetomorpha aerea* (Dillw.) Kuetz. and *C. Linum* (Fl. Dan.) Kuetz. on the one hand, and *C. melagonium* (Web. and Mohr) Kuetz. and the prostrate plant which has been known as *C. Picquotiana* Mont. on the other. This relationship, however, has not been demonstrated by cultures but is inferred from resemblances existing between the supposedly related states or species.

The attached species of *Chaetomorpha* possess an elongated basal segment which produces branched rhizoidal or even rhizome-like processes of attachment from its lower end. These elongated basal segments and organs of attachments distinguish these species of *Chaetomorpha* from those of both *Rhizoclonium* and *Hormiscia*.

The development of the prostrate species of *Chaetomorpha* has not been thoroughly worked out, but judging from what Okamura (1912, p. 163) has found in *Chaetomorpha spiralis* and from what we have observed in *C. torta,* the plants of these species are attached at first, but probably never assume the erect and tufted habit of the other group, very early assuming a prostrate and flexuous or crisped habit, intertwining and becoming detached. It may be that they are states of other species, but it also may be that they are distinct species or, at least, varieties of the erect species they most naturally resemble. Until more is certain along this line, it seems best to keep them distinct and this policy will be followed by us.

KEY TO THE SPECIES

1. Filaments erect, solitary or tufted... 2
1. Filaments soon declined, becoming horizontal and entangled.............................. 5
    2. Filaments reaching a diameter of 300μ or more.. 4
    2. Filaments 300μ or less in diameter.. 3

1. Chaetomorpha californica Collins

Filaments attached by a small disk formed of short, stout, coalescent rhizoids, erect, straight or flexuous, of uniform diameter throughout, not contracted at the nodes, about 20 cm. long, 20–40μ diam.; segments 1–2 times, rarely 3–4 times, as long as the diameter.

Growing on sandstone in shallow pools along high-tide level. Observed at Laguna Beach and at La Jolla, southern California.

Collins, *in* Collins, Holden and Setchell, Phyc. Bor.-Amer. (Exsicc.), 1900, no. 664; New species, etc., Rhodora, vol. 8, 1906, p. 106, Green Alg. N. A., 1909, p. 325.

Collins (1909, p. 326) says of this species that it is "the most slender erect marine species known" and that it is "not likely to be mistaken for any other." He has, however, recently described a still more slender species (*C. minima*) from Bermuda. In habit and habitat, *C. californica* resembles *Chaetomorpha aerea* as generally understood, but is much more slender than that species. The basal segment is very suddenly narrowed at the base and is up to 200μ long. The base, itself, is provided with a number of thick, short, very slightly if at all branched, stout rhizoids whose outer walls coalesce to form a sort of disk.

2. Chaetomorpha aerea (Dillw.) Kuetz.

Plate 14, figs. 9–11

Filaments rigid, erect, dark green becoming yellowish with age, nearly cylindrical throughout except at the tapering base, attached by delicate rhizoids issuing from the lower end of the basal segment and later coalescent into a more or less solid disk, 125–300μ diam.; segments at first cylindrical, later varying to almost spherical in the fertile segments, 0.5–2 times as long, basal segments much longer; membrane hyaline, thick, at times lamellate; chromatophore at first continuous, finely fenestrate.

Growing in rock pools along high-tide limit or even above. Common on the coast of California from Monterey to San Diego.

Kuetzing, Sp. Alg., 1849, p. 379; Collins, Green Alg. N. A., 1909, p. 324; Collins, Holden and Setchell, Phyc. Bor.-Amer. (Exsicc.), no. 76. *Conferva aerea* Dillwyn, Brit. Conf., 1809, pl. 80.

We feel reasonably safe in referring certain specimens from the central and southern coasts of California to this widespread species after carefully comparing them with specimens from the European and New England coasts. Our plants favor shallow rock pools, more or less lined with sand and situated high up in the littoral belt where the water is warmed by the sun. The specimens available for examination are rather young and range from 200μ to 300μ above. The filaments are erect and tufted, and collapse on being taken from the water. The plants are grass green below but lighter and yellowish above, the color dying away at the tips. This species does not seem to form entangled masses above and we have found no specimens corresponding to the *Chaetomorpha Linum* of the New England coasts and those of Europe, which is supposed to be a state or variety of *C. aerea*. We do not find plants over 300μ in diameter in any of our specimens, whether in those from our own coasts or those from New England or Europe. This is less than the extreme measurements given by some writers (e.g., $600–700\mu$ for zoosporangia by De-Toni, 1889, p. 273) but agrees well with those given by Farlow (1881, p. 46) for the New England plant. The basal segment is short, never ranging, in specimens examined, over 1.5 mm. in length. The upper segments in specimens approaching maturity are short, usually from 0.5 to once as long as broad, but the lower segments may be as much as twice longer than broad. The segments in fertile condition, or approaching it, are slightly swollen and short barrel-shaped.

3. **Chaetomorpha melagonium** (Web. and Mohr) Kuetz.

Filaments closely cespitose or, at times, scattered, attached by short, stout rhizoids coalescing with those of adjacent plants, erect, coarse, stiff, dark glaucous green, 2–6 dm. long, $300–700\mu$ diam.; segments 1–2 diameters long, basal segments slightly attenuated below, up to 2.5–2.75 mm. long.

Unalaska and Kadiak Island, Alaska.

Kuetzing, Phyc. Germ., 1845, p. 204; Collins, Green Alg. N. A., 1909, p. 323; Setchell and Gardner, Alg. N.W. Amer., 1903, p. 222. *Conferva melagonium* Web. and Mohr, Reise nach Schweden, 1804, p. 194, pl. 3, f. 2; Harvey, Phyc. Brit., 1846, vol. 1, pl. 99; Ruprecht, Tange, 1851, p. 396.

Comparatively little is known as to the occurrence of *Chaetomorpha melagonium* on the Pacific coast of North America. Ruprecht (1851, p. 397) credits *Conferva melagonium* to Unalaska and Kadiak Island in tufts over two feet long. Kjellman (1889, p. 55) credits it in typical form to St. Lawrence Island and to Port Clarence. Setchell (1899, p. 590) refers some fragments from the Pribilof Islands to the typical form and Sounders (1901, p. 413) credits Yakutat Bay, Alaska, with the f. *rupincola* (Aresch.) Kjellm. Kjellman unites *Chaetomorpha melagonium* (Web. and Mohr) Kuetz. and *C. Picquotiana* Mont. under *Chaetomorpha melagonium* f. *typica.* The type of *C. melagonium* is evidently a tall, rigid, erect, and attached plant and, judging from the description, so is *C. Picquotiana.* Whether or not the coarse, horizontal, and entangled plant of New England (cf. Farlow, 1881, p. 47) which has gone under the latter name is the true *C. Picquotiana* and whether or not it is also a state of *C. melagonium* are questions which we cannot decide at present; nothing like it, however, has been definitely recorded for our Alaskan waters.

Collins, 1909 (p. 324) states that *Chaetomorpha melagonium* f. *typica* Kjellm. is the unattached plant forming crisped and entangled masses about the roots of *Laminariae* in the sublittoral belt. We do not find that Kjellman so describes it anywhere, but he does distinctly cite the reference of Weber and Mohr, who describe and figure their *Conferva melagonium* as erect, attached, and tufted. The original description of *Conferva Picquotiana* Mont. (1849, p. 66) says "rigida, erecta" and below also "basi adnata," so that it does not seem as if either *Chaetomorpha melagonium* or *C. Picquotiana,* whether the two may be distinct or identical, is unattached and horizontal in characteristic form.

As to f. *typica* Kjellm. (1883, p. 312) and f. *rupincola* Aresch. (Alg. Scand. Exsicc., no. 275 a) the essential differences seem to be that the former has longer segments and the latter shorter segments. The prostrate plant of the New England coast has longer segments and, if it is to be regarded as a state of *C. melagonium,* ought, therefore, to be placed under f. *typica.*

Under *Chaetomorpha melagonium* may be mentioned the *Conferva confervicola* Ruprecht (1851, p. 397) reported as having been found epiphytic on, and attached basally to, *Chaetomorpha melagonium* at Sitka, Alaska. Harvey (1858, p. 88) states that he received a specimen of this plant from Ruprecht and suggests that it may be the young and attached state of *Chaetomorpha tortuosa.* It may possibly be the young attached state of what we have called *Chaetomorpha cannabina.*

4. **Chaetomorpha antennina** (Bory) Kuetz.

Filaments dark green, tufted, 4–9 cm. long, erect, rigid below, somewhat less so above, more or less clavate; basal segments 5–9 mm. long, emitting long, slender, intertwined, branching rhizoids from its base; segments 450–900μ in diameter, 1–4 times as long as broad, fertile somewhat swollen, walls finally thickened (up to 25μ thick) and stratified.

St. Augustin and Mazatlan, Mexico.

Kuetzing, Sp. Alg., 1849, p. 379; Collins, Green Alg. N. A., 1909, p. 324, Suppl. II, 1918, p. 79; Howe, Mar. Alg. Peru, 1914, p. 37. *Conferva antennina* Bory, Voy. quatre îles d'Afr., vol. 2, 1804, p. 161, Voy. Coquille, 1828, p. 227; Montagne, Voy. au Pol Sud, 1845, p. 4. *Chaetomorpha pacifica* Kuetzing, Sp. Alg., 1849, p. 379.

There are two coarse, rigid species of *Chaetomorpha* on the Pacific Coast of North America, viz., *C. melagonium* and *C. antennina*. The first is Alaskan, inhabiting the cold waters of the Boreal zones, while the second is found in tropical waters on the coast of Mexico. *C. melagonium* is a species reaching a length of several dm. while *C. antennina* is seldom over 1 dm. The basal segment in *C. melagonium* is short, not over 3 mm., while that of *C. antennina* is long, reaching a length of 5 to 9 mm.

Howe (*loc. cit.*) has examined the type of *Chaetomorpha antennina* (Bory) Kuetz. and also the Mexican type of *C. pacifica* Kuetz. He judges them to belong to the same species as does also the *Chaetomorphopsis pacifica* Lyon (*in* Tilden, Amer. Alg., Cent., 5, 1901, no. 458). He has also examined the Mazatlan plants collected by Dr. and Mrs. Marchant and, although there are some differences in diameters and length of basal segments, expresses his opinion that they, also, are best placed under *C. antennina*.

The type locality of *Chaetomorpha antennina* is Reunion or Bourbon Island. The specimens upon which *C. pacifica* is based come from Java and from St. Augustin, Mexico. Lyon's *Chaetomorphopsis pacifica* came from the Hawaiian Islands. *Chaetomorpha antennina*, then, in this interpretation is a widespread Indo-Pacific species, ranging from the tropical shores of Pacific North America to those of eastern Africa. The West Indian specimens formerly referred to *C. antennina* seem better referred to *C. media* (Ag.) Kuetz. (cf. Howe, 1914, p. 37).

5. **Chaetomorpha cannabina** (Aresch.) Kjellm.

Filaments unattached, except possibly in the early stage, entangled, soft and rather delicate, color light green, 45–150μ diam., narrow and wide together in the same mass, or even a single filament tapering from larger to smaller measurement; segments 3–8 diameters long, uniformly 500–600μ long.

Growing in tangled masses among other algae, and on wood between tides. From Alaska to Puget Sound, Washington.

Kjellman, Om Beringh. Algfl., 1889, p. 55; Collins, Green Alg. N. A., 1909, p. 325; Saunders, Alg. Harriman Exp., 1901, p. 413; Setchell and Gardner, Alg. N.W., Amer., 1903, p. 221; Collins, Holden and Setchell, Phyc. Bor.-Amer. (Exsicc.), no. 916. *Conferva cannabina* Areschoug, Alg. Sc. Ex., ed. 1, no. 14, 1840, ed. 2, no. 135, Alg. minus rite cog., Pug. II, 1843, p. 268, pl. 9, Phyc. Scand., 1850, p. 433.

What passes for *Chaetomorpha cannabina* seems to be fairly frequent along our coast from Puget Sound northward to Bering Sea. The habit is that of entangled prostrate masses, free or intertwined with other algae. No organs of attachment have been found in any of the specimens. The filaments are long, crisped and often intricately entangled. The diameters of the segments range from 44μ to 125μ in our specimens and a considerable range of variation is usually found in the same mass and even in the same filament.

In attempting to determine the exact status of the *Conferva cannabina* Areschoug, we are met at once with conflicting statements in the various descriptions of Areschoug. In 1843 Areschoug described the species as having filaments $\frac{1}{32}$–$\frac{1}{25}$ line in diameter which, assuming this to be the ''Paris'' line, means from 73μ to 90μ. In 1850 Areschoug, however, states that the more slender filaments are 0.10 mm. and the thickest are 0.20 mm., thus indicating 100μ to 200μ for the diameter of the filaments. An examination of no. 135 of the second issue of the Algae Scandinavicae Exsiccatae shows filaments ranging from 85μ to 145μ.

Our west American specimens agree fairly well with the specimens distributed by Areschoug under no. 135 of the second issue of his Algae Scandinavicae Exsiccatae (the first issue is not accessible to us) and it seems best to retain his name for our species, at least for the present. It seems also, judging from the specimen in the Herbarium of the Imperial Academy at St. Petersburg, to be the plant described as *Conferva tortuosa* var. *crassior* by Ruprecht (1851, p. 399) which was collected at Sitka, Alaska. Harvey (1858, p. 87) referred Ruprecht's plant to his *Chaetomorpha litorea* (cf. Harvey, 1849, p. 208 and 1851, pl. 333, under *Conferva*). The *Chaetomorpha litorea* Harv. itself, judging from descriptions and figures, is closely related to *C. cannabina* if not identical with it.

The *Chaetomorpha confervicola* (Rupr.) De-Toni (1889, p. 268) is a slender attached species growing on *C. melagonium* at Sitka (cf. Ruprecht, 1851, p. 397). This is suggested as being the young attached state of *Chaetomorpha tortuosa* (*Rhizoclonium tortuosum* of this account), but it seems fully as probable that it may be the young and attached state of *C. cannabina*.

6. **Chaetomorpha torta** (Farlow) McClatchie

Filaments rigid, attached when young, soon becoming declined, loosened and entangled among other algae, much coiled and contorted, 40–60 cm. long, up to 1 mm. in diam., nearly uniform throughout with segments slightly moniliform, 1–1.5 times as long as broad; color iridescent bluish green.

Growing in the sublittoral belt. Southern California.

McClatchie, Seedless plants, 1897, p. 351 (nomen nudum); *in* Yendo, Notes on algae new to Japan II, 1914, p. 264 (excl. syn. *C. spiralis*). *Chaetomorpha clavata* var. *torta* Farlow, *in* Farlow, Anderson and Eaton, Alg. Exsicc. Amer.-Bor., fasc. 5, 1889, no. 211 (nomen nudum); Collins, Holden and Setchell, Phyc. Bor.-Amer. (Exsicc.), fasc. XII, 1899, no. 571 (nomen nudum); Farlow, *in* Collins, Green Alg. N. A., 1909, p. 323 (descr.). *Chaetomorpha clavata* Collins, Holden and Setchell, Phyc. Bor.-Amer. (Exsicc.), no. 371 (not of Kuetzing).

Chaetomorpha torta is confined, so far as our knowledge goes, to the coast of southern California. It is so different from any of our other species as to be readily recognizable by its scattered or massed, long, coarse, contorted and even spirally coiled, rigid filaments of a peculiar light bluish green color. In searching for any trace of an attachment, our persistence was rewarded by the discovery of one among all the material examined and that, curiously enough, was found on no. 211, of Farlow, Anderson and Eaton's Algae Americae Borealis Exsiccatae. The rhizoids are slender, blunt, somewhat branched, and evidently were attached to a rock surface. The basal segment is slightly attenuated below and about 2 mm. long. The filament was bent over and declined just above the basal segment. The method of attachment and later growth of *Chaetomorpha torta* and *C. spiralis* Okam. seem to indicate the possibility of the prostrate species of *Chaetomorpha* being not later stages in growth of the erect species, but distinct species, declined almost from the first and resembling the species of *Rhizoclonium* in this respect as has been suggested above.

The status of the name is somewhat complicated. It was first published by Farlow in the form "*Chaetomorpha clavata* var. *torta* Farlow,*" in connection with a distributed specimen and printed label but without description. The description was appended first by Collins in 1909. The combination "*Chaetomorpha torta* Farlow" was first used in 1897 by McClatchie, but without description or citation. Yendo (1914, p. 264) used the combination "*Chaetomorpha torta* McClatchie,*" citing the various references to the California plant and assigning also as a synonym, the *Chaetomorpha spiralis* Okamura (1912, p. 95). Collins (1918, p. 78) has finally reviewed the situation and argued for the etiquette "*Chaetomorpha torta* (Farlow) Yendo.*" It seems to us, while agreeing in general with the principles enunciated by Collins in his later paper, better to write "*Chaetomorpha torta* (Farlow) McClatchie*" *in* Yendo, Notes on algae new to Japan, etc. Otherwise, and following out literally the contention of Collins, it should be written "*Chaetomorpha torta* (Collins) Yendo,*" since Collins was the first to make the varietal name valid. It must be borne in mind, however, that the situation is further complicated by the fact that the varietal name had not been validated when McClatchie published his combination.

Chaetomorpha torta has been associated with *C. clavata* (Ag.) Kuetz. and with *C. spiralis* Okam. The former is a plant described from the West Indies and has erect, straight filaments, tapering from below upward. *C. spiralis*, on the other hand, although attached at first, is soon declined, flexuous, and spirally twisted, with segments much narrowed towards the base, short and much swollen above and reaching a maximum diameter of 2.5 mm. In all these characters, *C. spiralis* differs from *C. torta* and should be kept distinct. From *Chaetomorpha moniligera* Kjellman (1897, p. 24, pl. 4, f. 17–23), an erect and delicate, though broad, species, both *C. spiralis* and *C. torta* seem amply distinct.

13. **Cladophora** Kuetz.

Plants composed of filaments of a single series of segments, the filaments branching, usually abundantly; branching lateral, but often coming to appear dichotomous in consequence of the pushing aside of the original filament by the branch; growth in length chiefly by division of the apical segment, subsequent division of segments being rather exceptional; branches all of the same type but different orders usually differing in diameter; segments multinucleate, the chromatophore either covering the segment wall, or forming a network on it, or in the form of numerous small disks; pyrenoids several in a segment; reproduction by 4-ciliated zoospores, and by 2-ciliated gametes, uniting and germinating immediately, or sometimes germinating without copulation.

Kuetzing, Phyc. Gen., 1843, p. 262.

The genus *Cladophora* was established by Kuetzing in 1843, the same year that Hassall (1843, p. 363, May) proposed the genus *Microspora* to include the branching species of the old genus *Conferva*. It is possible that *Microspora* Hassall antidates *Cladophora* Kuetzing, but Hassall (1845, p. 213) withdrew his genus and adopted that of Kuetzing. The name *Microspora* was later used by Thuret (1850, p. 221) to designate a later segregation from *Conferva* and this name is still current in the sense indicated by Thuret.

The type species of Kuetzing's genus is the *Conferva oligoclona* Kuetzing (1833, no. 62), a fresh-water species and there is little question as to the limits of the genus, except possibly towards some of its segregates. It is distinctly and readily separated from *Chaetomorpha* by its branching and from *Rhizoclonium* by the method of its branching. The difficulties of separating *Spongomorpha* Kuetz., *Aegagropila* Kuetz. and *Acrosiphonia* J. Ag., if it seems best to do so, are at times considerable. The species of all these genera are of more spongy habit than are those of the true or typical *Cladophora*. In *Spongomorpha* and *Acrosiphonia* the spongy condition is brought about by specialized rhizoidal or spinelike branches which hold the mass of branches together. In *Aegagropila*, there are no such specialized branches and it has seemed best to us not to separate the species of *Aegagropila* from those of *Cladophora*. The distinction between the two genera depends largely upon habit and is not always applicable with certainty and precision.

The determination of species of *Cladophora* presents unusual difficulties. The synonymy, especially of older specific names, is very much unsettled and intertangled. The identity between species of different coasts and oceans is very uncertain, and the recognition of species in their various ages and under varying conditions seemingly most difficult. We have adapted our account very largely from Collins (1909) and have accepted his determinations and limitations of species, since we ourselves have very little opportunity for the study of this genus and possess few facilities for the proper comparison of our west coast plants with authentically determined specimens from other coasts. It is desirable that collectors pay careful attention to these plants, to their seasonal conditions, their fertile states, their earlier stages of growth as well as adult, and attempt to make better known the entire life-history of our various species and forms.

KEY TO THE SPECIES

1. Low, pulvinate or matted, creeping at the base...2
1. Erect, tufted or loosely spreading...4
   2. Branching irregular, not at all dichotomous............1. **C. amphibia** (p. 209)
   2. Branching dichotomous or trichotomous, except above............................3
3. Main filaments 60–150μ in diameter.....................2. **C. hemisphaerica** (p. 209)
3. Main filaments 120–250μ in diameter................................3. **C. trichotoma** (p. 210)
   4. Main filaments usually over 150μ in diameter...5
   4. Main filaments usually under 150μ in diameter...9
5. Lower segments 10 or more diameters long...........4. **C. graminea** (p. 211)
5. Lower segments less than 10 diameters long..6
   6. Ramuli curved...5. **C. microcladioides** (p. 212)
   6. Ramuli straight...7
7. Ramuli only slightly less stout than branches................6. **C. Hutchinsiae** (p. 213)
7. Ramuli very much less stout than branches...8
   8. Main filaments 112–200μ in diameter..........................7. **C. ovoidea** (p. 214)
   8. Main filaments 135–300μ in diameter...............8. **C. MacDougalii** (p. 214)
9. Main filaments distinctly zig-zag or flexuous...10
9. Main filaments straight or nearly so...13
   10. Ramuli clustered at the tips..............................9. **C. laetevirens** (p. 216)
   10. Ramuli not clustered at the tips...11
11. Main filaments coarse, up to 150 (or 160) μ.......................10. **C. gracilis** (p. 216)
11. Main filaments slender, not over 80μ...12
   12. Segments long below (up to 6 diam.), shorter (as low as 2 diam.) above
              ....11. **C. flexuosa** (p. 217)
   12. Segments long throughout.....................12. **C. Rudolphiana** (p. 217)
13. Ramuli curved...........................13. **C. Bertolonii var. hamosa** (p. 218)
13. Ramuli straight...14
   14. Main filaments not over 30μ in diameter..................14. **C. albida** (p. 218)
   14. Main filaments considerably over 30μ in diameter..............................15
15. Ramuli acute...15. **C. glaucescens** (p. 219)
15. Ramuli blunt..16
   16. Branching dichotomous below the tips.............16. **C. Stimpsonii** (p. 219)
   16. Branching nowhere dichotomous..................17. **C. delicatula** (p. 220)

1. Cladophora amphibia Collins

Basal layer of densely branching, prostrate filaments, segments cylindrical, 40–70μ diam. and 2–5 diam. long, or fusiform, 1–2 diam. long, swollen to 100μ in the middle, emitting erect filaments with segments 30–50μ diam., 4–8 diam. long, cylindrical or irregular, terminal segments obtuse or truncate; slender descending rhizoids sometimes issuing from lower segments of erect filaments.

Growing on the ground among *Salicornia* and other salt marsh plants along high-tide line. Known only from a single locality, viz., Alameda, along the shores of San Francisco Bay, California.

Collins, *in* Collins, Holden and Setchell, Phyc. Bor.-Amer. (Exsicc.), no. 1284 (nomen nudum), and *in* Rhodora, vol. 9, 1907, p. 200 (description), Green Alg. N. A., 1909, p. 349.

Collins has correctly described this species as a "dull green unattractive plant." It is not readily mistaken for any other, both as regards its habitat and its habit. It is low and forms extended patches along the salt marsh covering of *Salicornia*. Its discovery in similar localities along the coast is to be expected. The species would ordinarily be grouped with those of the *Aegagropila* section; it is matted together below by the intertwining of the prostrate filaments and by the descending slender almost rhizoidal branchlets. The branching is irregularly alternate and not at all dichotomous. The turflike layer is not over a centimeter in thickness.

2. Cladophora hemisphaerica Gardner

Plants at first compact, hemispherical, of not over 1 cm. radius, profusely branching at the base, dichotomous, forkings narrow, segments 60–150μ diam., 3–6 diam. long, often distinctly clavate; later less regularly hemispherical, up to 3 cm. or more radius, branches more distant, with segments in the upper part 50–80μ diam., cylindrical; numerous slender tufts, up to 1 dm. long, on nearing maturity arise from the hemispherical mass, having cylindrical segments 50–80μ diam., with forkings very distant and narrow; substance firm, not adhering well to paper.

Growing in small, shallow pools on rocks in the upper littoral belt. Cypress Point, Monterey County, California.

Gardner, *in* Collins, Holden and Setchell, Phyc. Bor.-Amer. (Exsicc.), no. 2240 (nomen nudum), *in* Collins, Green Alg. Suppl. II, 1918, p. 83 (descr.).

Cladophora hemisphaerica is another species of the *Aegagropila* section, found thus far in a single locality, but growing on rocks where the small indentations retain water at the upper limits of the littoral belt. The filaments are bound together at the base by descending rooting branchlets but are entangled above. Tufts growing in deeper pools sometimes give rise to more slender and looser erect tufts more like individual plants of the *Eucladophora* section.

Cladophora hemisphaerica resembles somewhat *C. trichotoma*, as found on our coasts, but is more distinctly tufted, of different aspect and with more slender filaments. It is attached very firmly to the wave-swept rocks.

3. Cladophora trichotoma (Ag.) Kuetz.

Plate 16, fig. 2

Plants forming light or bright green, densely pulvinate masses, 2–5 cm. high in shallow pools, and up to 2.5 dm. in deep, quiet pools; filaments procumbent at the base, stiff, di-trichotomous, with rather few, short, alternate, rarely opposite branches, fastigiate at the tips; segments 120–250μ diam., 4–10 diam. long, nearly cylindrical below, above ovoid to pyriform; the branches about the same diameter as the filament.

Growing in rock pools near high water mark. Vancouver Island, British Columbia, to La Paz, Lower California.

Kuetzing, Sp. Alg., 1849, p. 414; Collins, Green Alg. N. A., 1909, p. 349; Howe, Phyc. Studies V, 1911, p. 492; Collins, Holden and Setchell, Phyc. Bor.-Amer. (Exsicc.), no. 820. *Conferva trichotoma* Agardh, Syst., 1824, p. 121. *Cladophora repens* Collins, Holden and Setchell, Phyc. Bor.-Amer. (Exsicc.), no. 727 (not of Harvey). *Cladophora columbiana* Collins, *in* Setchell and Gardner, Alg. N.W. Amer., 1903, p. 226. *Cladophora composita* Setchell and Gardner, Alg. N.W. Amer., 1903, p. 226, as to "N. L. G. no. 521." *Cladophora cartilaginea* Tilden, Amer. Alg. (Exsicc.), no. 376.

Cladophora trichotoma is the most common of all our species of *Cladophora* and is usually abundant on all rocky shores from Vancouver Island to La Paz in Lower California. It grows high up on the bare rocks or in small pools where its compact bright green cushions or expanded tufts form conspicuous objects. Our plant has been compared with a fragment of the type specimens and agrees so closely as to be referred here with considerable confidence. Its extensive distribution is to be explained, in all probability, by its occurrence so high up in the littoral belt that it is enabled to receive much heat from the sun and the air, while its dense tufts allow it to remain exposed at low tide without drying.

Cladophora trichotoma f. elongata Collins

A luxuriant form, sometimes up to 3 dm. high; fronds sparingly branched except near the surface of the water, where it forms characteristic dense tufts.

Growing on the margins of deep rock pools little affected by tides. Point Carmel, "Point Lobos," Monterey County, California.

Collins, *in* Collins, Holden and Setchell, Phyc. Bor.-Amer. (Exsicc.), no. 2141, Green Alg. Suppl. II., 1918, p. 81.

The typical form of *Cladophora trichotoma* is truly a member of the *Aegagropila* section of the genus *Cladophora* and occurs in dense cushions or tufts, low and spreading by prostrate branches which descend, attach themselves, and send off erect branches in turn. When growing in deep rock pools, and, therefore constantly immersed, the low cushions give off erect and luxuriant tufts of filaments, loosely entangled and sparing branched, until they reach the surface. These erect tufts simulate the plants of the *Eucladophora* section, and may pass for them unless care is taken to study the base. We feel fairly certain in referring here no. 127, Butler and Polley, collected at Port Renfrew, British Columbia and listed in Setchell and Gardner, Algae of Northwestern America (1903, p. 228) as *Cladophora Hutchinsiae* var. *distans*.

4. Cladophora graminea Collins

Plants loosely tufted, 10–15, cm. high, dark green, cartilagineous, very rigid, distantly di-trichotomous, all divisions erect; main filaments about 300μ diam., ultimate divisions about 150μ diam., tips blunt or slightly acute; segments very long below, up to 30 diam., shorter above, normally occupying the space from one forking to another; ultimate branches 4–6 diam. long; segment walls usually strongly lamellate.

Growing in rock pools usually in shaded habitats, in the upper littoral belt. Known along the California coast from Monterey to San Diego.

Collins, *in* Rhodora, vol. 11, 1909*a*, p. 19, pl. 78, f. 6. *Cladophora erecta* Collins, *in* Collins, Holden and Setchell, Phyc. Bor.-Amer. (Exsicc.), no. 1690.

Cladophora graminea is one of the coarsest of our west coast species of erect *Cladophoras*, superficially resembling a species of *Spongomorpha* in this and in its habit. It has been usually confused with *C. cartilaginea* which, properly, is a *Spongomorpha*. In its regular or unequal, dichotomous branching, where each arm is composed of a single segment, in its coarseness, as well as in the proportional length of the segments, it is to be distinguished from all our other species.

5. Cladophora microcladioides Collins

Plate 13, fig. 2

Plants more or less densely tufted, 10–20 cm. high; filaments about 200μ diam. at the base, segments 4–6 diam. long; stiff, straight or flexuous, distantly di-trichotomous, branches similar, erect or more or less recurved, bearing on the upper (inner) side numerous short branches, rarely with very short branches opposite one or more of them; this ramification continued, the ultimate ramuli of very few segments, 80–100μ diam., segments 1.5–2.5 diam. long.

Growing in the upper sublittoral belt. From Vancouver Island, British Columbia, to San Diego, California.

Collins, *in* Rhodora, vol. 11, 1909*a*, p. 17, pl. 78, f. 2, 3, Mar. Alg. Vancouver Island, 1913, p. 104.

Cladophora microcladioides f. stricta Collins

Similar to the species but differing in habit; branches of various orders virgate and little recurved.

Lower littoral belt. San Diego, California.

Collins, *in* Collins, Holden and Setchell, Phyc. Bor.-Amer. (Exsicc.), no. 1583.

In its typical form *Cladophora microcladioides* is one of our most readily recognized species of the *Eucladophora* section or erect *Cladophoras*. The regular recurved branches give the plant much of the habit of *Microcladia borealis* Ruprecht. There is much variation in height, flexuosity of principal axes, frequency of recurved pinnules, etc., but the pinnules are distinctive and usually readily recognizable. In the forma *stricta*, however, the pinnules are not recurved, but there is more or less of an indication of the close affinity with the species.

6. **Cladophora Hutchinsiae** (Dillw.) Kuetz.

Plants glaucous green, up to 40 cm. high; filaments 120–300μ diam., stiff, flexuous, sparingly branched; ramuli few, secund, blunt, with constricted nodes; cells 2–3 diam. long.

Vancouver Island, British Columbia.

Kuetzing, Phyc. Germ., 1845, p. 210; Collins, Green Alg. N. A., 1909, p. 345. *Conferva Hutchinsiae* Dillwyn, Brit. Conf., 1809, pl. 109.

Cladophora Hutchinsiae var. **distans** (Ag.) Kuetz.

Segments of primary branches longer than in the type, 280–400μ diam., with few secondary branches whose segments are 200–250μ diam., 3–4 times as long as the diameter.

West coast of Vancouver Island, British Columbia.

Kuetzing, Sp. Alg., 1849, p. 392; Collins, *loc. cit.;* Setchell and Gardner, Alg. N.W. Amer., 1903, p. 228. *Conferva distans* Agardh, Syst., 1824, p. 120.

Cladophora Hutchinsiae is one of the coarser species of the *Eucladophora* section, whose branchlets are only slightly reduced in diameter from the branches from which they spring. We have seen no specimens from our coast and Collins (1909, p. 345) gives only "Vancouver Island," stating (*loc. cit.,* p. 346) that it is likely to be found "on the west coast south of Vancouver." In his Marine Algae of Vancouver Island, however, Collins (1913, p. 104) lists only the var. *distans* as occurring at Port Renfrew, where it was collected by Butler and Polley. No. 127 of their collection, as distributed, which is the plant in question, seems clearly to be *Cladophora trichotoma* f. *elongata* Collins. In his Marine *Cladophoras* of New England, Collins says (1902, p. 126) that the type of *C. Hutchinsiae* has never been certainly known from New England, the one quoted by Farlow in the Marine Algae of New England proving to be wrongly marked and to belong to California. The status of the species on the West Coast, therefore, is not exactly established.

Cladophora Hutchinsiae is to be distinguished from both *C. ovoidea* and *C. MacDougalii* by the stout branchlets, the ultimate ramuli being, for the most part, of only one or two segments.

There is doubt, at least, as to the reference of var. *distans* to our coast, the Butler and Polley specimen (no. 127) seeming to be erroneously determined, as stated above. We are, however, inclined to refer here a specimen collected by one of us (Gardner, no. 4104) at Lands End, San Francisco, California. The specimens seem to agree well with Harvey's figure of *Cladophora diffusa* (1849, pl. 130) and fairly well with the figure of Dillwyn's *Conferva diffusa* (1803, pl. 21), and possibly also with the plate of Roth (1800, pl. 7). The San Francisco plant is more slender than the type of the species, with less profusion of branches and branchlets, and, in this, agrees with Harvey's *C. diffusa*. The synonymy, however, is too confused to be unraveled, at least at present.

7. Cladophora ovoidea Kuetz.

Plants 5–15 cm. high, stiff, rather dull green; filaments distantly dichotomous, 150–200μ diam. below, branches becoming more lateral and secund above; upper ramuli not over 60μ diam., tips rounded or slightly pointed; segments in lower part cylindrical, 4–8 diam. long; above ovoid, 1.5–3 diam. long.

Known only from Carmel Bay and Santa Cruz in California.

Kuetzing, Phyc. Gen., 1843, p. 266, Tab. Phyc. III, 1853, pl. 92, f. 1; Collins, Green Alg. N. A., 1909, p. 346.

We have seen only specimens identified by Collins and have been unable to compare them with European plants of *Cladophora ovoidea*. Our specimens are coarse, of the same general appearance as those of *C. Hutchinsiae*. The branchlets, however, are much more slender than the branches from which they spring and are, for the most part, composed of four to six segments. The upper segments are more or less ovoid. Our plants resemble reasonably closely the figures of Kuetzing (1853, pl. 92, f. 1).

8. Cladophora MacDougalii Howe

Plants rather stout, coarse and rigid, in strict tufts, dark or yellowish green, 10–17 cm. high; main filaments 135–310μ diam., sparingly dichotomous below the middle of the tufts; branching in median and upper parts lateral, the branches erecto-patent, secund, occasionally alternate, or very rarely opposite, becoming more or less secund-pectinate toward the apices, the main axes commonly excurrent beyond the last lateral branch as rather rigid tapering prolongations 10–40 segments long; the ultimate lateral branchlets 75–110μ diam., about one half the diameter of the filaments from which they spring, usually 3–7 segments long, in most cases gradually tapering from near the base, subacute or blunt, commonly rather rigid; segments in extreme basal parts 6–15 times as long as broad, in median and upper parts 1–4 (mostly 1.5–2.5) times as long as broad, usually a little constricted at the septa and appearing quite strongly constricted when dry.

San Felipe Bay, Lower California.

Howe, Phyc. Studies V, 1911, p. 491, pl. 33, f. 7.

Collins (Green Alg. Suppl. I, 1912, p. 96) appends the following note to *C. ovoidea,* "This species," referring to *C. MacDougalii,* "is compared by its author with *C. Hutchinsiae* (Dillw.) Kütz., and *C. Ovoidea* Kütz., but is considered distinct from both. A specimen kindly furnished by the author shows that it is amply distinct from the *C. Hutchinsiae* of England and France, N. J. and Barbados. It is, however, quite close to the California plant given as *C. ovoidea* by the writer, 1909, p. 346, and would have been placed there without question. It may well be that it is distinct from the *C. ovoidea* of Europe, which appears to be a little known species. The California plant is larger in all dimensions, less moniliform, and with long excurrent axes. A comparison of specimens from various stations shows much variation in the two last particulars, but all are much stouter than *C. ovoidea* as originally described. It often varies in the direction of *C. microcladioides* Collins, with denser and more fasciculate branching.''

Howe says (1911, p. 492) : "We have seen no European specimens of *C. ovoidea,* but are unwilling at the present time to identify with this species a Lower California plant with filaments and branches averaging twice as thick as those of the plant described and figured by Kützing, and with filaments so little constricted at the septa (in a soaked out condition, at least) that no one would think of describing any of the cells as 'ovoid' (Kützing, Phyc. Gen., 266). Also, according to Kützing's figure (Tab. Phyc., 3:26, pl. 92, f. 1, 1853), the branching in *C. ovoidea* is more fasciculate than in *C. MacDougalii,* the ultimate lateral branchlets are less tapering and less rigid, and the main axes do not show the long-excurrent prolongations of the Baja California species.''

It seems best to us to retain *C. MacDougalii* as a distinct species.

9. **Cladophora laetevirens** (Dillw.) Kuetz.

Plants 15–20 cm. high, erect, rigid, bright yellow green; filaments much branched, flexuous, 50–150μ diam.; branches erect, often opposite; ultimate ramuli short (usually 1–3 segments), obtuse or subacute, densely fastigiate at the tips of the branches; segments of main branches 6 diameters long, of ramuli 3 diameters.

Kuetzing, Phyc. Gen., 1843, p. 263; Harvey, Phyc. Brit., vol. 2, 1849, pl. 190; Notice of a collection of algae, etc., 1862, p. 177; Setchell and Gardner, Alg. N.W. Amer., 1903, p. 224; Collins, Green Alg. N. A., 1909, p. 345. *Conferva laetevirens,* Dillwyn, Brit. Conf., 1805, pl. 48.

The only reason for including this species beyond the record of Harvey (1862*a*, p. 177) for Fuca Strait, apparently based on two small and young specimens, is the collection of what seem to be characteristic specimens by Butler and Polley near Port Renfrew, B. C. These have been carefully compared with no. 143 of Wyatt's Algae Danmonienses both by F. S. Collins and by ourselves and they seem to be in close agreement with it.

10. **Cladophora gracilis** (Griff.) Kuetz.

Plants moderately rigid, forming somewhat slender, pyramidal fascicles, 15–30 cm. high, pale or glaucous green, di- trichotomous at the base; main filaments up to 160μ diam., segments 4–7 times as long as the diameter; main branches smaller and beset with numerous slender, tapering, secund ramuli constricted at the joints.

Growing on rocks in shallow tide pools, in the upper and middle littoral belts. Sitka, Alaska, and Neah Bay, near Cape Flattery, Washington.

Kuetzing, Phyc. Germ., 1845, p. 215; Collins, Green Alg. N. A., 1909, p. 342. *Cladophora vadorum* Kuetzing, Sp. Alg., 1849, p. 402, Tab. Phyc., vol. 4, 1854, p. 4, pl. 20, f. I. *Conferva gracilis* Griffiths, *in* Wyatt, Alg. Danm., no. 97.

The limits of *Cladophora gracilis* have not been clearly defined. Either it is an exceedingly variable species, or several closely related species have been grouped into one by different authors. Kuetzing (1845, p. 215) gives 113–124μ, De-Toni (1889, p. 322) gives 100–140μ, and Collins (1909, p. 342) gives up to 160μ as the diameter of the main filaments of this species. Collins (*loc. cit.,* p. 343) has recognized five forms of this species as abounding on the Atlantic coast of North America.

We are placing under *C. gracilis* two sets of plants differing considerably from one another, mainly in details of measurements, as apparently closely allied to the Atlantic coast forms, but they do not seem to be identical with any of them. No. 3954 Gardner, from Sitka, Alaska, has the main filaments up to 100μ in diameter. No. 3870a Gardner, from Neah Bay, Washington, has slightly smaller dimensions than no. 3954. The ultimate ramuli in both of the above mentioned collections, while being more or less secund, are not so much so as is usually recorded for the species. No. 3870a is fruiting abundantly at the tips of the ramuli, the segments becoming decidedly ventricose or even spherical in reproduction. These two collections are placed here provisionally awaiting further investigation into the status of the species.

11. **Cladophora flexuosa** (Griff.) Harv.

Plants 10–20 cm. high, light green; main filaments 80–120μ diam., regularly flexuous, with flexuous alternate branches, 40–80μ diam., which in turn have alternate or secund, curved and sometimes refracted ramuli; segments from 6 diam. long below to 2 in the ramuli.

Growing in rock pools in the lower littoral belt. Annettee Island, Alaska, to San Diego, California.

Harvey, Phyc. Brit., 1851, p. 353; Collins, Green Alg. N. A., 1909, p. 339; Setchell and Gardner, Alg. N.W. Amer., 1903, p. 224; Collins, Holden and Setchell, Phyc. Bor.-Amer. (Exsicc.), no. 2239. *Conferva flexuosa* Griffiths, *in* Wyatt, Alg. Danm., no. 227.

The status of the name *Cladophora flexuosa* is equivocal. The plant usually referred here does not seem to be at all certainly the *Conferva flexuosa* Dillwyn (1802, pl. 10) nor that of Mueller in the Flora Danica (1782, p. 5, pl. 882), but more probably that of Mrs. Griffiths (*in* Wyatt's, Alg. Danm., no. 227), later adopted by Harvey (1851, pl. 353) and referred to *Cladophora*. Our west coast plants have been referred here chiefly on the authority of Collins.

12. **Cladophora Rudolphiana** (Ag.) Kuetz.

Plants loose, soft, yellowish green, gelatinous, up to a meter in length; main filaments 40–80μ diam.; branches alternate or opposite, patent, flexuous, ramuli secund, tapering, about 20μ diam.; segments much longer than broad, up to 20 diam. below.

Kuetzing, Phyc. Gen., 1843, p. 268; Harvey, Phyc. Brit., 1846, pl. 86; Collins, Green Alg. N. A., 1909, p. 336.

The typical form has not as yet been detected on our coast. It is to be distinguished by its soft and gelatinous consistency and its segments which are long throughout the plant.

Cladophora Rudolphiana f. eramosa Gardner

Branches very long, subsimple.

Growing in a warm salt water pond. Key Route Power House, Oakland, California.

Gardner, *in* Collins, Green Alg. Suppl. II, 1918, p. 81; Collins, Holden and Setchell, Phyc. Bor.-Amer. (Exsicc.), no. 2241.

A curious plant, seemingly nearest to *Cladophora Rudolphiana,* has been found in flowing warm salt water. The branching is very slight. The main filaments are 55μ to 60μ diameter and the ramuli about 25μ diameter. The lower segments are from 4.5 to 6.5 diameters long.

13. Cladophora Bertolonii var. hamosa (Kuetz.) Ardiss.

Plants 3–10 cm. high, dark green; filaments rather stiff, 80–100μ diam. in the main divisions, 25–30μ in the ramuli; much branched, main divisions di- trichotomous, opposite or whorled branches, usually short and with densely set, secund, recurved ramuli, segments 1.3–3 diameters long, rarely more, terminal segments rounded, not tapering.

Central California (Pacific Grove).

Ardissone, Phyc. Med., 1886, part II, p. 242; Collins, Green Alg. N. A., 1909, p. 344. *Cladophora hamosa* Kuetzing, Phyc. Gen., 1843, p. 267, Tab. Phyc., 1854, vol. 4, pl. 8, f. 2.

We know of this plant only from the report of Collins (*loc. cit.*), who appends the following remarks: "The California plant seems to be more slender than the European, seldom exceeding 60μ in the main branches and 25μ in the ramuli. The dark color, short cylindrical cells and elegant feathery tips, with a long series of secund, usually slightly recurved ramuli on the similarly recurved branches, are fairly clear characters."

14. Cladophora albida (Huds.) Kuetz.

Plants soft, dense, pale green, filaments 20–30μ diam., segments 4–5 diameters long, delicate; branching irregular, ramuli long, patent, blunt.

Known in our region from East Sound, Washington, and San Pedro, California.

Kuetzing, Phyc. Gen., 1843, p. 267; Collins, Green Alg. N. A., 1909, p. 336. *Conferva albida* Hudson, Fl. Ang., 1732, Ed. I, 1778, Ed. II, p. 595; Dillwyn, Brit. Conf., 1809, p. 32, pl. E.

Cladophora albida is our most slender species as well as one of very soft and spongy consistency. By these characters it is usually to be distinguished, whether living or in dried specimens.

15. Cladophora glaucescens (Griff.) Harv.

Plants 10–40 cm. long, glaucous or yellowish green, loosely tufted, much branched, ending in long, erect, acute, alternate or sometimes secund, ramuli; segments at base 50–60μ diam., in ramuli 25–30μ; segments usually 4–6 diam. long, sometimes considerably longer.

Nanaimo, Vancouver Island, British Columbia, to Oakland, California.

Harvey, Notice of a collection of algae, etc., 1862a, pp. 160, 161, 176; Collins, Mar. Alg. Vancouver Island, 1913, p. 103; Setchell and Gardner, Alg. N.W. Amer., 1903, p. 224. *Conferva glaucescens* Griffiths, *in* Wyatt, Alg. Danm., no. 195.

The above description is from Collins' Green Algae of North America (1909, p. 336), and refers to the species on the Atlantic coast. Its occurrence on our coast rests largely on the authority of Harvey, who determined a plant collected by Dr. Lyall at Nanaimo, B. C. There is but a single plant in Harvey's herbarium to represent the species on our coast, and one of us (Setchell) has seen the specimen and expresses some doubt as to its identity with the Atlantic coast plant. We are inclined, however, to refer here a plant collected by one of us (Gardner, no. 2647) in the warm salt water pond at the Key Route Power House in Oakland, California.

16. Cladophora Stimpsonii Harv.

Plants loosely tufted, up to 30 cm. high, light green, of delicate and silky texture; filaments 100–150μ at the base, tapering gradually upward, di- trichotomously divided, branches continuously but distantly forking, successively smaller, ultimate branches lateral, secundly pectinate with long ramuli, 20–25μ diam., with rounded or slightly pointed tips; segments 5–8 diam. long, longest near the base.

On shells, etc. Ucluet Inlet, Vancouver Island, British Columbia, to southern California.

Harvey, Characters new alg., 1859*a*, p. 333; Collins, Green Alg. N. A., 1909, p. 338, Mar. Alg. Vancouver Island, 1913, p. 104; Collins, Holden and Setchell, Phyc. Bor.-Amer. (Exsicc.), no. 729.

As Collins remarks (1909, p. 338), this is "a soft, delicate, silky plant, reminding one of the more delicate forms of *C. gracilis*, but distinct in manner of branching, substance and cell dimensions." It seems possibly a not unusual plant about San Pedro, California. We have compared our plants with a specimen collected by Yendo at Hakodate, Japan, the type locality, and it certainly seems to come very close to it, although careful study of a larger series of specimens might show some essential differences.

17. Cladophora delicatula Mont.

Plants loosely tufted, soft, dull green, about 10 cm. high; filaments 40–60μ diam. below, 4–6 diam. long; loosely branching, branches virgate, erect; ramuli in short, secund series, seldom over 8 segments in length, segments 20–30μ diam., 1–2 diam. long, joints somewhat constricted.

Growing on rocks along high water limit exposed to the surf. San Pedro, California.

Montagne, Crypt. Guyan., 1850, p. 302; Collins, Green Alg. N. A., 1909, p. 337; Collins, Holden and Setchell, Phyc. Bor.-Amer. (Exsicc.), no. 1582.

The reference of these delicate plants to *Cladophora delicatula* is on the authority of Collins. He is also inclined to refer here, with some doubt, plants collected on shaded, sandstone rocks in quiet water at Coos Bay, Oregon (Gardner, no. 2747).

14. Spongomorpha Kuetz.

Plants composed of profusely branched, monosiphonous filaments, with terminal segments frequently larger and longer than the segments below, and usually blunt and rounded, though acute in some species; intercalary divisions predominate, though divisions of the apical segments may occur; in age the filaments are usually bound tightly together in cushion-like or rope-like tufts by descending rhizoidal branches below, or by special, short, spine-like or hooked branches, segments uni- or plurinucleate; chromatophore either a thin, closed band, or a finer or coarser reticulum with few to many pyrenoids.

Kuetzing, Phyc. Gen., 1843, p. 273. *Cladophora* (*Spongomorpha*) Kuetzing, Spec. Alg., 1849, p. 417.

As originally founded by Kuetzing, *Spongomorpha* included two species, *Conferva uncialis* Ag. and *Conferva aggregata* Ag., the latter now considered to be a synonym of *Conferva lanosa* Roth. Since these two species have generally been regarded as varieties of one, we may say that *Spongomorpha* Kuetzing has for its type *S. lanosa* (Roth) Kuetzing. In the Phycologia Germanica (1845, p. 237), Kuetzing credited *Spongomorpha* with nine species, six of which are now referred under *Spongomorpha lanosa* or its variety *uncialis,* while three are members of the Ectocarpaceae. He also placed the genus near *Ectocarpus* and well removed from his *Cladophora* and *Aegagropila.* In the Species Algarum (1849, p. 387 *et seq.*), Kuetzing placed both *Spongomorpha* and *Aegagropila* as subdivisions (subgenera?) under *Cladophora.* In this arrangement he extended *Spongomorpha* to embrace twenty-one species, but did not include under this subdivision his section *"Comosae"* (1849, p. 389) of six species now generally referred to *Spongomorpha arcta.* Kuetzing's extension, however, is sufficient to indicate a conception decidedly advanced beyond his first proposal. His first conception (1843) of the character of *Spongomorpha* was the dense habit, but later (1849) he called attention to the slender, descending rhizoidal branches. In 1854, Kuetzing reviewed *Spongomorpha* as a genus, illustrating the various species belonging to it (1854, p. 16 *et seq.*, and 1855, p. 29). Kuetzing's idea, even then, was founded chiefly on habit.

J. G. Agardh (1846, p. 12) founded a genus *Acrosiphonia* (*"Acroliphonia"* as printed) which was based largely upon *Conferva lanosa* Roth and *C. arcta* Dillw. The chief character given is that of having the upper segments long, while the lower are very short.

We find, therefore, two genera, or generic conceptions, of practically the same content, but which have usually been included under *Cladophora.* Kjellman was the first to recall attention to *Spongomorpha* (1883, p. 304 *et seq.*), but in 1893, he reviewed J. G. Agardh's genus *Acrosiphonia* and substituted it for *Spongomorpha.* A later proposal is that of Wille (1899, p. 281, 1900, p. 238) who would restrict *Spongomorpha* to the original type species, *S. lanosa,* on account of its uninucleate segments, and apply *Acrosiphonia* to the other species because of their multinucleate segments. To this distinction he adheres in a later article (1909, pp. 117, 118). It seems to us desirable to defer judgment on the value of the nuclear characters until more extensive investigations have been undertaken. None of our species has been subjected to an examination of this sort, although they all seem to be made up of multinucleate segments. We shall consequently adopt *Spongomorpha* for our species instead of following the revision of Wille and adopting *Acrosiphonia.*

Kjellman (1893) has laid great stress on the variation in the position of the fertile segments. We have not been able, as yet, to study this character in many of our species and can not, therefore, judge properly of distinctions founded on such characters. It is very desirable that a study of the forms, arrangement and dates of formation of the fertile segments be made for our various species, and in widely separated localities.

We have followed Collins (1909) very largely in the descriptions and limitations of our species, but have examined all our west coast plants carefully. As a result, we have found it necessary to depart from his account in some minor details and to differ from him in referring some of our plants. Of the seven species recognized by us, five are supposedly restricted to the North Pacific Ocean and originally named by Ruprecht (as species of *Conferva*). Two, however, are assumed to be identical with North Atlantic species. Possibly some of our species may be identical rather with some of the species distributed by Kjellman under *Acrosiphonia*, but we have not been able to make sufficiently certain of this to place any under his names.

The species of our coast are most numerous along the northern range, i.e., from Puget Sound northward. *Spongomorpha coalita*, however, ranges from Sitka, Alaska to just north of Point Conception, California. A form referred to *Spongomorpha arcta* has been found in one locality in this same range, but we are inclined to associate it rather with *S. Mertensii.*

KEY TO THE SPECIES

1. Branches all blunt and rounded at the apex 2
1. Branches sometimes rounded, sometimes acute at the apex.................................. 5
    2. Hooked branchlets present...5. **S. Mertensii** (p. 227)
    2. Branchlets never hooked.. 3
3. Slender, not over 100μ in diam. at tip.........................1. **S. arcta** (p. 223)
3. Stouter, 200μ, or over, in diam. at tip.. 4
    4. Walls of segments thin, not striate.............................2. **S. Hystrix** (p. 224)
    4. Walls of segments thick, striate............................3. **S. duriuscula** (p. 225)
5. Hooked branchlets absent..4. **S. saxatilis** (p. 226)
5. Hooked branchlets present... 6
    6. Hooked branchlets always simple..........................6. **S. spinescens** (p. 229)
    6. Hooked branchlets usually compound or branched......7. **S. coalita** (p. 230)

1. **Spongomorpha arcta** (Dillw.) Kuetz.

Plants rich green, in dense fastigiate tufts, up to 15 cm. high; filaments erect, stiff, 60–100μ diam. above, below smaller, much branched; branches erect or appressed, obtuse or clavate at tips; segments above 4–6 diam. long, below 1.5–3 diam. long; rhizoidal descending branches 40–60μ diam. with segments 2–6 diam. long, firmly matting together the lower part of the tuft.

On rocks and on *Fucus* in the middle and lower littoral belts. Alaska (Bering Sea) to Washington (Puget Sound).

Kuetzing, Sp. Alg., 1849, p. 417; Collins, Green Alg. N. A., 1909, p. 359 (excluding varieties). *Conferva arcta* Dillwyn, Brit. Conf., 1809, p. 67, pl. E. *Cladophora arcta* Kuetzing, Phyc. Gen., 1843, p. 263; Harvey, Phyc. Brit., vol. 2, 1849, pl. 135; Setchell and Gardner, Alg. N.W. Amer., 1903, p. 224 (in part and excluding varieties except in part). *Cladophora lanosa* var. *uncialis*, Tilden, Amer. Alg. (Exsicc.), no. 372 (not of Thuret).

We have referred a number of specimens to *Spongomorpha arcta*, but without any knowledge of the exact nature of the type. The type specimens came from Bantry Bay, on the southwest coast of Ireland, and seem to be plants all of whose filaments are slender, obtuse or clavate at the tips, decidedly matted together below by descending rhizoidal branches but free and spreading in the upper portions. Neither hook nor spine branchlets are present. The plants referred here from our territory correspond in these general details and are, at least, closely related to this species. We are left with the impression, however, that this may be the conception of a group of species rather than of a single specific entity.

We have been unable to divide the *Spongomorpha arcta* of our coast into the varieties given by Collins in The Green Algae of North America (1909, pp. 359, 360). His f. *conglutinata*, based chiefly on the separation of groups of terminal filaments into "Symploca-like tufts" and said to have "acute branches occasionally found at the base of the older plants," proves to have, on careful examination of the plants on which it is based, and which show the characteristic habit, not only spiny branchlets but simple hooked branchlets as well. We feel compelled, therefore, to refer these plants rather to *Spongomorpha spinescens*, at least as we understand that species in this account. We find a similar combination of spiny branchlets and hooked branchlets in no. 918 of Collins, Holden and Setchell's Phycotheca Boreali-Americana referred to f. *pulvinata*. This plant also is referred by us to *S. spinescens*. While specimens of what we retain under *S. arcta* have also been referred to these varieties, they differ in habit from that described for either variety and we consequently refer them to *S. arcta* without attempting any varietal distinction.

It has been customary to refer under *Spongomorpha arcta*, at least in the broader sense in which we are, at present, compelled to use the name, *Conferva cohaerens* of Ruprecht (1851, p. 402). Ruprecht, himself, says that it is probably only a subspecies of *C. arcta*. The type specimens of *C. cohaerens* came from the North Pacific Ocean, probably from "Awatschabai" in the Sea of Ochotsk. Ruprecht states that his *Conferva cohaerens* is to be only slightly distinguished from the British *C. arcta* (of Mrs. Griffiths) by having the thicker variegated lateral branchlets which are abruptly attenuated to a point of $\frac{1}{200}$ line (about 11.3μ) in diameter (main filaments 75–80μ). *C. cohaerens* also differs particularly from *C. saxatilis* Rupr. in its intertwined filaments and its lower, more slender and shorter branchlets. The type specimen of *C. cohaerens* in the Herbarium of the Imperial Academy of St. Petersburg has the appearance and habit of *Spongomorpha arcta*, but it is very desirable that it be carefully examined microscopically before being finally referred.

2. Spongomorpha Hystrix Stroemf.

Plants rich green, in rather dense tufts, filaments straight, very erect, except those at the base of the tuft, which are somewhat more open; about 100–300μ diam. at the base, 200–500μ diam. at the tip ultimate segments blunt or clavate, not attenuate; segments up to 4 diam. long at the tip, 0.5–1.5 diam. long below; rhizoidal branches fairly common in the older parts, 40–70μ diam., segments 3–10 diam. long.

Forming a dense mass on rocks in the littoral belt. From Agattu Island to Sitka, Alaska.

Stroemfelt, Om Algvegt. vid Island Kuster, 1887, p. 54; Collins, Green Alg. N. A., 1909, p. 358; not Setchell and Gardner, Alg. N.W. Amer., 1903, p. 226.

Spongomorpha Hystrix seems to be a high northern species of the *arcta* group, differing primarily from *Spongomorpha arcta* in the greater diameter of the filaments. The specimens referred by us in our final disposition of the species, have much larger filaments than

those of *S. arcta* (up to 200μ or over at the tips), but we have seen
none approximating the maximum diameter (500μ at the tips) given
for the Atlantic-Arctic plants. The Tilden specimens (American
Algae, no. 374), issued as *Cladophora arcta* form *b*, and assigned to
C. Hystrix by Setchell and Gardner (1903, p. 226), prove on examination to have compound hooked branchlets and have been assigned,
consequently, to *Spongomorpha coalita* in this account. Specimens
from Karluk and from Uyak Bay, Alaska, assigned by Setchell and
Gardner to *Cladophora arcta* and its form *conglutinata* are now
referred here, after careful examination. The status of this species
on our coast, however, is subject to careful revision as soon as more
abundant material for study and comparison is available.

Spongomorpha Hystrix might be compared with *S. duriuscula*,
so far as description goes, but the species are very different in aspect,
the former being closely matted together below, while the latter shows
so loose an intermingling of the filaments as to suggest the possibility
of its belonging rather to *Cladophora* than to *Spongomorpha*. The
segments also in *S. duriuscula* are usually shorter and those of the
main filaments are horizontally striate.

3. **Spongomorpha duriuscula** (Rupr.) Collins

Tufts 15–25 cm. high, erect, loose, main filaments firm, straight,
with thick, horizontally striate walls, 200–250μ diam. below, 300μ at
the tip; segments 0.5–1.5 diam. long, 2–3 diam. at the blunt tips;
branches similar, erect, scattered or in secund series of two or more;
near the base of the tuft more slender, 150–200μ diam., with thinner
walls, not striate, with numerous short, patent ramuli, scattered or
secund.

Growing on rocks in the upper sublittoral and lower littoral belts.
Alaska (Pribilof Islands to Karluk).

Collins, Green Alg. N. A., 1909, p. 357. *Cladophora alaskana*
Collins, *in* Collins, Holden and Setchell, Phyc. Bor.-Amer. (Exsicc.),
no. 917 (nomen nudum), *in* Setchell and Gardner, Alg. N.W. Amer.,
1903, p. 228 (description). *Conferva duriuscula* Ruprecht, Tange,
1851, pp. 401–404. *Conferva cartilaginea* Ruprecht, Tange, 1851,
p. 404 (fide Yendo). *Acrosiphonia duriuscula* Yendo, Notes on algae
new to Japan, V, 1916, p. 246.

The type locality of *Conferva duriuscula* Ruprecht is Unalaska,
where it was collected by Wosnessenski (no. 108) as the type specimen

shows. The type specimen is of exactly the same habit and appearance as the specimens distributed under no. 917 of Collins, Holden and Setchell's Phycotheca Boreali-Americana, which are topotypes, probably collected at the very same spot whence Wosnessenski obtained the type.

Spongomorpha duriuscula is a coarse, lax species whose filaments are so slightly bound together that the plant seems much less like a *Spongomorpha* than either of the preceding species. There are some slender rhizoidal branches, however, and the main filaments increase in diameter towards their summits. The walls of the segments of the main filaments and longer branches are thick and horizontally striate as Ruprecht has stated. It seems to be a very distinct species of the Upper Boreal Zone.

Yendo (1916, p. 246) has placed this species under *Acrosiphonia* and, after examining the types, has united with it *Conferva cartilaginea* Ruprecht (1851, p. 404) whose type locality is also Unalaska. He also refers here Tilden's no. 373 (Amer. Alg.) under *Cladophora arcta*. The two (''a'' and ''b'') specimens in our copy are, however, clearly *Spongomorpha coalita*, both as to habit and as to the possession of compound hooked branchlets. Yendo states that Kjellman (1889, p. 55) included plants of this species under his *Cladophora diffusa* from Bering Island, Siberia.

4. **Spongomorpha saxatilis** (Rupr.) Collins

Plants dense but not much matted together; filaments 80–120μ diam., about the same diameter throughout, segments below 1–3 diam. long, above 3–6 diam., terminal segments sometimes 10–12 diam.; branching di- trichotomous, with occasional lateral branches, divisions erect, somewhat acute or tapering, but usually with rounded tip; older parts with descending rhizoidal filaments, about half the diameter of the filaments from which they spring, and with longer segments sometimes 10–12 diam. long.

On rocks in the lower littoral belt. Alaska to San Francisco, California.

Collins, Green Alg. N. A., 1909, p. 360, Mar. Alg. Vancouver Island, 1913, p. 104. *Conferva saxatilis* Ruprecht, Tange, 1851, p. 403. *Cladophora saxatilis* (Rupr.) De-Toni, Syll. Alg., vol. 1, sect. I, 1889, p. 311; Setchell and Gardner, Alg. N.W. Amer., 1903, p. 223; Collins, Holden and Setchell, Phyc. Bor.-Amer. (Exsicc.), no. 921. *Cladophora arcta* Tilden, Amer. Alg. (Exsicc.), no. 279 (not of Kuetzing).

Spongomorpha saxatilis var. **Chamissonis** (Rupr.) Collins

Filaments 40–60μ diam., cells 3–4 diam. long, nodes constricted; cells slightly shorter towards the base.

Alaska to Washington.

Collins, Green Alg. N. A., 1909, p. 360. *Cladophora Chamissonis* (Rupr.) De-Toni, Syll. Alg., vol. 1, sect. I, 1889, p. 333; Collins, Holden and Setchell, Phyc. Bor.-Amer. (Exsicc.), no. 920. *Conferva Chamissonis* Ruprecht, Tange, 1851, p. 403.

Spongomorpha saxatilis resembles looser conditions of *S. arcta*, but is even more lax, as a rule, than any condition of that species. It resembles *S. spinescens* in having very much attenuated branches and even moderately sharp branchlets, but it never has hooked, acute, tipped branchlets. It is also more lax and less split up into symplocoid tufts than is usual with the *S. spinescens* of our coast.

The type of *Conferva saxatilis* Ruprecht was found at "Cap Nichta" in the Sea of Ochotsk, and the other specimens referred to it seem to differ from it and among themselves in coarseness and proportions of segments. Yendo, who has examined the type specimens of Ruprecht's species of *Conferva,* refers (1916, p. 245) *Conferva Chamissonis* and *C. saxatilis* to the same species, as Collins had already done. Collins, however, was inclined to add *Conferva Mertensii* and *C. viminea* as well, but Yendo says that while the type specimens in St. Petersburg show that the two latter are identical they also show that they are quite distinct from the two former. We, also, have felt it necessary to keep *C. Mertensii* separate from *C. saxatilis.*

5. **Spongomorpha Mertensii** (Rupr.) S. and G.

Plants up to 11 cm. high, erect, moderately rigid, lax, bright green; branching alternate, branches erect, angles acute; main filaments and branches 110–160μ diam. above, 80–110μ diam. below; segments 0.5–2.5 times as long as the diameter, terminal rounded or gradually tapering, even prolonged into long spinous rhizoids, up to 5 times as long as the diameter; spiny branchlets absent; short, blunt, branched, hooked branchlets generally present.

Alaska (Unalaska and Sitka) to California (San Francisco).

Setchell and Gardner, Phyc. Cont. I, 1920, p. 280. *Conferva Mertensii* Ruprecht, Tange, 1851, p. 403. *Conferva viminea* Ruprecht, *loc. cit.* (fide Yendo). *Cladophora Mertensii* De-Toni, Syll. Alg., vol. 1, 1889, p. 317. *Cladophora viminea* De-Toni, *loc. cit.*, p. 318. *Spongomorpha arcta* var. *limitanea* Collins, Green Alg. N. A., Suppl. I, 1912, p. 97, *in* Collins, Holden and Setchell, Phyc. Bor.-Amer. (Exsicc.), no. 1736. *Acrosiphonia Mertensii* (Rupr.) Yendo, Notes on algae new tó Japan, V, 1916, p. 246.

Collins (1909, p. 360), on evidence furnished by authentic specimens in Herbarium Farlow, unites not only *Conferva Mertensii* Rupr. with *C. viminea* Rupr., but also unites with them *C. Chamissonis* Rupr. and *C. saxatilis* Rupr., placing all these names as synonyms under *Spongomorpha saxatilis.* Yendo (1916, p. 245), however, as a result of study of the type specimens preserved in the Herbarium of the Imperial Academy of Sciences of St. Petersburg, is of the opinion that *Conferva saxatilis* and *C. Chamissonis* are simply forms of one species and that *C. Mertcnsii* and *C. viminea* are also forms of one species, but that the two latter are quite distinct from the two former. Yendo does not, however, point out in what this difference consists.

We are inclined to refer here provisionally and with considerable doubt, a specimen (no. 3288 of Setchell and Lawson) collected at Amaknak Island in the Bay of Unalaska. It is somewhat coarser than specimens of *Spongomorpha saxatilis* and of even more lax habit. It has no spines, but has occasional inrolled, circinate, yet blunt, branchlets, very different, however, from the hooked branches of either of the two succeeding species. The general aspect, the dimensions, and these inrolled branchlets are all indicated in Ruprecht's description of *Conferva Mertensii* and assist in distinguishing this species from *C. saxatilis.*

Very similar to the Unalaska plant is that distributed under no. 1736 · of the Phycotheca Boreali-Americana. It has no spiny branches or branchlets but it usually shows circinate, often branched, branchlets. The terminal segments vary from very blunt and dilated to extremely attenuated and in some plants are prolonged into curious long and sinuate rhizoids. The last character has been noticed only in summer plants. We may add that the description of this species has been drawn up largely from our Californian specimens.

The type locality for *Conferva Mertensii* Rupr. is given as Sitka, while *C. viminea* Rupr. is given as occurring at both Sitka and Unalaska.

6. **Spongomorpha spinescens** Kuetz.

Plants orbicular in outline, with short symplocoid divisions; filaments about 80μ diam. below, 100μ diam. at tip; segments 0.5–1 diam. long below, 2 diam. long at tip; normal, erect, somewhat obtuse branches abundant; also patent and acute branches, either short and spine-like, or long, hooked, revolute and circinate, uniting the filaments into branching symplocoid tufts; descending rhizoidal branches slender and abundant.

Growing on the tips of algae and sponges along the upper tide limit in exposed places. Bay of Unalaska, Alaska, to Coos Bay, Oregon.

Kuetzing, Tab. Phyc., vol. 4, 1854, p. 16, pl. 75, II; Collins, Green Alg. N. A., 1909, p. 360. *Cladophora spinescens* Kuetzing, Sp. Alg., 1849, p. 418; Setchell and Gardner, Alg. N.W. Amer., 1903, p. 227. *Cladophora saxatilis* Setchell and Gardner, *loc. cit.*, p. 223 (in part). *Cladophora arcta* f. *pulvinata* Collins, *in* Setchell and Gardner, *loc. cit.*, p. 225. *Spongomorpha arcta* f. *pulvinata* Collins, Green Alg. N. A., 1909, p. 360 (not of Foslie). *Cladophora arcta* f. *conglutinata* Collins, *in* Setchell and Gardner, *loc. cit.*, p. 225 (in large part). *Spongomorpha arcta* f. *conglutinata* Collins, Green Alg. N. A., 1909, p. 359. *Cladophora arcta* Harvey, List N.W. Alg., 1862a, p. 176 (at least in part). *Cladophora scopaeformis* Setchell and Gardner *loc. cit.*, p. 227 (in part).

Spongomorpha spinescens was founded on a specimen sent by Lenormand to Kuetzing from the coast of Morbihan, a department of France on the northern short of the Bay of Biscay. A topotype of this species, if not possibly a cotype, exists in the Herbarium of the University of California (no. 145264), i.e., it is from the same locality and sent out by Lenormand. It also answers in full to Kuetzing's description.

Spongomorpha spinescens is characterized by abundant acute branches and branchlets with some of the short, simple, acute branchlets curved into definite hooks. In looking over our Alaskan and Puget Sound materials, we find a number of specimens which correspond well in habit and other characters with the descriptions and figures of *S. spinescens* as well as with the specimen mentioned above. We feel fairly certain in referring them to Kuetzing's species. We have also compared our specimen with the descriptions of *Acrosiphonia albescens* Kjellman and *A. hamulosa* Kjellman as well as with specimens of the former from both Iceland (Jónsson) and the Faeröes (Börgesen). Our plants do not agree completely with the description of either of Kjellman's species nor do they correspond to either of the specimens of *A. albescens* which specimens, in their turn, seem to us to belong to different species.

We find that our specimens, seeming to belong to this species, have been variously referred in our previous paper (1903) as will be seen from the synonymy given above. It is to be easily distinguished from the next, both in habit and in the unbranched, simple, hooked branch-lets. The fertile segments are all intercalary, usually one or two together, and rarely three to four in a series, and generally distributed over the plant.

7. Spongomorpha coalita (Rupr.) Collins

Plate 16, fig. 4, and plate 32

Plants elongated, at first loosely tufted, but soon forming long, dense, ropelike, branching tufts, up to 30 cm. long: at first bright, later dull or yellowish green; filaments 100–250μ diam. in the terminal segment; branching dichotomous below, irregularly alternate above, all branches of this class erect, with blunt, truncate, or, at times, acute ends; also present, except in very young plants, abundant, patent, tapering, very acute, compound, or branched, sharply and abruptly hooked branches by which all the older parts are densely matted together; segments 0.3–1 diam. long in the lower part of the older plants, 2–3 diam. in the younger plants, and even 6–25 diam. in the active terminal segment.

Growing on rocks and on other algae in the middle and lower littoral belts. Very abundant along the entire Pacific Coast from southeastern Alaska (Sitka) to central California (San Luis Obispo County).

Collins, Green Alg. N. A., 1909, p. 361. *Cladophora scopaeformis* Setchell and Gardner, Alg. N.W. Amer., 1903, p. 227; Collins, Holden and Setchell, Phyc. Bor.-Amer. (Exsicc.), nos. 819 and 922; Farlow, Anderson and Eaton, Alg. Exsicc. Amer.-Bor., no. 203. *Cladophora coalita* Setchell and Gardner, *loc. cit.*, p. 227. *Cladophora Hystrix* Setchell and Gardner, *loc. cit.*, p. 226 (not of Stroemfelt). *Conferva coalita* Ruprecht, Tange, 1851, p. 404. *Cladophora arcta* form *a*, Tilden, Amer. Alg., no. 373, and form *b*, no. 374; Collins, Mar. Alg. Vancouver Island, 1913, p. 104, as to no. 374.

Spongomorpha coalita, as it usually occurs, is readily recognizable, both by its habit and by the possession of compound, strongly hooked branchlets. It seems reasonably certain that the *Conferva scopae- formis* Ruprecht represents only a younger stage of *C. coalita* Ruprecht, with somewhat more slender filaments and longer terminal segments (cf. Collins, 1909, p. 361). The plants are distinctly elongated and their filaments are combined into ropelike masses. The recurved branchlets usually show either three or four sharply recurved and pointed tips or a less number, sometimes only one, with one or more longer straight branchlets arising from them. We have seen nothing like this in any other of our species of *Spongomorpha,* nor is any branching of this sort described for any of the species of this genus, or of *Acrosiphonia,* except that Kjellman figures compound recurved branchlets in his *Acrosiphonia hamulosa* (1893, pl. 1, f. 5). This species, however, is of very different habit, and has more slender filaments as well as more slender recurved branchlets than *Spongo- morpha coalita,* resembling more closely *S. spinescens,* but seemingly distinct from it.

There is considerable variation in the dimensions of the filaments and in the proportions of the segments in different individuals. The walls of the segments also vary much in thickness, up to as great as 40μ, e.g., in a specimen collected on the west coast of Whidbey Island, Washington (Gardner, no. 467).

A seemingly young, but unusual plant, collected on June 3 at Carmel Bay, California (no. 5418 Setchell), is about 200μ in diameter at the tips, but the terminal segment is often as much as 3 mm. long. Below, the hooked branchlets are often simple, but some are com- pound. If separated, these specimens might be referred under *Con- ferva scopaeformis* Rupr.

15. **Microdictyon** Dec'ne

Plant a sessile, membranaceous network, formed of monosiphonous filaments, densely branching in one plane in a radiate manner, the tips of the branches attaching themselves to the sides of other branches by a terminal thickening, producing irregular, angular, open spaces between the segments; reproduction by zoospores formed in any segment.

Decaisne, Pl. de l'Arab., 1841, p. 115.

The genus *Microdictyon* comprises species which are totally different in habit from any of the other Cladophoraceae, and possesses a
thallus of a leaflike form and general appearance. The species may,
however, be considered to represent opposite branched Cladophorae
whose branchlets anastomose to produce a plane, reticulate thallus,
the main filaments and branches giving the appearance of veins, while
the branchlets form the general groundwork. The species are all
tropical or subtropical, one species only being credited to our coast.

Microdictyon Agardhianum Dec'ne

Plants delicately membranaceous, filaments 50–200μ diam., main
veins rather distinct, radiate, branches patent; segments usually 2–4
diam. long.

Guadalupe Island, Lower California.

Decaisne, Pl. de l'Arab., 1841, p. 115.

Collins (1909, p. 366) refers under the name *Microdictyon
Agardhianum* a plant from Guadalupe Island off the coast of Mexico.
Since we have not seen specimens of this, we have not attempted to
consider whether the Guadeloupe plant is the same as *M. Agardhianum*
Decaisne of the Red Sea or *M. umbilicatum* (Velley) Zanard. from
southwest Australia, but have left our reference under the name used
by Collins.

16. **Boodlea** Murray and De-Toni

Boodlea composita (Harv.) Brand (1904, p. 187, pl. 6, f. 28–35)
is a tropical species, the type locality of which is the island of Mau-
ritius. This species has been credited to our coast (cf. Setchell and
Gardner, 1903, p. 226; Collins, 1909, p. 367; Yendo, 1916, p. 247;
Collins, Holden and Setchell, Phyc. Bor.-Amer., no. 722). An examination of the specimens upon which this representation is based shows,
however, that none of them is really the *Conferva composita* Harv.
(1834, p. 157). Collins (1918, p. 85) states that what has been said
concerning *Boodlea composita* in his Green Algae of North America
(1909, p. 367) is to be cancelled. *Boodlea,* consequently, can not be
said to have been found on our coast.

·ORDER 4. ULVALES BLACKMAN AND TANSLEY

Fronds membranaceous of one or two layers of cells, or tubular with wall of a single layer of cells, or filamentous of two or more vertical rows of cells, simple or branched, attached by rhizoids either free or united into a disk; cells with a single nucleus and a single parietal chromatophore containing usually one, but, occasionally, two or three pyrenoids; multiplication vegetative, non-sexual and sexual; vegetative, by abscission of proliferous shoots, by accidental rupture, by gemmae or by akinetes; non-sexual, by 2- or 4-ciliated zoospores; sexual by 2-ciliated isoplanogametes forming a zygote capable of germinating at once.

Blackman and Tansley, Rev. Class. Green Algae, 1902, pp. 20 and 136; West, Algae, 1916, vol. 1, p. 275.

FAMILY 9. ULVACEAE GREVILLE

Characters of the order.

Greville, Alg. Brit., 1830, p. 168 (lim. mut.). *Ulvacées* Lamouroux, Essai, 1813, p. 59 (in part); Thuret, Note sur la Syn., 1854, p. 27.

KEY TO THE GENERA

17. **Capsosiphon** Gobi

Plants filamentous, hollow, gelatinous, the cells mostly in twos and fours, enclosed within the walls of the mother cell, and arranged in distinct longitudinal series, the series loosely connected laterally.

Gobi, Ber. Alg. Forsch. im Finn. Meer., 1877, etc., 1879, p. 88. *Ilea* Fries, Syst. orb. veg., pt. 1, pl. homon, 1825, p. 336 (in part); J. G. Agardh, Till Alg. Sys., pt. 3, 1883, p. 114.

We have preferred to use the generic name *Capsosiphon* of Gobi (1879) rather than *Ilea* J. Ag. for several reasons. Fries founded the genus *Ilea* in 1825 (p. 336) to contain various species which were later found to belong to *Enteromorpha* Link (1820). The genus *Ilea* was refounded by Fries in 1835 (p. 321) to contain two species, viz., *I. Fascia* (Muell.) Fries and *I. foeniculaceus* (Huds.) Fries. If the genus *Ilea* is to be retained at all, it ought to be retained for *Ilea Fascia* (*Phyllitis Fascia* Kuetz.), since Fries states (1835) that the character of *I. Fascia* was the one upon which the genus was really founded. Finally the genus *Capsosiphon* Gobi (1879) was founded two to three years before J. G. Agardh (1883, p. 114) resurrected the name *Ilea* to confer it upon the *Ulva aureola* Ag. (1835, no. 29, pl. 29).

Capsosiphon fulvescens (Ag.) S. and G.

Fronds 1–5 cm. up to 8 cm. high, thread-like, later becoming tubular, up to 2 cm. diam., cylindrical or somewhat compressed, with an occasional swelling, unbranched, or in age slightly proliferating; cells roundish or oval with a thick membrane resembling the cells of *Gloeocapsa*, 4–5μ diam., arranged in long rows, 2–4 rows grouped together.

Growing on muddy rocks. St. Michael, Alaska.

Setchell and Gardner, Phyc. Cont. I, 1920, p. 280. *Ulva fulvescens* Agardh, Sp. Alg., 1821, p. 420. *Ilea fulvescens*, J. Agardh, Till Alg. Syst., part 3, 1883, p. 115, pl. 4, f. 95–99; Collins, Green Alg. N. A., 1909, p. 206, f. 71. *Enteromorpha aureola* Kuetzing, Tab. Phyc., vol. 6, 1856, p. 14, pl. 40; Setchell and Gardner, Alg. N.W. Amer., 1903, p. 214. *Capsosiphon aureolum* (*sic!*) Gobi, *loc. cit.* *Ulva aureola* Agardh, Icon. Alg. Eur., 1835, no. 29, pl. 29.

Capsosiphon fulvescens, better known perhaps as *Enteromorpha aureola*, was included in our previous account (1903, p. 214) as having been found at St. Michael, Alaska by one of us (Setchell) intermixed with *Rhizoclonium riparium*. The specimen, on reëxamination, fails to show a single filament of the *Capsosiphon*, hence its proper inclusion in our flora must remain, for the present, a matter of doubt.

We have chosen the specific name *fulvescens* rather than that of *aureolus*, because J. G. Agardh (1883, p. 115) has stated that it belongs to this species, *Ulva fulvescens* Ag. being a younger condition, while *U. aureola* Ag. is more developed.

18. **Monostroma** Thur.

Frond at the beginning a closed sack or tube, at times splitting very early or again retaining the saccate or tubular shape until late, in almost all cases, however, finally becoming a flattened or crisped membrane of a single layer usually parenchymatous but occasionally of gloeocapsoid cells, except at the base where thickening occurs by the descent of elongated rhizoidal cells forming several layers; vegetative multiplication by gemmation or proliferation, non-sexual reproduction by 2- or 4-ciliated zoospores and sexual reproduction by 2-ciliated isoplanogametes all originating in unchanged cells; zygote usually germinating immediately.

Thuret, Note sur la Syn. des Ulv., 1854, p. 13.

The genus Monostroma comprises those members of the Ulvaceae which, at maturity, form an expanded membrane of a single layer of cells. Certain species of *Monostroma* develop nearly to full size as sacks which then split open by one or more slits and become expanded membranes. There are still other species, however, which seem to consist of expanded membranes of a single layer of cells almost, if not quite, from the beginning. We have some reason to suppose that these species are tubular or saccate only in their very youngest stages, splitting early and becoming, therefore, one-layered almost from the beginning.

In some of the species of *Monostroma* the cells are closely placed, with thin, or even thicker, firm walls, giving a parenchymatous appearance, while in others the intercellular substance is ample and more or less gelatinous causing the cells to stand off from one another, usually in small groups (2–4 or more) after the fashion of *Gloeocapsa* or *Chroococcus*. The relative sizes of the cells in different dimensions, as well as the abundance or scarcity of intercellular substance, added to the size, development, and shape of the frond, furnish characters which may be used for the separation of the species.

The species of *Monostroma* are marine, and also found in brackish and fresh water. At times what appears to be the same species may be found in both salt and fresh water. The thirty-five species credited to the genus are known only with certainty from the Northern Hemisphere, where they occupy, for the most part, the colder waters, intruding into warmer zones only in winter and spring when the temperature of the water is lowered.

1. **Monostroma Grevillei** (Thur.) Wittr.

Frond attached, at first saccate, then opening at the top, and ultimately splitting to the base; soft and delicate, pale green; membrane 15–20μ thick, cells quadrate with rounded angles, closely set, horizontally oval in cross section, 12–14μ high; sporiferous cells enlarged, vertically elongate in cross section; cell wall dissolving after emission of spores.

On stones in the upper sublittoral and lower littoral belts, Alaska.

Wittrock, Monostr., 1866, p. 57, pl. 4, f. 14; Collins, Green Alg. N. A., 1909, p. 209; Setchell and Gardner, Alg. N.W. Amer., 1903, p. 208. *Enteromorpha Grevillei* Thuret, Note sur la Syn. Ulv., 1854, p. 25.

Certain of the species of *Monostroma* retain the saccate habit until late, or rather until the sack has reached considerable size, before splitting. Two species, at least, on our coast do this; one is *Monostroma Grevillei* and the other is *M. arcticum*. In *M. Grevillei*, especially if obtained in position, the saccate habit is usually easily observed or inferred, even when the plant is split to the base into segments. When detached fragments, especially of some size, are collected, the saccate habit may not be in evidence. *Monostroma Grevillei* is of delicate consistency and lubricous, differing in both these characters from any of the forms of *M. arcticum*. It is also decidedly thinner than *M. arcticum*. Otherwise the two species are much alike. Rosenvinge (1893, p. 949 and 1894, p. 152) has united them as varieties of one species. We are inclined, however, to follow

Collins (1909, pp. 209, 210) and keep them separate, at least for the present, for the reasons given above.

On the Atlantic coast of North America, *Monostroma Grevillei* is a summer plant in the Greenland waters (Upper Boreal Zone), but invades the North Temperate Zone in spring-time. On the Pacific Coast, our only specimens are from Bering Sea but Collins (1903, p. 13) states that it descends to Monterey, California, which is much above its accustomed temperature.

The reference of Collins to Monterey is based on two small specimens found attached to *Gloiosiphonia verticillaris* Farl., collected by Mrs. J. M. Weeks. Reëxamination of these specimens, although not convincing, leads as to the opinion that they may be nearer *Monostroma zostericola* than to any forms of *M. Grevillei*.

Collins gives, in addition to the typical form, the two following varieties as found on our coast. Since we have access to very scanty material we follow him and other authorities as to their disposition.

Monostroma Grevillei var. lubricum (Kjellm.) Collins

Frond up to 15 cm. long, pale or whitish green, delicate, very lubricous and flaccid, of irregular outline, laciniate, plicate, margin often crisped and lacerate; frond 18–22μ thick; cells seen superficially, circular or rounded angular, often in twos or fours, cell wall thick; in cross section horizontally ovate or oblong, 4.5–8μ high.

Floating in shapeless masses in quiet waters. Alaska.

Collins, Green Alg. N. A., 1909, p. 209. *Monostroma lubricum* Kjellman, Spetsb. Thall., 1877, p. 48, pl. 4, f. 8, 9; Setchell and Gardner, Alg. N.W. Amer., 1903, p. 207.

Monostroma Grevillei var. Vahlii (J. Ag.) Rosenv.

More slender in form, often cylindrical, retaining its saccate shape longer, and with cells arranged in more or less distinct longitudinal series. An early spring plant.

Alaska (Kukak Bay, *Saunders*, Sitka, *Gardner*).

Rosenvinge, Groenl. Havalg., 1893, p. 949; Collins, Green Alg. N. A., 1909, p. 209. *Monostroma Vahlii* J. G. Agardh, Till Alg. Syst., VI, 1883, p. 109, pl. 3, f. 84–89; Saunders, Alg. Harriman Exp., 1901, p. 410; Setchell and Gardner, Alg. N.W. Amer., 1903, p. 208.

We have seen only a fragment collected by one of us (Gardner) at Sitka. This seems, however, to be clearly the *Monostroma Vahlii* of J. G. Agardh.

2. **Monostroma arcticum** Wittr.

Frond attached, at first saccate, later splitting into a few broad laciniae; subradiately plicate, with crisped margin; pale green, becoming yellowish in drying; membrane 25–45μ thick; cells 4–6 angled, closely set, irregularly placed; in cross section either vertically or horizontally oval, 10–30μ high.

On stones, in shallow pools of the middle littoral belt. Known only from Alaska.

Wittrock, Monostr., 1866, p. 44, pl. 2, f. 8; Collins, Green Alg. N. A., 1909, p. 210; Setchell and Gardner, Alg. N.W. Amer., 1903, p. 208; Collins, Holden and Setchell, Phyc. Bor.-Amer., (Exsicc.), no. 910. *Monostroma latissimum* Setchell and Gardner, Alg. N.W. Amer., 1903, p. 207 (in part).

Rosenvinge (1893, p. 949, 1894, p. 152) unites *Monostroma arcticum* with *M. Grevillei*, keeping it as a variety. Much may be said in favor of such a disposition, especially in view of the treatment usually accorded *M. fuscum, M. splendens* and *M. Blyttii.* Collins (1909, p. 210), however, decides to keep them separate, although acknowledging the close relationship, on the ground that *M. arcticum* is a somewhat tougher and thicker plant than typical *M. Grevillei.* It seems best to us, also, to keep them separate. We follow Rosenvinge, however, in uniting with *Monostroma arcticum, M. angicava, M. cylindraceum* and *M. saccodeum* of Kjellman (1883, pp. 295–297). These last species seem to differ only in general habit and this is probably due to earlier or later splitting of the saccate frond.

We find in carefully examining some specimens from Bering Sea and northwestern Alaska, previously referred (cf. Setchell and Gardner, 1903, p. 207, as to nos. 4020 and 5077) to *Monostroma latissium,* that they agree better with *M. arcticum.*

3. **Monostroma zostericola** Tilden

Plate 14, figs. 12, 13

Frond more or less cucullate, cuneate-obovate or divided into segments of that form; cells angular, in more or less distinct series, longitudinal and transverse; margins plane, often ragged; membrane 7–10μ thick, cells quadrate to vertically oblong in cross section, 5–8μ high.

In the sublittoral belt, growing on *Zostera*. Known definitely only from the waters of the Puget Sound region.

"*Monostroma zostericolum*" Tilden, Amer. Alg. (Exsicc.), no. 388, 1900. *Monostroma leptodermum* Collins, Green Alg. N. A., 1909, p. 213; Setchell and Gardner, Alg. N.W. Amer., 1903, p. 209 (probably not of Kjellman). ·

Monostroma zostericola is known thus far from the original collection by Tilden from the waters between Brown and San Juan Islands, Washington; from Port Renfrew, Vancouver Island, British Columbia, collected by Butler and Polley (cf. Collins, 1913, p. 103, under *M. leptodermum*); and from near Victoria, British Columbia, collected by one of us (Gardner). There is some reason for referring here, although doubtfully, young specimens from Monterey previously placed under *M. Grevillei*. The *specimens* of all collections agree in showing sessile plants, attached and split in such a way as to suggest their earlier saccate form and growing on *Zostera*. The species is very delicate and its membrane very thin. The only species of this genus on our coast with anything like so thin a membrane is our *Monostroma areolatum* which differs decidedly in size, undulate and crisped lobes, areolate surface and cells in definitely delimited and separated groups. Collins, in various papers, has been inclined to refer the Tilden plant to *Monostroma leptodermum* of Kjellman (1877a, p. 52, f. 23, 24). Kjellman's plants, unfortunately, were not found attached, but were floating fragments, the largest of which was 10 cm. long and about 6 cm. wide. We have seen no entire plants among the considerable number examined from our coast which approach these dimensions. Kjellman also states that his larger fragment had the margins undulate and crisped which is not the case in any of the specimens from our coast. The cells in ours seem to agree fairly well with the description and figures of Kjellman except in being more elongated vertically than horizontally in cross section.

Rosenvinge (1893, p. 944, f. 49, 1894, p. 148, f. 49) and Jónsson (1904, p. 63) refer a very different plant from ours to the *Monostroma leptodermum* Kjellm. Their plant has a long, slender, tubular stipe and the membrane is undulate and ruffled. It seems best to us, therefore, to consider that their plant is more likely to represent Kjellman's species and to place ours under the name bestowed upon it by Tilden.

4. **Monostroma areolatum** S. and G.

Plate 25 and Plate 26, fig. 2

Frond very delicate, lubricous, 20–35 cm. high, sessile, saccate when young, soon splitting and forming numerous, long, broadly ovate or obovate, undulate, plicate and much crisped lobes, pale green; membrane distinctly and finely areolate, 9–12μ thick; cells with rounded angles, 6–7μ diam., subspherical in cross section, grouped within each areole.

Growing on *Zostera* in quiet waters. Sitka, Alaska.

Setchell and Gardner, Phyc. Cont. I, 1920, p. 281, pl. 30 and pl. 31, fig. 2.

This species of *Monostroma* is exceedingly beautiful and among the most delicate and flaccid of the genus. The frond remains saccate for a brief period only, attaining a height of but a millimeter or two. The sack then breaks and the membrane spreads out at once, early developing small lobes. Finally a few primary lobes are established and these develop numerous secondary lobes. The growth on the whole margin greatly exceeds that of the interior, which results in the production of a great number of folds, making the margin very much crisped. In the thickness of the frond and shape of the cells *M. areolatum* closely approximates *M. zostericola* Tilden. The cells of the latter are, however, more angular and more closely placed, and the frond is not divided into areolae. There is a marked difference in the size of these two species as well as in their method of development. *M. zostericola* is diminutive, remains saccate for some time, and then splits longitudinally, forming several lobes broadening outward. *M. areolatum* very closely resembles the genus *Prasiola* in the grouping of the cells as seen in surface view.

5. **Monostroma quaternarium** (Kuetz.) Desmaz.

Frond at first attached, soon becoming free, soft and delicate, irregularly lobed and folded, 20–23μ thick; cells rounded, when actively dividing set closely in threes and fours within the mother cell wall; in cross section semicircular or oval, 15–17μ high.

Floating in brackish and in fresh water. Washington to southern California.

Desmazières, Plantes Crypt. de France, 3 Sér., no. 603, 1859; Collins, Green Alg. N. A., 1909, p. 212; Setchell and Gardner, Alg. N.W. Amer., 1903, p. 207. *Ulva quaternaria* Kuetzing, Tab. Phyc., vol. 6, 1856, p. 6, pl. 13, f. 2.

Monostroma quaternarium resembles *M. latissimum* in habit and habitat. Both are usually found in either brackish or fresh water and are not strictly marine. Both are usually found floating and are more or less indefinitely expanded. In *M. quaternarium* the cells are rounded, segregated in small groups of three or four in surface view, while in *M. latissimum* the cells are angular and closely placed, although at times appearing somewhat grouped. It is by no means a certain matter to place some specimens definitely in one species or the other. Such a case is presented by no. 218 of Farlow, Anderson and Eaton's Algae Americae Borealis Exsiccatae, collected near Santa Cruz, California, probably in an estuary.

6. **Monostroma latissimum** (Kuetz.) Wittr.

Frond at first attached, afterwards floating; thin and soft, glossy, of irregular shape, more or less plicate near the even or undulate margin; membrane 20–25μ thick, cells 4–6 cornered or roundish, closely set, without order or more or less distinctly in twos, threes and fours; in cross section vertically oval or nearly circular, 14–18μ high.

Attached to various objects in the lower littoral belt when young, but soon becoming free and floating in quiet waters, salt marshes, ditches, etc. Washington to central California.

Wittrock, Monostr., 1866, p. 33, pl. 1, f. 4; Collins, Green Alg. N. A., 1909, p. 211; Setchell and Gardner, Alg. N.W. Amer., 1903, p. 207. *Ulva latissima* Kuetzing, Phyc. Gen., 1843, p. 296, pl. 20, f. 4.

As stated under the preceding species, *Monostroma latissimum* is usually to be found floating in shallow warmer waters, either brackish or fresh, seldom, if ever, in true marine localities. It is to be distinguished from the last (*M. quaternarium*) by the appearance of the cells both in surface view and in section as indicated in the descriptions.

The propriety of adopting the specific name *latissimum* for this species may be questioned. It seems to be the *Ulva latissima* of Kuetzing, but probably not the *Ulva latissima* of Linnaeus. We are not in a position to discuss this question, and simply follow later usage.

7. **Monostroma orbiculatum** Thur.

Frond membranaceous, attached by fibrils, or later free; soft and flaccid, sub-orbicular or irregular in outline, often radially plicate, with undulate margin, 30–40µ thick; cells angular, varying much in size and arrangement, often irregularly elongate, closely set, but with chromatophore not occupying the whole cell; in cross section vertically oval, 25–30µ high.

In brackish water attached to various objects in ditches of salt marshes. Central California.

Thuret, Note sur la Syn. Ulv., 1854, p. 388; Collins, Green Alg. N. A., 1909, p. 212; Wittrock, Monostr., 1866, p. 39, pl. 2, f. 6.

The present species resembles very closely the preceding and is, in fact, to be distinguished from it chiefly by its greater thickness. It may be a question as to whether it ought to be united with *Monostroma latissimum* or not. So far as our experience goes, however, the *M. latissimum* plants are definitely not over 25µ in thickness, and those of *M. orbiculatum* seldom less than 35µ. This seems to indicate sufficient difference for keeping them distinct.

8. **Monostroma fuscum** (Post. and Rupr.) Wittr.

Frond membranaceous, at first tubular, soon splitting, dull green, more or less lobed but not divided to the base; membrane 20–35µ thick; cells 4–6 angled, very closely set, in cross section quadrate, with only slightly rounded corners, occupying nearly the entire thickness of the frond.

On stones in the middle littoral belt, and floating in salt marshes. From Alaska to Puget Sound.

Wittrock, Monostr., 1866, p. 53, pl. 4, f. 13; Collins, Green Alg. N. A., 1909, p. 213; Saunders, Alg. Harriman Exp., 1901, p. 409; Setchell and Gardner, Alg. N.W. Amer., 1903, p. 208. *Ulva fusca* Post. and Rupr., Illust. Alg., 1840, p. 21. *Ulva Lactuca* var. *rigida* Setchell and Gardner, Alg. N.W. Amer., 1903, p. 209 (in part).

Monostroma fuscum var. **splendens** (Rupr.) Rosenv.

Frond deep green, glossy, 50–55µ thick, more deeply parted than in the other forms; cells similar to those of var. *Blyttii* or more rounded.

From Alaska to Vancouver Island.

Collins, Ulvaceae of N. A., 1903, p. 12, Green Alg. N. A., 1909, p. 213; Setchell and Gardner, Alg. N.W. Amer., 1903, p. 209; Collins, Holden and Setchell, Phyc. Bor.-Amer. (Exsicc.), no. 911. *Monostroma splendens* Wittrock, Monostr., 1866, p. 50, pl. 3, f. 12; Setchell, Alg. Prib., 1899, p. 591. *Ulva splendens* Ruprecht, Tange Och., 1851, p. 410.

Monostroma fuscum var. Blyttii (Aresch.) Collins

Frond deep green, blackish in drying, 60–70μ thick; cells "palisade-form" in cross section.

Growing in tide pools and on pebbles in the sublittoral belt. Vancouver Island and Washington.

Collins, Ulvaceae of N. A., 1903, p. 12, Green Alg. N. A., 1909, p. 213. *Monostroma Blyttii* Wittrock, Monostr., 1866, p. 49, pl. 3, f. 11. *Ulva Blyttii* Aresch., *in* Fries, Sum. Veg. Scand., 1846, p. 129.

We are puzzled, when it comes to the discussion of *Monostroma fuscum*, whether to treat of it as a species or as a group of species. The typical *Monostroma fuscum* is comparatively thin (20–25μ according to Wittrock, 1866, p. 53), turning only a light brown on drying instead of black, and with cells quadrate or only slightly vertically elongated in cross section. The var. *splendens* is a thicker plant (49–53μ according to Wittrock, 1866, p. 51), turning black, adhering even less well to paper than the typical form and with cells vertically much elongated in cross section. The var. *Blyttii* differs from var. *splendens* chiefly in being thicker (65–72μ according to Wittrock, 1866, p. 49), but is otherwise essentially the same.

Rosenvinge (1893, p. 940, 1894, p. 146), relying upon the observations of Kleen (1874, p. 42), reduces the three species of Wittrock to two varieties of one, viz., var. *typica* and var. *splendens* of *Monostroma fuscum*. The arrangement of Rosenvinge has been generally followed and we feel that we can not do better than adopt the general opinion. We have, however, followed Collins (1909, p. 213) in retaining the varieties (or forms?) of *splendens* and *Blyttii* as well as the typical form. The typical form is perhaps a younger or less developed form and seems more distinct from both var. *splendens* and var. *Blyttii*, than they do from one another. We may expect any, or all, of the three forms anywhere along the coast from the Bering Sea to the Puget Sound region.

We have had no opportunity of observing the earlier stages of growth of any of the forms of this species, but Postels and Ruprecht

(1840, p. 21) describe the young specimens as provided with a short stipe, and Rosenvinge (1893, p. 942, f. 48, 1894, p. 148, f. 48) confirms this and adds figures of the young plants with stipes which remain tubular. No. 387 of Tilden's American Algae is, in our copy at least, a light colored plant, provided with a distinct tubular stipe about 3 cm. long and 3–4 mm. in diameter (in pressed specimen). This specimen previously referred by us to *Ulva Lactuca* var. *rigida* (cf. Setchell and Gardner, 1903, p. 210), has the structure, although not the exact color, of *Monostroma fuscum*.

Rosenvinge (1893, p. 942, f. 17 C, D, 1894, p. 147, f. 17 C, D) calls attention to the fact that each cell in this species contains two chromatophores, one at each end, and Jónsson (1904, p. 631) emphasizes this as characteristic of this species in distinction from all other species of *Monostroma*. Collins (1909, p. 25), however, finds only one chromatophore in fresh material of *M. fuscum* from Revere Beach, Massachusetts, and also calls attention to Wittrock's figure (1866, pl. 3, f. 11) which he says "shows a perfectly uniform chromatophore quite like that of the Revere Beach plant." We have seen what appear to be two distinct chromatophores in some of our specimens of varieties *splendens* and *Blyttii* where the cells are much elongated vertically, but have failed to find them in plants seemingly to be more of the type of the species.

19. **Enteromorpha** Link

Frond persistently tubular, usually slender, but often ample, simple, proliferous or branched, its wall consisting of a single layer of cells, commonly, but not always, arranged parenchymatously; all the cells of the membrane, except the very lowest, capable of producing zoospores or gametes, which are discharged through an opening in the outer cell wall.

Link, Epistola, 1820, p. 5.

The name *Enteromorpha* has been for so many years practically agreed upon among writers as the name for this genus that there is little need, perhaps, for anticipating any change. The only name which seems strictly and definitely to antedate it is that of *Tubularia* of Roussel (Flore du Calvados, ed. II, 1806) which is said to have been founded upon *Ulva intestinalis* (cf. Desvaux, Journ. Bot., 1813, p. 144). It is not possible for us to verify this reference at present and we follow the weight of authority in retaining *Enteromorpha* for the accepted generic name.

Enteromorpha is a genus of the Ulvales characterized by its tubular frond which is usually narrow, although much dilated at times in the case of certain varieties of *Enteromorpha intestinalis.* The species are usually branched, although some are normally simple or, at times, proliferous. This definition or characterization leads us to exclude *E. percursa* and its allies, *E. aureola* and its allies, and *E. Linza,* the two former being referred each to its own genus, viz., *Percursaria* and *Capsosiphon* respectively, and the last to the genus *Ulva.* The particular reasons for thus excluding from the genus *Enteromorpha* plants heretofore commonly referred to it will be given under the respective genera or species, while the general reason is that *Enteromorpha* thus reduced is more readily defined and apparently more natural. In addition to these changes, we have added to the genus *Enteromorpha* the species commonly known as *Monostroma groenlandicum,* because in detail of habit, at least, it is very much closer to *Enteromorpha* than to *Monostroma* and not so very diverse from some species of *Enteromorpha* even in structure.

The more characteristic species of *Enteromorpha* have the cells arranged closely set, and parenchymatous in appearance, but certain species show, at times, the cells separated considerably from one another by an intercellular jelly and this condition is normal in *E. groenlandica,* which, for this reason, usually has been placed under the genus *Monostroma.* The cells are arranged in longitudinal rows in many species. Some species show a continuance of this arrangement even on into the adult condition, while others soon lose it. The chromatophore generally fills the cell as seen in surface view, but in a few species it occupies only a small portion of the surface of the cell.

Much is to be determined in *Enteromorpha,* as is also the case in *Monostroma* and *Ulva,* from the thickness, shape and proportions of the cells, and the disposition of the enclosing intercellular substance in cross sections of the membrane. Such sections also show the degree of compression of the frond and whether the walls are completely separated from one another or not.

In branching there is great variation, not only in the genus, but within many of the species. It has seemed best, nevertheless, to consider the method of branching as one of the chief characters in separating the species.

The species of *Enteromorpha* present difficulties which have not yet been entirely overcome. The habit is of importance, but varies considerably either under environmental changes or with age in such a fashion as to be insufficient in itself for diagnosis. The anatomical characters, on the other hand, seem to be more reliable, but, of themselves, present difficulties and seeming abnormalities or departure from type. Very few of the older species have been carefully described anew from the type specimens, so that uncertainty holds in many cases as to the exact application of specific names. Different writers, also, differ decidedly as to their views of specific limits and groupings of forms, so that a student of this group finds much variance of opinion, and consequent resulting confusion.

We have followed the accounts of Collins (1903 and 1909, p. 195 *et seq.*) very largely, but with due attention to the revision of J. G. Agardh (1883, p. 115 *et seq.*) and the critical remarks of Reinbold (1889, p. 113 *et seq.*), Börgesen (1902, p. 487 *et seq.*), Jónsson (1903, p. 343 *et seq.*) and Kylin (1907, p. 4 *et seq.*). Very considerable assistance has been obtained also from the earlier monograph of the group by Ahlner (1877) especially as interpreted by J. G. Agardh. Study has been made of the living plant so far as possible, and dried specimens have been boiled in water before examination to restore cell outlines and chromatophores. This method of preparation has given very satisfactory results, the dried specimens swelling up and assuming a form and structure closely approximating that of the living material. Even the chromatophores are fairly well restored to size, shape and position.

The species of *Enteromorpha* inhabit brackish water and strictly fresh water as well as strictly salt water and that which passes for the same species may be found in all three. The amount of salinity of the water may, seemingly at least, have very different effects particularly upon the branching and the thickness of the membrane. Careful cultures, however, are needed to establish this fact clearly and convincingly. It may be stated that the culture of various species and forms of the different species of *Enteromorpha* is very necessary before a definite basis for many of the distinctions now employed may be assured.

The species of *Enteromorpha* are probably more nearly "cosmopolitan" than those of any other genus of marine algae, although the term cosmopolitan can not, in all probability, be used in the strictest sense, even in connection with them. The ranges of the different species along our western coast of North America are more extensive than those of the species of most other genera. This is to be explained, we think, by the fact that the species of *Enteromorpha* are of rapid development and early maturity, so that their effective season of growth may be short, and on the basis of their being, for the most part, essentially tropical or subtropical species. Such warm water species of rapid development may extend into zones of colder surface waters by growing in the upper littoral belt where they may take advantage of the temperatures of the air, or by inhabiting shallow pools or lagoons the temperature of which is raised by the influence of the atmosphere and sunlight.

KEY TO THE SPECIES

1. **Enteromorpha groenlandica** (J. Ag.) S. and G.

Frond filiform, tubular, cylindrical, up to 15 cm. long, from a very slender base expanding to 1 mm. diameter; apex broken only at exit of spores; cells in the lower part loosely arranged in twos and fours, roundish angular; cells in the upper part more evenly distributed, more or less loosely set; in cross section the membrane 25–35μ thick; the cells radially elongate, 2–4 times as long as broad; in the younger parts the central cavity filled with a gelatinous substance which disappears as the plant becomes older; spores or gametes forming first at the summit of the frond, and developing successively in lower cells.

On small boulders in the middle littoral belt. Alaska (Bay of Unalaska and Kukak Bay).

Setchell and Gardner, Phyc. Cont. I, 1920, p. 280. *Monostroma groenlandicum* J. Agardh, Till Alg. Syst., part III, 1883, p. 107, pl. 3, f. 80–83; Collins, Green Alg. N. A., 1909, p. 208; Saunders, Alg. Harriman Exp., 1901, p. 410; Setchell and Gardner, Alg. N.W. Amer., 1903, p. 208.

Enteromorpha groenlandica has always been puzzling as to its proper placing. While technically it may seem to belong to the genus *Monostroma,* under which it was originally described, more properly than to any other genus of the Ulvaceae, yet its slender, filiform habit certainly more closely resembles that of some species of *Enteromorpha.* From *Enteromorpha,* however, it differs in not having its cells set sufficiently closely together to be parenchymatous in appearance. It is at first solid, becoming hollow only late, but never rupturing longitudinally and opening out into a membrane as do the characteristic species of *Monostroma.* Certain species of *Enteromorpha* show a tendency towards abundance of intercellular jelly at times, while certain species of *Monostroma* are parenchymatous. It seems best to us, therefore, to transfer this species to *Enteromorpha.*

The plants of the North Pacific Ocean, as Collins (1909, p. 209) states, have decidedly smaller cells than those of the North Atlantic, measuring 8–10μ in diameter, as against 12–16μ as seen superficially. It seems, therefore, to constitute a different form.

The species is a summer plant of the Upper Boreal Zone, invading the Lower Boreal and North Temperate Zones only as a short-lived plant of the springtime when the waters are colder than in the summer. This intrusion happens, so far as we have evidence, only on the eastern coasts of North America where the Massachusetts coast experiences a much colder winter and spring season than do western coasts of North America of the same zones. On the Pacific Coast, so far as our knowledge goes, the species is confined to the Bering Sea and adjacent portions of the Alaskan Peninsula where the summer temperature of the surface waters seldom, if ever, rises above 10° C.

2. **Enteromorpha micrococca** Kuetz.

Frond 1–5 cm. long, 1–5 mm. wide, tubular or compressed, simple or slightly proliferous at times, much curled and twisted; cells angular, 4–5μ diam., in no definite order; membrane 15–20μ thick, with distinct inner hyaline layer.

Growing in the upper littoral belt, on rocks and on woodwork. From Alaska (Dutch Harbor) to Mexico (*fide* Collins, *loc. cit.*).

Kuetzing, Tab. Phyc., vol. 6, 1856, p. 11, pl. 30, f. 2; Collins, Green Alg. N. A., 1909, p. 204; Setchell and Gardner, Alg. N.W. Amer., 1903, p. 211; Saunders, Alg. Harriman Exp., 1901, p. 411.

2a. **Enteromorpha micrococca** forma **subsalsa** Kjellm.

Plate 16, fig. 1

Frond compressed, much contorted, with numerous, patent or uncinate, shorter or longer branches from the margin, the latter again branched, all broad at the base and tapering to a point.

Growing on stones in the littoral belt. Alaska (Skagway) to Washington (Puget Sound).

Kjellman, Alg. Arctic Sea, 1883, p. 292, pl. 31, f. 1–3; Collins, Green Alg. N. A., 1909, p. 204, Mar. Alg. Vancouver Is., 1913, p. 102; Setchell and Gardner, Alg. N.W. Amer., 1903, p. 211; Collins, Holden and Setchell, Phyc. Bor.-Amer. (Exsicc.), no. 1068 (fresh water). *Enteromorpha minima* Setchell and Gardner, Alg. N.W. Amer., 1903, p. 213 (in part); Collins, Holden and Setchell, Phyc. Bor.-Amer. (Exsicc.), no. 912.

Enteromorpha micrococca is, in typical form, a low plant forming a layer on rocks and woodwork high up in the littoral belt. It is to be distinguished by its small cells, not arranged in longitudinal rows, and its thicker membrane which is reinforced by a hyaline layer on the inside. In its ordinary marine habitat, it seldom shows any tendency towards branching, but plants agreeing with it in anatomical characters, but growing in brackish or fresh waters, branch abundantly from the margins, the branches tapering at the tips and being usually curved. Such plants are referred under the forma *subsalsa*. This variety occurs at times, especially in fresh water, up to 10 cm. or more long. The var. *bullosa* which Collins distributed under no. 1067 of the Phycotheca Boreali-Americana seems to differ decidedly from *Enteromorpha micrococca* in the size and shape of the cells and approaches more nearly *E. intestinalis* in structure. This variety is known, thus far, only from fresh water (San Leandro, California) and does not strictly come under our consideration.

3. **Enteromorpha minima** Naeg.

Frond 1–10 cm. long, 1–5 mm. broad, simple or slightly proliferous, dilated or collapsing, soft and delicate, cells angular, 5–7μ diam., arranged in no definite order; membrane 8–10μ thick, equally thickened on both surfaces.

Growing on stones and on wood in uppermost littoral belt.

From Alaska (Unalaska) to Mexico.

Naegeli, *in* Kuetzing, Sp. Alg., 1849, p. 482, Tab. Phyc., vol. 6, 1856, p. 16, pl. 43, f. i-m; Collins, Green Alg. N. A., 1909, p. 201, Mar. Alg. Vancouver Is., 1913, p. 102; Setchell and Gardner, Alg. N.W. Amer., 1903, p. 213.

Enteromorpha minima resembles *E. compressa* but is, in normal form, less likely to show any branching, is more commonly dilated, and of softer and more delicate texture. The cells are also slightly smaller in surface view. From *E. micrococca*, this species is to be distinguished by the larger cells and thinner membrane which is seldom noticeably thickened on the inside, although Kuetzing (1856, pl. 43, f. m) so represents it. *Enteromorpha minima* resembles *E. micrococca* rather than *E. compressa* in size, but is found in typical form up to 10 cm. high. In fresh water forms attributed to this species we find plants up to 20 cm. long and sometimes with the cells decidedly separated from one another as in *Monostroma*. In a form from dripping rocks above high water mark on San Juan Island, Washington, distributed under no. 912 of the Phycotheca Boreali-Americana, the plants are large and the cells small, with the membrane thickened on the inside. This seems to approach very closely to *E. micrococca*.

4. Enteromorpha compressa (L.) Grev.

Plate 14, figs. 7, 8; plate 16, fig. 3

Frond tubular, more or less compressed, sometimes constricted, varying much in dimensions; branches usually simple, cylindrical or expanding above, in either case narrowed at the base, similar in appearance to the main axis; cells in no definite order; membrane rather thin.

Growing in the middle and lower littoral belts. From Alaska (Bering Sea) to Mexico (Magdalena Bay).

Greville, Alg. Brit., 1830, p. 180, pl. 18; Collins, Green Alg. N. A., 1909, p. 201, Mar. Alg. Vancouver Is., 1913, p. 101; Setchell and Gardner, Alg. N. W. Amer., 1903, p. 213. *Enteromorpha prolifera* Setchell and Gardner, Alg. N.W. Amer., 1903, p. 221 (as to no. 5687 only). *Enteromorpha fascia* Postels and Ruprecht, Illust. Alg., 1840, p. 21; Setchell and Gardner, Alg. N.W. Amer., 1903, p. 211; Collins, Green Alg. N. A., 1909, p. 204. *Ulva compressa* Linnaeus, Fl. Suec., Ed. II, 1755, p. 433.

We have followed the opinion of J. G. Agardh (1883, p. 137) as to the nature and limits of *Enteromorpha compressa*. The habit of typical plants is well illustrated in our figures (plate 14, figs. 7, 8, and plate 16, fig. 3). The branches may arise from the very base or at different heights along the axis. They are uniformly constricted at the base and usually expanded to a rounded tip. The broader portions of the frond are almost always flattened and the layers may be very imperfectly separated. Sections of such imperfectly tubular fronds often bear a striking resemblance to those of *Ulva Linza*, especially when separation is present only on the margins as may happen in spots.

The differences between *Enteromorpha compressa* and *E. minima* have already been noticed under the latter species. The resemblance to narrow forms of *Ulva Linza* is sometimes puzzling, especially in unbranched specimens (var. *subsimplex* J. Ag.) or in specimens only slightly branched.

We refer under *Enteromorpha compressa* the *E. fascia* of Postels and Ruprecht (1840, p. 21) since the habit (plate 16, fig. 3) is the same and the size and shape of the cells, both in surface view and in cross section, are identical. The cell contents, however, in the type specimens of *E. fascia* are disorganized in such a way as to seem almost as if there were groups of small cells within the larger ones. The color of the type specimens is also somewhat brownish. We ascribe this as well as the peculiar appearance of the cell contents to some abnormal state or unusual treatment of the specimens. The type specimens of *E. fascia,* as they were found in the Imperial Academy of St. Petersburg in 1903, were labelled as having been collected by the Luetke Expedition in Kamtschatka.

Of specimens distributed from our coast we find in our copy of the American Algae that Tilden's no. 265, under the name of *E. compressa* var. *complanata* is *E. crinita* and no. 264, under the name of *E. compressa* var. *subsimplex* is *E. plumosa.*

Most of the specimens available to us for examination are close to the typical form of *E. compressa* but certain plants collected by one of us (Gardner) at Coos Bay, Oregon, seem referable rather to var. *subsimplex* J. Ag. (1883, p. 137).

5. **Enteromorpha intestinalis** (L.) Link

Frond simple or having at the base a few branches similar to the main frond, or occasionally a few proliferations above; length varying from a few centimeters to several meters; diameter from 1–10 cm.; at first attached by a short cylindrical stipe, but often later detached and floating; cylindrical or expanding above, more or less inflated, often much crisped and contorted, and irregularly and strongly constricted; cells 10–16μ diam., in no regular order; thickness of membrane varying from 50μ below to 20μ above, generally thickened on the inside; cells in cross section from 12–30μ.

Common in its various forms from Alaska (Kukak Bay) to Mexico (La Paz).

Link, Epistola, 1820, p. 5; J. Agardh, Till Alg. Syst., part 3, 1883, p. 131; Collins, Green Alg. N. A., 1909, p. 204, Mar. Alg. Vancouver Is., 1913, p. 102; Setchell and Gardner, Alg. N.W. Amer., 1903, p. 212; Saunders, Alg. Harriman Exp., 1901, p. 411. *Ulva intestinalis* Linnaeus, Flo. Suec., Ed. II, 1755, p. 418.

Howe (1911, p. 490) has referred doubtfully a plant from La Paz, Mexico, to this species.

The following forms have been detected on our coast:

Forma **cylindracea** J. Ag.

Frond long and slender, of uniform diameter; usually floating unattached.

J. Agardh, Till Alg. Syst., part 3, 1883, p. 131; Collins, Green Alg. N. A., 1909, p. 205; Mar. Alg. Vancouver Is., 1913, p. 102; Saunders, Alg. Harriman Exp., 1901, p. 411; Setchell and Gardner, Alg. N.W. Amer., 1903, p. 212.

Forma **maxima** J. Ag.

Frond large, up to 4 cm. diam., inflated and bullate, producing small, scattered branches.

J. Agardh, Till Alg. Syst., part 3, 1883, p. 132; Collins, Green Alg. N. A., 1909, p. 205, Mar. Alg. Vancouver Is., 1913, p. 102; Saunders, Alg. Harriman Exp., 1901, p. 411; Setchell and Gardner, Alg. N.W. Amer., 1903, p. 212; Collins, Holden and Setchell, Phyc. Bor.-Amer. (Exsicc.), no. 1182.

Forma **clavata** J. Ag.

Frond always attached, filiform below, enlarging more or less abruptly upwards, open at the upper end, 1–5 cm. wide, 1–5 dm. long.

J. Agardh, Till Alg. Syst., part 3, 1883, p. 131; Collins, Green Alg. N. A., 1909, p. 205, Mar. Alg. Vancouver Is., 1913, p. 102. *Enteromorpha intestinalis* f. *genuina* Hauck, Meeresalg., 1885, p. 426; Setchell and Gardner, Alg. N.W. Amer., 1903, p. 212; Tilden, Amer. Alg. (Exsicc.), no. 263.

There appears to be no authenticated type specimen of the *Ulva intestinalis* of Linnaeus in existence (cf. Jackson, B. D., 1912, p. 147), so that we must follow general tradition as to the nature of this species. As generally agreed upon, it possesses a tubular frond usually inflated, of varying length and diameter, often twisted or constricted, usually branched from a slender base, occasionally slightly proliferous above, and with the rather large, angular or slightly elongated cells not arranged in longitudinal rows, at least not in the adult frond. Under this conception are arranged many and seemingly diverse forms both as to shape and size. The most slender plants of f. *cylindracea*, e.g., may not be much over 1 or 2 mm. in greatest diameter, while the largest of f. *maxima*, on the contrary, may be 10 cm. through. In regard to amount of inflation, smoothness or rugosity of surface, constriction or lack of it, and even of thickness or thinness of the membrane itself, there is much difference between specimens seemingly correctly referred to *Enteromorpha intestinalis*. What these very diverse forms indicate needs cultural experimentation to demonstrate. At present, we assume an identical genetic constitution for all and hold the varying environmental conditions responsible for transformations of form. This method is very unsatisfactory, but it is the best that can be done at present. Even with the wide range of characters we have given to the species, we have, nevertheless, followed the more restricted point of view rather than the more ample conception. The form names appended to the species, as given above, are simply for the purpose of giving some idea of the amplitude of variation of the species even as most narrowly delimited.

6. **Enteromorpha acanthophora** Kuetz.

Frond more or less proliferously branched, the branches usually constricted at the base, beset with numerous short, spinelike ramuli, with somewhat narrowed base and acute tip; cells 11–13μ diam., angular, showing no longitudinal arrangement except indistinctly at the tips of the ramuli and in the spinelike branchlets.

Guaymas, Mexico.

Kuetzing, Sp. Alg., 1849, p. 479, Tab. Phyc., vol. 6, 1856, pl. 34, f. 1; Collins, Green Alg. N. A., 1909, p. 200 (in part).

The figure and description of Kuetzing (1849, p. 479 and 1856, pl. 34, f. 1) provide our chief knowledge of this species, the type specimens of which are from New Zealand. We have not seen the type but have studied New Zealand specimens which seem to belong here. We are encouraged to refer to this species, although doubtfully, specimens collected by T. S. Brandegee at Guaymas, Mexico, which have a general resemblance to *Enteromorpha intestinalis,* but which are beset with short, spinelike branches. We do not think that no. 515 of the Phycotheca Boreali-Americana, a plant of fresh water, is properly to be referred to *E. acanthophora.* Its membrane is too thin (about 13μ, instead of 30–45μ) and its cells (4–5μ diam., instead of 10–13μ) and cross section, as well as its branching more closely, resemble those of *E. micrococca* f. *subsalsa,* although not strictly in agreement with them.

7. **Enteromorpha prolifera** (Muell.) J. Ag.

Frond up to several meters long and 2 cm. diam., tubular or compressed, with more or less abundant proliferous branches, which are usually simple, but sometimes also proliferous; branches varying much in length and diameter; cells 10–12μ in diameter, in the younger parts always arranged in longitudinal series, which become somewhat less distinct in the older parts; membrane 15–18μ thick, not much exceeding the dimensions of the cells in cross section.

Growing on sticks and stones, sometimes floating, in quiet waters and sheltered bays. From Alaska (Sitka) to central California.

J. Agardh, Till Alg., Syst., part 3, 1883, p. 129, pl. 4, f. 103, 104; Collins, Green Alg. N. A., 1909, p. 202; Saunders, Alg. Harriman Exp., 1901, p. 411; Setchell and Gardner, Alg. N.W. Amer., 1903, p. 211; Collins, Holden and Setchell, Phyc. Bor.-Amer. (Exsicc.), no. 913; Tilden, Amer. Alg. (Exsicc.), no. 385. *Ulva prolifera* Mueller, *in* Fl. Dan., vol. 5, fasc. 13, 1778, pl. 763, f. 1.

We must necessarily adopt the idea of J. G. Agardh as to the nature and limits of *Enteromorpha prolifera*, but neither he, nor any other writer, as far as we know, has examined the type. The illustration in the Flora Danica (*loc. cit.*) simply shows the habit and might represent either this species or some form of *E. intestinalis*. The fronds of forms of *Enteromorpha prolifera* resemble those of forms of *E. intestinalis* in habit, but are generally more proliferous. They vary from those of *E. intestinalis* in having slightly smaller cells, but differ particularly in having the cells arranged in longitudinal rows in the lower portions, at least, and in the branches. In size and shape, as well as in extent and variety of branching, there is great variation.

Enteromorpha prolifera also resembles *E. compressa* at times, when the tube is collapsed, but may generally be distinguished from that species by the longitudinal arrangement of cells in the branches. It is closely related to *E. tubulosa* and *E. flexuosa,* but in these species the cells are more regularly and uniformly arranged in longitudinal rows than they are in *E. prolifera. Enteromorpha tubulosa* is more or less branched and the membrane is not thickened within, while *E. flexuosa* is typically unbranched and with the membrane inwardly thickened. *E. prolifera* may be simple at first, but is usually branched later and is destitute of a thickening of the inner surface of the membrane.

8. **Enteromorpha flexuosa** (Wulf.) J. Ag.

Frond cylindrical, tubular, simple, tapering to a filiform stipe below, inflated above, flexuous and intestine-like; cells 8–12μ long, 6–8μ wide, roundish polygonal, in longitudinal series; membrane somewhat thickened on the inside; chromatophore filling the thick-walled cell.

Growing on rocks and on other plants. Santa Barbara, California.

J. Agardh, Till Alg. Syst., part 3, 1883, p. 126; Collins, Green Alg. N. A., 1909, p. 203. *Conferva flexuosa* Wulf., *in* Roth, Cat. Bot., II, 1800, p. 188.

J. G. Agardh (*loc. cit.*) is responsible for separating this species from among the forms previously referred to *E. intestinalis*. No recent examination of the type is reported. As taken by J. G. Agardh it seems to be a simple, more or less slender tube with the cells arranged in longitudinal rows and with the membrane somewhat thickened on the inside. It is generally regarded as being an inhabitant of warmer waters (cf. J. G. Agardh, 1883, p. 127, and Collins, 1909, p. 203). We have not seen it on our coast, but J. G. Agardh credits to this species a specimen collected by Mrs. Bingham at Santa Barbara, California.

9. **Enteromorpha tubulosa** Kuetz.

Plate 14, figs. 4, 5

Frond simple or with short proliferations, usually near the base, but with occasional longer proliferations some distance above the base, tubular and nearly cylindrical throughout, or enlarging upward from a delicate cylindrical stipe and becoming compressed above; cells squarish, 11–15μ diam., arranged in longitudinal series throughout, less distinctly so in the upper mature parts; membrane 15–24μ diam., walls equally thickened on both sides, with cells squarish or slightly elongated radially, chromatophore filling the outer end of the cell.

Growing attached to rocks in the lower littoral belt, or floating in intertwined masses in pools in salt marshes. Central California.

Kuetzing, Tab. Phyc., 1856, p. 11, pl. 32, f. 2; Ahlner, *Enteromorpha*, 1877, p. 49, f. 9a, 9b. *Enteromorpha prolifera* var. *tubulosa* Collins, Green Alg. N. A., 1909, p. 203; Collins, Holden and Setchell, Phyc. Bor.-Amer. (Exsicc.), no. 462 (Key West, Florida).

Kuetzing's figure of *Enteromorpha tubulosa* represents a simple plant, but J. G. Agardh (1883, p. 128) states that it branches. The main frond is tubular and slender, of nearly uniform diameter throughout. Our specimens are all branched more or less, but usually from near the base. The membrane may be thickened on both sides or not at all. There has been some difference of opinion among writers as to the proper relationship of this plant, but it seems best to us to retain it as an independent species.

10. **Enteromorpha marginata** J. Ag.

Frond filiform, compressed, simple or with a few proliferous branches; cells 4–8μ diam., squarish, arranged in longitudinal series, very distinctly in the two or three rows at each side, less so in the middle portion.

Vancouver Island (Departure Bay) to California.

J. G. Agardh, Algae Med., 1842, p. 16; Collins, Green Alg. N. A., 1909, p. 202, Mar. Alg. Vancouver Is., 1913, p. 102.

Enteromorpha marginata is a very slender plant, usually of salt springs or salt marshes. It is most commonly simple and of low stature. It is credited to our coast by Collins (*loc. cit.*), but we have had no specimens for examination.

11. **Enteromorpha salina** var. **polyclados** Kuetz.

Frond small, tubular, with occasional branches similar to the main filaments, all beset with short, spinelike, patent ramuli ending in a single series of cells and varying from few in some specimens to very numerous in others; cells squarish, arranged in longitudinal series.

Floating in tangled masses in salt-water ponds. Central California.

Kuetzing, Phyc. Germ., 1845, p. 248; Collins, Green Alg. N. A., 1909, p. 202. *Enteromorpha polyclados* Kuetzing, Tab. Phyc., vol. 6, 1856, pl. 36, II.

We have encountered floating in sun-heated pools in the salt marshes or in the artificially warmed water of the Key Route Pool, but all in the neighborhoods of Oakland and Alameda, California, what seems to be the above listed variety of *Enteromorpha salina.* Our plants are slender, with few or no main branches, cells squarish and in distinct longitudinal rows, and beset with branchlets consisting of one or two longitudinal rows of cells throughout. The main portions of the broader fronds sometimes show as many as twenty cells across, but there are only four to eight to be seen in the more usual slender fronds. We have referred all our plants to the variety but some show details of structure exactly corresponding to those of the species (cf. Kuetzing, 1856, pl. 36, f. I).

12. **Enteromorpha torta** (Mert.) Reinb.

Frond small, 1–3 cm. long, filiform, simple or with occasional proliferations consisting of two rows of cells; cells rectangular, always in longitudinal series throughout the filaments and more or less in cross series; chromatophore thin, covering the greater part of the cell.

Growing attached to rocks, in shallow pools in the upper littoral belt. San Diego, California. December.

Reinbold, Rev. Juergens' Alg. aquat., 1893, p. 201 (p. 14, Repr.). *Conferva torta* Mertens, msc., *in* Juergens, Dec. 13, no. 6.

We are inclined to refer here a slender plant from San Diego, California, collected by one of us (Gardner, no. 3574) although it shows no branches in any of the samples we have examined. It agrees fairly closely with the unbranched plants in the specimen distributed by Reinbold under no. 624 of the Phykotheka Universalis. Our plant may possibly be a short, capillary form of *E. tubulosa,* with cells more regularly arranged in longitudinal rows.

13. **Enteromorpha crinita** (Roth) J. Ag.

Frond filiform, cylindrical or compressed, much and repeatedly branched, the branches tapering towards the tips, the smallest, as well as the tips of the larger, usually of a single series of quite short cells; cells almost always in longitudinal series, often rounded, quite or nearly filled by the chromatophore.

Growing on wood or floating in the littoral belt. From Alaska (Valdes) to California (San Diego).

J. G. Agardh, Till Alg. Syst., part 3, 1883, p. 144; Collins, Green Alg. N. A., 1909, p. 199, Mar. Alg. Vancouver Is., 1913, p. 101; Saunders, Alg. Harriman Exp., 1901, p. 412; Setchell and Gardner, Alg. N.W. Amer., 1903, p. 214; Collins, Holden and Setchell, Phyc. Bor.-Amer. (Exsicc.), no. 965. *Enteromorpha compressa* f, *complanata* Tilden, Amer. Alg. (Exsicc.), no. 265 (not of J. G. Agardh). *Enteromorpha prolifera* Tilden, Amer. Alg. (Exsicc.), no. 385 (not of J. G. Agardh). *Conferva crinita* Roth, Cat. Bot., 1797, I, p. 162, pl. 1, f. 3.

Enteromorpha crinita is usually a slender, much branched plant, with tapering branches of several orders, the tips of the branches ending in a single series of cells. The chromatophore, as seen in surface view, seems to fill the cell and thereby distinguishes this species from

the very similar *E. plumosa*. This interpretation of *E. crinita* is that
of J. G. Agardh (*loc. cit.*) who founds his opinion upon a specimen
determined by Roth.

14. **Enteromorpha erecta** (Lyng.) J. Ag.

Frond filiform, with numerous long, usually erect branches, more
slender than the main filament; the ultimate ramuli of varying length,
polysiphonous, the cells being symmetrically arranged in successive
segments, similar to those of *Polysiphonia;* cells in the main axes and
branches in longitudinal and usually in transverse series; chromatophore filling the cell.

Rare on the Pacific Coast, known only from Vancouver Island
(Comox) and California (Santa Rosa Island).

J. G. Agardh, Till Alg. Syst., part 3, 1883, p. 152; Collins, Green
Alg. N. A., 1909, p. 200. *Scytosiphon erectus* Lyngbye, Hydr. Dan.,
1819, p. 65, pl. 15 C.

Enteromorpha erecta is to be distinguished from *E. crinita* by
having the tips of the branches polysiphonous instead of monosiphonous (ending in a single row of cells). It is to be distinguished
from *E. plumosa* in the same way as well as by having the chromatophores filling the cells. By the latter character it is also to be distinguished from *E. clathrata*.

J. G. Agardh (1883, p. 152) states that his *Enteromorpha erecta*
is the same as a specimen sent out by Dillwyn under the name of
Conferva paradoxa Dillwyn, but, since the description of Dillwyn
seems to him to have been made from another plant and Dillwyn's
figure ill drawn, he does not adopt Dillwyn's specific name. It is
impossible for us to do more than call attention to this statement,
but it seems to us very probable that Dillwyn's specific name may be
the proper one to be used for this plant.

15. **Enteromorpha plumosa** Kuetz.

Frond slender, filiform or later compressed, repeatedly branched,
branches tapering from the base and ending in a single series of cells;
cells in longitudinal series, less distinctly so in the older parts of the
frond, 8–10μ wide in the monosiphonous parts; chromatophore not
filling the cell.

Floating in slightly brackish water. Washington (Puget Sound)
and California (San Francisco).

Kuetzing, Phyc. Gen., 1843, p. 300, pl. 20, f. 1; Collins, Green Alg. N. A., 1909, p. 198. *Enteromorpha compressa* f. *sub-simplex* Tilden, Amer. Alg. (Exsicc.), no. 264 (not of J. G. Agardh).

In *Enteromorpha plumosa,* we find the counterpart of *E. crinita,* as mentioned under that species, except that the chromatophore, as seen from surface view, does not fill the cell. This is a character sometimes difficult to determine in dried specimens and may possibly be variable under different conditions of metabolism. Thus far it has been generally accepted since its suggestion by J. G. Agardh (1882, p. 151), but cultural experimentation to determine its constancy is desirable. J. G. Agardh (*loc. cit.*) prefers *Enteromorpha Hopkirkii* McCalla (cf. Harvey, 1849, p. 215, 1851(?), pl. 263) for reasons which we can not comprehend. Kuetzing's figure (1843, pl. 20, f. 1) seems to us definite and exact.

16. **Enteromorpha clathrata** (Roth) Grev.

Frond filiform, cylindrical or compressed, much branched in all directions, the branches tapering from base to summit, but not ending in a single series of cells; cells rectangular, usually longer than broad, always in longitudinal series, the chromatophore noticeably smaller than the cell.

Forming large, floating masses in warm, quiet waters. Alaska (Sitka) to central California.

Greville, Alg. Brit., 1830, p. 181; Collins, Green Alg. N. A., 1909, p. 199, Mar. Alg. Vancouver Is., 1913, p. 101. *Conferva clathrata* Roth, Cat. Bot., 1806, III, p. 175.

This species is the counterpart of *Enteromorpha erecta,* but with the chromatophore not filing the cell. *Enteromorpha clathrata* and *E. erecta* bear a similar relation to one another to that borne by *E. crinita* to *E. plumosa.* The identification of Roth's species seems to rest with Mertens, who is followed by J. G. Agardh (1883, p. 153) and, in turn, by all recent writers.

20. **Ulva** L.

Frond membranaceous, flat, consisting of two layers of cells usually closely applied throughout, but in some species separating at the base and margins; zoospores or gametes formed from any cell except those of the thickened, or hollow, stipe and escaping through an opening in the surface of the frond.

Linnaeus, Gen. Plant., 1737, p. 326, Sp. Plant., vol. 2, 1753, p. 1163 (in part).

The name *Ulva* goes back into classical Latin and was used to designate some marsh plant. It was adopted by some of the botanists to designate the expanded or gelatinous algae of any color. Linnaeus used it, at first, for a combination of species now referred to *Ulva*, *Monostroma*(?), *Enteromorpha*, *Porphyra*, *Botrydium* and *Nostoc*, later extending it to other expanded or non-*Fucus* species of all four groups of algae. If we are to follow some weighty authorities, we may be compelled to believe that Linnaeus included in his original list no one of the species generally referred to *Ulva*. If *Ulva Lactuca* L. is really a *Monostroma*; if *Ulva latissima* L. (at least of the Species Plantarum) was founded on a portion of the blade of *Laminaria saccharina*; and if we refer *Ulva Linza* to the genus *Enteromorpha*, then there is no species of *Ulva*, in the sense in which it is now used, left in the original list of Linnaeus.

As genera came to be more strictly delimited, *Ulva* came more and more to be reserved for membranous or tubular forms and ultimately for those belonging to the Chlorophyceae. Finally, in 1854, Thuret gave it its final description and content by the separation of the species of *Monostroma*. J. G. Agardh (1883, p. 160), although differing from Thuret as to some details, followed Thuret's segregations, and since that time the general concept has been the same for all writers. The genus *Ulva*, therefore, may be defined as including those species of the Chlorophyceae which have a parietal chromatophore and with the cells arranged, in large part at least, in a two-layered membrane. We have arranged our plants under *Ulva* in accordance with this idea, including even those like *Ulva Linza* which, in habit, seem to belong to *Ulva* and the greater portion of whose fronds remain as two layers closely applied to one another.

The species of *Ulva* are generally regarded as being not readily separable from one another, and the universal tendency has been to divide the genus into a few widely distributed and variable species. There is, however, great need of careful and extensive monographic work on this genus. In our attempt to distinguish the species and forms of our own coast, we have come to the conclusion that the plants fall into certain groups, fairly readily to be distinguished from one another. We have been able, also, to refer these narrower groups of forms to described species with plausible certainty. We trust that we may be able to stimulate, at least, careful scrutiny of these species and to refer observations as to the behavior of the plants referred to them. Cultural studies are needed and, although possibly difficult, may, even for certain more readily grown forms, yield results of extended application.

From our own experience, we feel convinced that general habit and size do not, as a rule, vary within extensive limits in the adults of the same species, while the details of cell structure are fairly uniform for any particular part of an adult plant within the species. Both habit and cell structure are sufficiently variable among the species, however, to afford intelligible diagnostic characters.

KEY TO THE SPECIES

1. Frond lanceolate with tubular stipe..............:..............................1. **U. Linza** (p. 262)
1. Frond variously shaped, stipe when present, solid... 2
    2. Cells square, or nearly so, in cross section.. 3
    2. Cells distinctly vertically elongated in cross section................................... 6
3. Frond seldom over 2 cm. high..2. **U. californica** (p. 264)
3. Frond usually over 2 cm. high.. 4
    4. Frond narrowly lanceolate...3. **U. angusta** (p. 264)
    4. Frond broad in proportion to length... 5
5. Frond attached, orbicular or ovate, often deeply split........4. **U. Lactuca** (p. 265)
5. Frond usually soon free, ample and exapnded.....................5. **U. latissima** (p. 266)
    6. Frond broad in proportion to height... 7
    6. Frond narrow in proportion to height or with long, narrow laciniae........... 9
7. Frond apparently regularly and abundantly perforated....6. **U. fenestrata** (p. 267)
7. Frond not perforate or occasionally showing a few irregular holes....................... 8
    8. Frond ample, usually with deep ruffled margins..........7. **U. expansa** (p. 268)
    8. Frond moderate, deeply lobed, with slightly ruffled or plane margins.........
    ....8. **U. lobata** (p. 268)
    8. Frond short ovate, plane, usually deeply split...............9. **U. rigida** (p. 269)
9. Frond long, usually simple, more or less ruffled...........10. **U. stenophylla** (p. 271)
9. Frond very short, simple, plane...............................11. **U. vexata** (p. 271)
9. Frond deeply divided into long, narrow, crisped laciniae.......................10
    10. Laciniae borne on a short, broad, basal portion, not dentate below...........
    ....12. **U. dactylifera** (p. 272)
    10. Laciniae split to the very base, dentate below........13. **U. taeniata** (p. 273)

. 1. **Ulva Linza** L.

Plate 12, figs. 1–4

Frond lanceolate or linear-lanceolate, simple, 1–5 dm. long, 1–20 cm. broad; stipe longer or shorter, hollow; upper part of the frond flat, the two layers of cells completely united or remaining free along the whole or part of the margins, which are plane or more or less undulate; membrane 25–70μ thick; cells usually vertically elongated in section, up to twice as high as broad.

Growing on wood, rocks, and on other algae, in the lower littoral belt. Alaska (Orca) to Mexico (La Paz).

Linnaeus, Sp. Plant., vol. 2, 1753, p. 1163. *Enteromorpha Linza* J. Agardh, Till Alg. Syst., part 3, 1883, p. 134, pl. 4, f. 110–112; Collins, Green Alg. N. A., 1909, p. 206, Mar. Alg. Vancouver Is., 1913, p. 102; Setchell and Gardner, Alg. N.W. Amer., 1903, p. 212; Howe, Phyc. studies, V, 1911, p. 490; Collins, Holden and Setchell, Phyc. Bor.-Amer. (Exsicc.), no. 967 b; Tilden, Amer. Alg. (Exsicc.), no. 384. "*Ulva Lactuca* forma *genuina*" Tilden, Amer. Alg. (Exsicc.), no. 260 (not of Hauck).

The figure of Dillenius (1741, pl. 9, f. 6) quoted by Linnaeus (1753, p. 1163) under *Ulva Linza,* seems sufficiently characteristic to distinguish this species. We find, however, that it becomes necessary to include under the name a very considerable variety of forms. Some of these forms are very narrow, while others are comparatively broad. In some, the hollow stipe gradually expands into the blade, while in others the passage from one to the other is extremely abrupt and the blade is broad, even slightly cordate, at the base. Many plants of *Ulva Linza* are short (a few cm. long) while some are very long (up to 1 M. or more). The margins, in turn, may be perfectly flat and plane while, in others, they are decidedly, even conspicuously, undulate or deeply ruffled. The hollow stipe and greater or less extent of hollow margin, however, distinguish all forms of *Ulva Linza* from any other species of *Ulva,* and the considerable expanse of two-layered blade distinguishes them from any species of *Enteromorpha.*

While *Ulva Linza* is an *Enteromorpha* at the base and on the lower margins, it is decidedly an *Ulva* so far as the expanded blade is concerned. It might, with justice, be placed in either genus, but since the habit in general is that of an *Ulva* and the greater portion of any plant of the species is ulvoid, it seems to us that the novice, at least, is more likely to arrange it with *Ulva* than with *Enteromorpha.* We have decided, therefore, to restore it to the genus *Ulva.*

It has been customary since the account of J. G. Agardh (1883, p. 134) to distinguish two forms of *Ulva Linza,* the one (f. *lanceolata*) with the margins plane or undulate and the other (f. *crispata*) with the margins crisped. Since all degrees of ruffling or absence of it occur in plants seemingly to be referred to the species, it does not seem practicable to distinguish sharply between them.

2. **Ulva californica** Wille

Frond 1.5–2 cm. long, up to 1.5 cm. wide, triangular or reniform with wavy edge, sometimes with proliferations of a few cells each, passing abruptly into a slender, filiform stipe; cells of the stipe, which on the inner side form rhizoidal prolongations, are in cross section about quadrate; membrane 30–35µ thick; the cells in the upper part of the frond are rather irregularly polygonal with rounded corners; no noticeable arrangement in longitudinal series.

Growing in profusion on rocks near high-tide line. California (region about San Diego).

Wille, *in* Collins, Holden and Setchell, Phyc. Bor.-Amer. (Exsicc.), no. 611; Collins, Green Alg. N. A., 1909, p. 215.

Our knowledge of *Ulva californica* is derived from the type and other specimens collected by Mrs. M. S. Snyder at La Jolla, near San Diego, California. It seems to be a very small species with a comparatively long stipe and having the cells small and nearly isodiametric. In this combination of characters it seems amply distinct from all other species of the genus.

3. **Ulva angusta** S. and G.

Plate 22, and plate 26, fig. 1

Frond simple or very rarely lobed, lanceolate to oblanceolate, 8–15 cm. long, 0.5–1.5 cm. wide, 35–45µ thick (occasionally about 53µ), tapering either gradually or abruptly at the base to a delicate, solid stipe with discoidal holdfast, color of fronds pale green, margins varying from almost plane to very much crisped; cells in surface view 3- 6-sided, with rounded angles, 5–12µ diam. in section, quadrate to one and a half times longer than broad, with rounded angles; chromatophore filling the outer half of the cell.

Growing in shallow pools along high-tide level. California (region of San Francisco).

Setchell and Gardner, Phyc. Cont. I, 1920, p. 283, pl. 27, and pl. 31, fig. 1.

We find at several places along the coast of central California, a rather short and narrow *Ulva* which does not seem to belong to any of the hitherto described species. We have felt compelled, therefore, to give it a name. It resembles the *Phycoseris lapathifolia* of Kuetzing (1856, pl. 25) but is shorter and narrower. It also resembles, even more closely, Kuetzing's figure of *Phycoseris Linza* (1856, pl. 16, f. I), but is a smaller plant than that also. The short, flattened stipe is solid. The narrow blade varies from plane to undulate or even crisply ruffled on the margins. The cells are oblong or rounded in section, each provided with more or less of a distinct wall. Although we have only recently become acquainted with it, this seems to be a vernal species. It has been observed in fertile condition in April.

4. Ulva Lactuca L.

Fronds short, usually broader than long, attached by a disk from a broad or attenuate base, generally deeply and irregularly split, light to dark green in color, delicate in texture, margins plane or ruffled; membrane 35–50μ thick (usually about 40μ); cells in section nearly square with rounded angles or slightly elongated, seldom one and one half times higher than broad even in fertile condition.

Growing attached at first in the upper littoral belt. From Alaska (St. Michael) to Mexico (Gulf of California).

Linnaeus, Sp. Plant., vol. 2, 1753, p. 1163 (in part?); Thuret, Note Syn. *Ulva Lactuca*, 1854, p. 24; Thuret and Bornet, Études Phyc., 1878, p. 5, pl. 2, 3; Collins, Green Alg. N. A., 1909, p. 214, f. 75 (in part). *Ulva Lactuca* f. *rigida* Tilden, Amer. Alg. (Exsicc.), no. 386 (not of Le Jolis).

Ulva Lactuca, as described by Linnaeus, is not so definitely delimited as to be intelligible, so that different views have been held as to its nature. A specimen exists in the Linnaean Herbarium (cf. Benjamin Daydon Jackson, 1912, p. 147) labeled in the handwriting of the younger Linnaeus. We have no knowledge of this specimen and have seen no account as to its examination. It is certain that a number of the earlier writers agree in referring here a plant which, tubular at first, finally splits. On this account, the *Ulva Lactuca* L. is a *Monostroma* or perhaps included both a *Monostroma* and an *Ulva*. With all due respect to these authorities, it has seemed best to us to follow a considerable number of recent writers who have followed the opinion of Thuret (1854, p. 24) as to the real nature of *Ulva Lactuca*. The plant described by Thuret and Bornet in their Études Phycologiques (1878, p. 5, pl. 2, 3) is the one we have in mind in the reference of the specimens of our coast. This species has a frond broadly lanceolate, attenuated to a very short stipe below, broadened and more or less deeply and coarsely lobed above. The membrane is about

40μ thick at about the middle and the cells in section are circular or oblong and but very slightly elongated vertically in section even when fertile. The substance of the membrane is soft and slightly lubricous. Taken in this narrower sense, we find specimens both from our own coasts and those of Europe in agreement. This leads us to include this species as defined above, but not in the broad sense, as including *Ulva rigida* Ag., *U. latissima* L., and other species as has been the customary usage of recent writers. On this account, it is impossible, at present, to give an extended synonymy under the species. Vickers (1908, pl. 1) has figured what seems to be a tropical form of true *Ulva Lactuca* and Hauck has distributed a specimen from Trieste (cf. Hauck and Richter, Phyk. Univ., no. 17).

5. Ulva latissima L.

Frond ample, broader than long, usually soon free and expanded, often reaching a considerable size, yellowish green; membrane 35–40μ thick; cells, in section, nearly square or elongated horizontally.

On mud flats and in quiet bays sometimes completely covering extensive areas. Alaska (Juneau).

Linnaeus, Sp. Plant., vol. 2, 1753, p. 1163 (in part?). *Ulva Lactuca* var. *latissima* De Candolle, Flore Française, ed. 3, vol. 2, 1805, p. 9; Collins, Green Alg. N. A., 1909, p. 215 (in part), Mar. Alg. Vancouver Is., 1913, p. 103 (in part?); Setchell and Gardner, Alg. N.W. Amer., 1903, p. 210 (in part). *Ulva Lactuca myriotrema* Saunders, Alg. Harriman Exp., 1901, p. 410 (not of Le Jolis).

Ulva latissima of the Species Plantarum (Linnaeus, 1753, p. 1163) may or may not be the same as that of the Flora Suecica (Linnaeus, 1755, p. 433) and the questions as to whether either of these is the *U. latissima* of other authors, whose conceptions also have varied, are not to be solved by us. We infer that the specimen of *Ulva latissima* in the Linnaean Herbarium is not labeled either in Linnaeus' own hand or those of any known amanuensis (cf. Benjamin Daydon Jackson, 1912, p. 147). It is apparently (cf. Turner, Fuci, vol. 3, 1811, p. 72 and English Botany, vol. 22, 1806, text under pl. 1551) a fragment of the blade of *Laminaria saccharina*. The plant upon which the name given in the Species Plantarum is based was collected by Linnaeus on his trip into West Gothland, Sweden (Linnaeus, 1753, p. 1163, "iter w. gotl. 160").

We have followed the conception of J. G. Agardh (1883, pp. 164–166) as to the proper nature of the Linnaean species and have referred here those ample floating forms, with thin membranes, and cells which are cubical or horizontally elongated in section. We find very few such plants on our coast, their place as ample, expanded, floating membranes being taken by *U. expansa.* One specimen, however, collected near Douglas, Alaska, by Mr. Eldred Jenne seems clearly to belong here, and other specimens may be expected in quiet waters. The specimen from Douglas, Alaska, agrees fairly well with no. LXXVI of the Phycotheca Boreali-Americana, but is somewhat thinner.

6. **Ulva fenestrata** P. and R.

Frond ample, usually soon free and expanded, yellowish green, completely, and more or less uniformly, perforated with larger and smaller round or elongated openings with undulate edges, margins often wavy; membrane up to 60µ thick; cells in section nearly square or slightly vertically elongated (about 20µ high by 16µ broad in thicker sections); chromatophore cucullate at outer end of cell.

Growing on rocks in the lower littoral and upper sublittoral belts. Alaska (Sitka) to Puget Sound, Washington.

Postels and Ruprecht, Ill. Alg., 1840, p. 21, pl. 37.

The question as to the origin of the perforations found apparently regularly in some species of *Ulva* and occurring more or less sporadically in many or all species, has not, as yet, been at all carefully investigated. Greville (1830, p. 172) speaks of the frond of *Ulva latissima* as being "frequently much perforated by marine animals." On the other hand, J. G. Agardh (1882, p. 171) in speaking of *U. rigida* and its forms, states that the holes found in this species, as well as in *U. reticulata* Forsk., are not the work of animals, but due to inequalities of growth. We find holes frequently in considerable numbers in various of the *Ulvae,* as well as of the *Porphyrae,* of our coast and we feel certain that in many cases, at least, they are the work of mollusks, but we also find specimens of a large species of *Ulva,* which seems likely to be *U. fenestrata,* in which the holes are so numerous and so regular, and so constantly found that we are inclined to believe them to be the results of growth of the uninjured frond. We have, consequently, referred them here, although with much doubt, since other than as to perforations, they agree well with specimens of *Ulva expansa.* Some of the specimens in our possession are as long as 4 meters and up to 13 decimeters wide. As to details of perforation, our specimens agree with those figured by Postels and Ruprecht.

7. **Ulva expansa** (Setchell) S. and G.

Frond ample, pale green, orbicular or broadly elongated, margin deeply ruffled; frond 60–70μ thick in the middle, 38–45μ on the margins; cells, in section, vertically elongated in the middle of the frond (up to 28–30μ long, 10–12μ wide), nearly square in the margins.

Growing on rocks in the lower littoral belt. Washington (Puget Sound) to Mexico (La Paz).

Setchell and Gardner, Phyc. Cont. I, 1920, p. 284. *Ulva fasciata* f. *expansa* Setchell, *in* Collins, Holden and Setchell, Phyc. Bor.-Amer. (Exsicc.), no. LXXVII; Collins, Green Alg. N. A., 1909, p. 216.

We find along the coast of central California a broad species of *Ulva*, often also long, something·like *Ulva latissima* in appearance, yet of a more vivid green color, thicker in the center of the frond and with distinct, broad, ruffled margins. The cells of the thicker center of the frond are distinctly palisade-like in section, while on the thinner margins they are nearly square. A younger specimen of this plant was distributed by one of us as *Ulva fasciata* f. *expansa* (Phyc. Bor.-Amer., no. LXXVII), but it has seemed, on further study, to belong neither to *Ulva fasciata* Delile nor to the *Ulva fasciata* f. *taeniata* also distributed by one of us (Phyc. Bor.-Amer., no. 809), but described later in this account as *Ulva taeniata*. We have therefore described it (Setchell and Gardner, 1920, p. 284) as an independent species under the name of *Ulva expansa*.

Ulva expansa, so far as we have observed it, remains attached only for a short time. It soon becomes free and floats or drifts, increasing in size, becoming at times at least 3 meters long and varying in width from 18 cm. to 75 cm. In form and structure it differs from *Ulva latissima* and from all the other species of *Ulva* of our coasts. It comes nearest to *Ulva fenestrata* as we have described that species, but is little, if at all, perforate. Plants of what appears to be the same species have been found in the Puget Sound region and Howe (1911, p. 490) is inclined to credit here some from La Paz, Mexico.

8. **Ulva lobata** (Kuetz.) S. and G.

Frond dark green, of moderate size (up to about 30 or more cm. long and 10–15 cm. broad), more or less deeply lobed or divided, attenuate to a cuneate, crisped, often more or less twisted base, margins plane or slightly undulate; membrane 45–90μ thick, thicker in the center and thinner near the margins; cells elongated vertically, up to two and one half times as high as broad in the thicker central portion, nearly square at the margins.

On rocks in the lower littoral belt. Central California (San Francisco) to southern California (Pacific Beach).

Setchell and Gardner, Phyc. Cont. I, 1920, p. 284. *Phycoseris lobata* Kuetzing, Spec. Alg., 1849, p. 477, Tab. Phyc., vol. 6, 1856, p. 10, pl. 27. *Ulva fasciata* f. *lobata* Setchell, *in* Collins, Holden and Setchell, Phyc. Bor.-Amer. (Exsicc.), no. 863; Collins, Green Alg. N. A., 1909, p. 216.

Among the *Ulvae* of the Californian coast is one of moderate size (up to 30 cm. or more long and to 15 or more cm. broad) which is distinct in general appearance. It is attenuate at the crisped base, broadening above and usually lobed or divided into several broad divisions. The margins are either plane or slightly undulate. Like *U. expansa* it is thicker in the center with palisade-like cells (in section) and thinner on the margins where the cells are nearly square (in section). It bears a striking likeness in every way to Kuetzing's figure (1856, pl. 27) of his *Phycoseris lobata* from Chili. We have, therefore, referred it to his species with some doubt.

Ulva lobata belongs to the same group of species as *U. expansa* but is generally firmer in substance, slightly thicker, never reaches a great size, and is less deeply or conspicuously ruffled. It is well represented by the specimens distributed in the Phycotheca Boreali-Americana (under no. 863).

The most typical plants are those of the central Californian coast (San Francisco to Monterey). We have referred here also one plant from southern California, but with some doubt.

9. **Ulva rigida** Ag.

Frond low, at first lanceolate or ovate-lanceolate, firm and stiff, with distinct stipe, later broader and irregularly deeply divided; membrane 60–110μ thick, varying with age and position in the frond; cells, in section, vertically elongated, one and one half to three times as high as broad.

Growing attached to rocks and other algae, upper littoral belt. Alaska (Uyak Bay) to Mexico (La Paz).

Agardh, Sp., vol. 1, part 2, 1822, p. 410. *Ulva Lactuca* var. *rigida* Le Jolis, Alg. Mar. Cherb., 1863, p. 38; Collins, Green Alg. N. A., 1909, p. 215 (in part), Mar. Alg. Vancouver Is., 1913, p. 103 (in part); Saunders, Alg. Harriman Exp., 1901, p. 410 (in part); Setchell and Gardner, Alg. N.W. Amer., 1903, p. 209 (in part); Howe, Phyc. Studies, V, 1911, p. 490.

Ulva rigida must needs be carefully studied and redescribed from the type specimen before any exact knowledge is possible as to the nature of the species. Agardh (1822, p. 410) describes it as from 3 or 4 up to about 9 inches long, split to the base into curved and crisp laciniae. J. G. Agardh (1883, p. 168) describes it as having cells vertically elongated in section to 2 to 3 times their width. He, however, refers as typical the figure of *Ulva Lactuca* of Thuret and Bornet's Études Phycologiques (1878, pl. 2, e). Yendo (1916, p. 244) says that J. G. Agardh has taken a broader view of the species than did C. Agardh. Yendo refers to *U. rigida*, in the sense of its founder, the *Ulva conglobata* Kjellm. and its f. *densa* as well as the *Ulva fasciata* f. *caespitosa* Setchell (Phyc. Bor.-Amer., no. 809). Yendo does not, however, state definitely just the characters of the type of *U. rigida* Ag.

The *Ulva conglobata* Kjellm. seems to us to agree well with the *U. fasciata* f. *caespitosa* Setchell (Phyc. Bor.-Amer., no. 809, nom. nud.), both as to habit and as to structure. The cells in each are only slightly, if at all, elongated vertically in section. We are inclined to refer both of these plants to *Ulva Lactuca* as small forms. The *Ulva conglobata* f. *densa*, however, seems different in its structure. Judging by Kjellman's figure (1897a, pl. 3, f. 15), the cells are decidedly vertically elongated in cross section, and this form probably, therefore, belongs to *U. rigida* in the sense of J. G. Agardh.

We have, in the light of what has been written, considered the *Ulva rigida* to be a low plant, rigid, deeply divided, rather thick and with cells vertically elongated in section. The membrane usually shows a rather thick hyaline layer under each surface, and another between the layers of cells. We have found that certain of our specimens conform to these characters and are to be distinguished by them from any other species of *Ulva*.

10. **Ulva stenophylla** S. and G.

Plate 21, fig. 2, and plate 24

Frond simple, linear-lanceolate, tapering abruptly at the base to
a very short, flattened, cuneate stipe, 5–8 dm. high, 5–10 cm. wide,
plane in the middle with undulate margins; membrane 60–110μ thick;
cells squarish in surface view, 14–20μ diam., 1.5–2 times as long as
the diameter in section, chromatophore a thin parietal layer, covering
a part or the whole of the cells; pyrenoids absent.

Growing on rocks in the lower littoral belt. Central California.

Setchell and Gardner, Phyc. Cont. I, 1920, p. 282, pl. 26, fig. 2, and
plate 29.

The plants described under this name are quite distinct from the
other species of *Ulva* in shape, texture and anatomical details. They
are dark green, tough and harsh to the touch. The usually simple,
long, lanceolate shape serves to distinguish them from other species
at a glance.

11. **Ulva vexata** S. and G.

Plate 17, figs. 4–7

Frond small, unbranched, rigid, linear to oblanceolate or spatulate,
plane or slightly undulate, more or less bullate with cuneate base and
small, solid stipe, 1–3 cm. long, 3–10 mm. wide, dark green, black on
drying; membrane 45–55μ, up to 100μ thick, cells vertically elongated,
11–15μ, up to 18μ long, 3.5–5μ wide, with thick walls and very blunt
angles in surface view; chromatophore filling the cell, pyrenoids
absent.

Growing on rocks along high-tide level. In the vicinity of San
Francisco, California.

Setchell and Gardner, Phyc. Cont. I, 1920, p. 282, pl. 22, figs. 4–7.
Ulva californica Reed, Two Ascomycetous Fungi, etc., 1902, p. 149
(not of Wille).

Ulva vexata has been observed only in the vicinity of San Francisco, as mentioned above, where it grows in considerable profusion.
It seems quite probable that it may be much more widely distributed
both north and south of San Francisco. It might be suspected of
being a malformation due to the parasite always found more or less
infesting it, but the size and proportions of the cells of the less parasitized portions seem to mark it as a distinct species.

12. **Ulva dactylifera** S. and G.

Plate 21, fig. 1

Frond sessile or with a very short stipe; basal portion orbicular or reniform, much crisped, 2–4 cm. high, giving rise from the upper margin to 1–6 lanceolate, simple or occasionally branched lobes or laciniae with plane midrib and much crisped margins, 5–15 cm. high, . 0.5–1.5 cm. wide; membrane of basal portion 50μ thick at the margin, up to 100μ thick in the middle, with cells 16–20μ diam. in the surface view, quadrate to 2 times as long as wide in section, membrane of the laciniae 40–50μ thick on margin, up to 190μ thick in the middle, with cells 12–16μ diam. in surface view, quadrate to 5 times as long as wide in section; chromatophore filling the outer half of the cell.

On exposed rocks, uppermost littoral belt. Southern California to Mexico (San Roque?).

Setchell and Gardner, Phyc. Cont. I, 1920, p. 285, pl. 26, fig. 1.

We have along the Californian coast two species related to *Ulva fasciata,* neither of which seems to be exactly like the Mediterranean species. Both are characterized by long, narrow fronds or laciniae, much thicker along the middle and with thinner, very much crisped margins. One of these, *Ulva dactylifera,* possesses a comparatively broad, though short, undivided basal portion from which arise the several narrow, elongated, crisped laciniae. Neither the basal portion nor the laciniae show distinctly toothed margins. The other species, *Ulva taeniata,* is either simple, long, slender, plane and dentate below, but with crisped margins above, or divided to the very disk itself into two or three such divisions. The "midrib" portions differ slightly in thickness in the two species and the cells of the "midribs" differ in proportions.

Ulva dactylifera has been distributed under no. 221 b (sub *"Ulva fasciata"*) of the Phycotheca Boreali-Americana. Unfortunately the plants under this number are not uniform. We have examined no. 221 b in two copies. In one, the plant is certainly, although not typically, *U. dactylifera.* In the other it seems rather to be a form of *Ulva Lactuca.*

Ulva dactylifera is nearest to *U. fasciata* f. *costata* Howe (1914, p. 20, pls. 1, 2, f. 10–23), but differs as to the basal portion, thickness, and possibly also in proportions of cells. It differs from *U. fasciata*

Delile, so far as descriptions and figures indicate, in branching, in ruffling, and probably in thickness. It is a very much thinner plant than *U. nematoidea* Bory, judging from the dimensions given by Bornet (1892, p. 36 or 196).

13. **Ulva teniata** (Setchell) S. and G.

Plate 23

Frond elongated, up to 1 or 2 M. long, simple or split to the very base into long, narrow segments, plane below and coarsely dentate, densely crisped and ruffled on the margins above, with a plane, thicker midrib; membrane up to 140μ thick as to the "midrib" and down to 40μ thick on the margins; cells of the "midrib" vertically elongated in section up to two and one half times as high as broad, but becoming nearly square towards the margins.

On rocks in lowermost littoral or upper sublittoral belts. Central California (Tomales Bay to Monterey).

Setchell and Gardner, Phyc. Cont. I, 1920, p. 286, pl. 28. *Ulva fasciata* f. *taeniata* Setchell, *in* Collins, Holden and Setchell, Phyc. Bor.-Amer. (Exsicc.), no. 862; Collins, Green Alg. N. A., 1909, p. 216.

Ulva taeniata has been found thus far on the coasts of central California only, while *U. dactylifera* has been found only on those of southern California. The differences between the two species have been enumerated under the latter species. From *U. fasciata* f. *costata* Howe, it differs particularly in its basal portion. *Ulva fasciata* Delile seems to be a species nearly if not absolutely plane, while *U. taeniata* is always crisply ruffled. No. 862 of the Phycotheca Boreali-Americana represents this species very well.

21. **Percursaria** Bory

Frond slender, filamentous at first, of a single series of cells, later becoming two longitudinal rows of cells placed symmetrically side by side throughout, or only in portions, of the filament; cells rectangular, with thick walls and a single chromatophore.

Bory, Dict. class. d'hist. nat., vol. 13, 1828a, p. 206. *Tetranema* Areschoug, Phyc. Scand. Mar., Sect. II, 1850, p. 418 (not *Tetranema* Bentham, 1843). *Diplonema* Kjellman, Norra Ishaf. Algfl., 1883a, p. 371 (not *Diplonema* Don, 1838, nor *Diplonema* De Notaris, 1846).

The genus *Percursaria* was founded by Bory in 1828 to receive the *Ulva percursa* Ag. Bonnemaison founded a genus of the same name in 1822 (p. 178) quoted as *"Percussaria"* by Leman (Dict. d'hist. nat., vol. 38, 1825, p. 425) which is a mixture of filamentous Myxophyceae with no type designated. It certainly seems to us that the *Ulva percursa* Ag. is a very distinct plant belonging in no other genera of the Ulvaceae as properly limited. We follow Rosenvinge and others in keeping it separate. It is, in reality, a very narrow membrane, seldom, if ever, more than two cells wide. It may possibly be looked upon as a very primitive form among the Ulvaceae.

Percursaria percursa (Ag.) Rosenv.

Plate 14, fig. 6

Frond several cm. long, flexuous and contorted, generally irregularly and frequently contracted to a single row of cells or expanded to a double row; cells 10–15µ wide and from once to twice as long.

In entangled masses with other filamentous algae in upper tide pools, in ditches in salt marshes and similar places where the water is warmed by the sun. From Alaska (Bay of Unalaska) to central California (San Francisco Bay).

Rosenvinge, Groenl. Havalg., 1893, p. 963, Alg. Mar. du Groenl., 1894, p. 160. *Conferva percursa* Agardh, Syn. Alg. Scand., 1817, p. 87. *Tetranema percursum* Areschoug, Phyc. Scand. Mar., Sect. II, 1850, p. 418. *Diplonema percursum* Kjellman, Norra Ishafv. Algfl., 1883*a*, p. 371, Alg. Arct. Sea, 1883, p. 302. *Enteromorpha percursa* Setchell and Gardner, Alg. N.W. Amer., 1903, p. 214; Collins, Green Alg. N. A., 1909, p. 197; Collins, Holden and Setchell, Phyc. Bor.-Amer. (Exsicc.), no. 968.

This interesting plant, so unlike a member of the Ulvaceae in general appearance, seldom occurs pure, but is generally mixed with those species of *Cladophora, Enteromorpha* and filamentous Myxophyceae which delight in the same conditions of life. It is readily to be told from its associates by the very symmetrically placed double row of cells.

ORDER 5.   SCHIZOGONIALES WEST

Thallus filamentous or membranaceous, deep green; cells dividing in one, two or three planes strictly at right angles to one another, chromatophores axile; multiplication by proliferation (or gemmation) of the thallus, by akinetes, or by the latter directly forming aplanospores; zoospores and gametes unknown.

West, G. S., Brit. Freshw. Algae, 1904, p. 98.

This order differs from that of the Ulvales by the cells possessing axile instead of parietal chromatophores, and by the absence of zoospores and gametes. Otherwise, the members in general appearance closely resemble those of the Ulvales. The order comprises one family and one to four genera according to the views of different writers. The species are largely inhabitants of fresh water or of damp, usually foul, earth, but a few are marine.

FAMILY 10.   SCHIZOGONIACEAE CHODAT

The characteristics are the same as those of the order.

Chodat, Alg. vertes de la Suisse, 1902, p. 341.

In choosing the name of this family, all depends upon whether *Schizogonium* is to be retained as a distinct genus. Until more definite cultural work has been carried through, it seems best to us to retain *Schizogonium* distinct from *Prasiola* and, consequently, we have adopted Chodat's name in preference to the Prasiolaceae of West (1904) or the Blatosporaceae of Wille (1909).

KEY TO THE GENERA

1. Adult frond an expanded membrane...22. **Prasiola** (p. 275)
1. Adult frond a solid filament...23. **Gayella** (p. 279)

22. **Prasiola** Menegh.

Frond membranaceous, monostromatic, attached by short filiform prolongations, by the edge of the membrane, or by a thickened stipe; cells with stellate, axile chromatophore and one pyrenoid, dividing to form groups of fours, these groups forming similar larger groups, the spaces between the groups of various orders constituting narrower or wider spaces, running in definite directions through the frond; asexual reproduction, (1) by the breaking off of smaller portions of the frond which attach themselves and grow independently; (2) by akinetes, formed from individual cells assuming thick walls, and developing either directly into a filament or a membrane, or indirectly by aplanospores, several in each akinete; (3) by aplanospores, formed 4–512 in a cell, by walls in 2 or 3 directions; sexual reproduction and zoospores unknown.

Meneghini, Cenni Organo. e Fisiol., 1838, p. 36. Tribe 4, *Ulvae* (*Prasiola*) Agardh, Sp. Alg., 1822, p. 416.

The genus *Prasiola* was first described as a tribe of the genus *Ulva* by C. A. Agardh (*loc. cit.*) and later raised by Meneghini (*loc. cit.*) to independence. There are three monographs of *Prasiola*, viz., by Jessen (1848), by Lagerstedt (1869) and by Imhäuser (1889). De-Toni (1899, p. 140 *et seq.*) enumerates twelve species, of which three are reckoned doubtful. Three of the remaining nine species are marine and the rest terrestrial or inhabitants of fresh waters. Several additional species have been described, one of which is a marine species from our own coast.

We find four species along our coast which have claims to be considered marine and there are probably as many more terrestrial or fresh-water species known from western North America. Much more study of seasonal forms and developmental stages is necessary, and the number of aplanospores produced by the aplanosporangia needs more careful determination. The formation of the "aplanospores" resembles very closely the formation of the male cells or sperms in *Porphyra* and may, according as the mother cell separates completely or incompletely into two or four at the first two divisions, be a certain number or four times that number as Hus has shown takes place in antheridial formation in *Porphyra* (cf. Hus, 1902, p. 190). Complete aplanospore formation has been followed in only one species, viz., *Prasiola delicata*.

KEY TO THE SPECIES

1. **Prasiola calophylla** (Carmich.) Menegh.

Fronds linear to narrowly cuneate, with truncate apex, many from the same holdfast, seldom over 1 cm. long, 1 mm. wide; cells near the base in a single series, about 10μ long by 3–5μ broad; farther up in two rows, the number increasing towards the upper part of the frond or as the frond grows older; the series of cells and the intercellular lines nearly parallel throughout; cells near the apex of the frond about 3–5μ diam., square; thickness of frond about 15μ; cells 8–10μ high in cross section.

Growing in brackish water. Washington (Whidbey Island).

Meneghini, Cenni sull' Organografia, 1836, p. 36; Setchell and Gardner, Alg. N.W. Amer., 1903, p. 215; Collins, Green Alg. N. A., 1909, p. 219. *Bangia calophylla* Carmichael, *in* Greville, Scottish Crypt. Flora, vol. 4, 1826, p. 220.

This species has been observed only once in anything like a marine locality. One of us (Gardner) found it at the head of Penns Cove on Whidbey Island, growing within reach of pure salt water. It is a small, slender species and is easily recognized by having a long tapering stipe ending below in a single row of cells. It is probably not to be expected at all frequently in marine localities.

2. **Prasiola borealis** Reed

Plate 10, figs. 1–3

Fronds cuneate to obovate, stipitate or sessile, margin crenulate, crisped or entire, soft membranaceous, 33–45μ thick, 5–10 mm. high, in tufts of several from one holdfast; cells in distinct tetrads, areolar arrangement manifest; cells 4–9μ diam., seen superficially; in cross section oblong or palisade-form, 11–14μ high; aplanospores probably numerous within each aplanosporangium, in groups of 8.

Growing on rocks just above high water mark. Alaska.

Reed, Two new ascomycetous Fungi, 1902, p. 160, pl. 15, f. 7; Collins, Green Alg. N. A., 1909, p. 220; Setchell and Gardner, Alg. N.W. Amer., 1903, p. 215; Trelease, The Fungi of Alaska, 1904, p. 34, pl. 7, f. 1, 5.

Prasiola borealis has been found only on the Alaskan coast (Unalaska, Kadiak and Baranof Island) and always infested with a fungus (*Guignardia alaskana* Reed), thus resembling the *Mastodia tessellata* Hook. and Harv. (cf. J. D. Hooker, 1847, p. 499, pl. 194, II) whose algal portion was later referred to *Prasiola tessellata* by Kuetz-ing (1849, p. 473). *P. borealis* differs from *P. tessellata* at least in habit and in being less regularly areolate. *Prasiola borealis* also resembles *P. furfuracea* (Mert.) Menegh. except in that its cells are larger, the membrane thicker, and the areolae and intercellular lines more distinct (cf. also Reed, *loc. cit.*, p. 156).

A similar composite to those of *Prasiola borealis* Reed with *Guignardia alaskana* Reed and *Prasiola tessellata* Kuetz. with *Guignardia Prasiolae* (Winter) Reed is the *Dermatomeris georgica* Reinsch (1890, p. 425, pl. 19, f. 1–6), which has not been, so far as we are aware, examined carefully with regard to its two components.

3. **Prasiola delicata** S. and G.
Plate 17, fig. 3; plate 19, fig. 8; plate 20, fig. 1

Frond 1–1.5 mm. high, broad and shortly stipitate, expanding directly and abruptly to broadly oblong or cordate, margins crisped and inrolled, dark bluish green; membrane 17–20μ thick, cells not grouped into distinct areolae and not separated by intercellular lines; akinetes not seen; aplanospores up to 512 from a single cell ($8 \times 8 \times 8$) but often only 128; cells palisade-like and 10–12μ in vertical diameter, in section twice or more as high as broad.

Growing at or near the upper tide mark on rocky islets. Sitka, Alaska.

Setchell and Gardner, Phyc. Cont. I, 1920, p. 291, pl. 22, fig 3; pl. 24, fig. 8; pl. 25, fig. 1.

Prasiola delicata has a decidedly thinner membrane than any other of our marine species, shows little areolation, and has a large number of aplanospores formed within a single aplanosporangium. These characters seem to indicate its just claim to be considered a distinct species. It is known to us, as yet, from a single collection, although in considerable quantity.

4. **Prasiola meridionalis** S. and G.
Plate 20, fig. 2

Frond up to 7 mm. high, with short and broad stipe, soon expanded into a broad, cordate, rosulate or cucullate blade, dirty green; cells neither arranged in distinct areolae nor separated by intercellular lines; membrane 40–45μ thick, section showing cells 14–18μ high and 7–8μ wide with broad, hyaline margins (up to 13μ thick); akinetes scattered, large thick walled; aplanospores probably 128–512 ($4 \times 4 \times 8$ or $8 \times 8 \times 8$) from a single aplanosporangium.

On exposed rocks or rocky islets above the high water mark but exposed to the force of the waves. Washington (Friday Harbor and Neah Bay) to central California (entrance to Tomales Bay).

Setchell and Gardner, Phyc. Cont. I, 1920, p. 291, pl. 25, fig. 2.

The specimens taken as the type of this species were collected by one of us (Gardner, no. 3824) at Neah Bay, Washington. We are also inclined to refer here specimens collected on ''Minnesota Reef'' at Friday Harbor, Washington, and at the entrance to Tomales Bay, California. In both of the last two localities, the species is associated with *Gayella constricta*. Cultures, however, definitely indicate the independence of the two plants of one another.

Prasiola meridionalis comes near to *P. borealis* Reed, but the frond of the latter is areolate and with more or less distinct intercellular lines. *P. borealis,* so far as found, is infested with a fungus (*Guignardia alaskana* Reed) while none of the three collections of *P. meridionalis* shows any trace of such a parasite.

23. Gayella Rosenv.

Frond filiform, simple or very slightly branched, at first of a single series of cells, later dividing longitudinally into many series, but always remaining filiform, not flat; cell structure as in *Prasiola*.

Rosenvinge, Groenl. Havalg., 1893, p. 936.

Gayella may be a genus of doubtful autonomy, but the cylindrical rather than flattened character of its fronds seems to mark it off distinctly from *Prasiola*. The behavior of *Gayella constricta* in cultures, in retaining its characters and never showing a *Prasiola*-stage, induces us to retain the genus. Our experience of both species of our coast does not confirm that of Börgesen (1902, p. 482 *et seq.*) and we, therefore, feel unwilling to consider *G. polyrhiza* as a subspecies of *Prasiola crispa*. Certainly our *G. constricta* shows no distinctly transitional forms to the *Prasiola* found associated with it.

KEY TO THE SPECIES

1. Up to 175μ diam. above, frequently constricted................2. **G. constricta** (p. 280)
1. Up to 70μ diam. above, without constrictions...................1. **G. polyrhiza** (p. 280)

1. Gayella polyrhiza Rosenv.

Frond at first a simple filament of a single series of disk-shaped cells, 10–12µ diam., attached to the substratum by a rhizoidal projection from the lower cell; later attached at various parts of the filament by rhizoidal growths, one or two from a cell; increasing in diameter by growth and division of cells, up to 70µ diam.; terete or somewhat irregular in surface, but not flattened; cells with parietal chromatophore and one pyrenoid; in the mature plant showing superficially an arrangement in longitudinal and transverse lines; in cross sections an arrangement by 2–4–8–16, etc., in somewhat *Gloeocapsa*-like form; asexual reproduction by aplanospores, arranged in longitudinal and horizontal series.

Known only from the west shores of Amaknak Island, Alaska.

Rosenvinge, Groenl. Havalg., 1893, p. 937, f. 45, 46; Collins, Green Alg. N. A., 1909, p. 221; Setchell and Gardner, Alg. N.W. Amer., 1903, p. 217; Collins, Holden and Setchell, Phyc. Bor.-Amer. (Exsicc.), no. 914.

The specimens distributed under no. 914 of the Phycotheca Boreali-Americana agree so closely in every way with the description and figures of Rosenvinge that we feel justified in referring them to *Gayella polyrhiza*. They were not associated with anything resembling a true *Prasiola*.

2. Gayella constricta S. and G.

Plate 12, figs. 5–10

Filaments small, dark green, somewhat tufted, 1–4 mm. high, 18–20µ diam. at the base, up to 175µ at the apex, cylindrical-clavate, uncinate, constricted at frequent intervals, sparingly branched at the base; cells disk-shaped, 10–15µ diam., at first in a single series throughout, remaining so for some distance at the base, but dividing into groups longitudinally in two or more planes above, increasing the diameter of the filament and preserving its cylindrical form in general through gradually becoming larger, the terminal group of cells having the greatest diameter; at frequent intervals groups of 2–6 cells remain undivided vertically, giving a constricted appearance to the mature plants; cell walls hyaline, homogeneous; cross-walls very thin; chromatophore single, occupying nearly the entire cell; pyrenoid obscure.

Growing in depressions and crevices in rock, above high-tide level, kept moist by dashing salt spray. Discovered at Tomales Point, Marin

County, California. Since its first publication it has been observed in a similar habitat near Friday Harbor, San Juan County, Washington.

Setchell and Gardner, *in* Gardner, New Pac. Coast Mar. Alg. I, 1917, pp. 384, 385, pl. 33, f. 5–9, and pl. 32, f. 5.

Gayella constricta differs from *G. polyrhiza* in having fewer rhizoids and these usually much longer and multicellular. It differs also in having deep constrictions in the mature filaments caused by the failure of certain cells to divide vertically. It differs finally in the much greater diameter of the upper portions of the filaments and in their uncinate tips.

Gayella constricta has been observed in cultures in the laboratory for over eighteen months as well as the *Prasiola* (*P. meridionalis*) found growing with it. In pure cultures no transformation from one to the other was observed. New plants of *Gayella* which were constantly arising retained the *Gayella* form. No plants of *Gayella* arose in pure cultures of the *Prasiola*. It seems to us that these cultures justify keeping the two genera separate as well as establishing the independence of both *Gayella constricta* and *Prasiola meridionalis*.

Order 6. ULOTRICHALES Blackman and Tansley

Frond of branched or unbranched filaments, typically of a single series of cells; cells uninucleate with one or few parietal chromatophores with or without one or more pyrenoids; multiplication by fragmentation, akinetes or aplanospores; reproduction by zoospores and by isogamous or oogamous gametes.

Blackman and Tansley, Class. Green Algae, 1902, p. 137; West, Algae, vol. 1, 1916, p. 281. *Chaetophorales* Wille, *in* Engler and Prantl, Natürl. Pflanzenfam., Nachtr. zum I Theil, 2 Abt., 1909, p. 3.

The Ulotrichales form a fairly compact group of families having the frond branched or simple, and composed of uninucleate cells. The cells are usually in a single series. There are a few exceptional genera, but these usually occur among the fresh water species which constitute the greater portion of the group.

Key to the Families

Filament simple, or very rarely branched, fixed to the substratum by a usually specialized basal cell; cells in a single series, or double by concrescence, uninucleate, with a single complete or broken annular chromatophore having one to several pyrenoids; multiplication by fragmentation, akinetes, or aplanospores; zoospores 2- 4-ciliated; isogametes 2-ciliated.

Borzi, Studi Algologici, 1883, p. 25 (*"Ulothriciaceae"* sic!) (in part); Blackman and Tansley, Class. Green Algae, 1902, p. 137. *Ulothricheae* Kuetzing, Phyc. Gen., 1843, p. 251.

The Ulotrichaceae was first designated as a family by Kuetzing (*loc. cit.*). The name in its present form was given by Borzi (*loc. cit.*) although the orthography was incorrect. The contents of Borzi's family was greater than used later by Blackman and Tansley (*loc. cit.*) whose idea we follow. The majority of the Ulotrichales are branched, but there are some with simple filaments and these are all contained in the Ulotrichaceae. Very few of the members of this family show any branching at all, and when they do it is slight compared with that of the members of the Chaetophoraceae or of the Trentepohliaceae. The great majority of the species of the Ulotrichaceae are inhabitants of fresh waters but a few species of *Ulothrix* are marine.

24. **Ulothrix** Kuetz.

Filaments simple or rarely branched, of a single series of uninucleate cells, all similar, and, with the exception of the attached basal cell, capable of division and of producing spores; chromatophore band-shaped, with one or more pyrenoids; asexual reproduction by aplanospores and akinetes, also by 4-ciliated zoospores, with red stigma, formed 1–4 in a cell, germinating immediately; sexual reproduction by 2-ciliated gametes formed 8 or more in a cell, germinating after conjugation; external conditions may induce many modifications of the normal process; akinetes may be formed, ultimately producing zoospores; filaments may break up into individual cells, and these by copious formation of gelatine pass into a *Palmella* or a *Gloeocystis* condition.

Kuetzing, Algolog. Mitth., 1833, p. 517.

The genus *Ulothrix* contains both fresh water and marine species and is to be distinguished from the genera *Hormiscia* and *Chaetomorpha* of the Cladophoraceae by its uninucleate cells. The pyrenoids,

also are less numerous in each chromatophore in *Ulothrix*, the majority
of species containing only one.

The species of *Ulothrix* still need careful study especially in cultures. These cultures, however, are not easy to carry on. The filaments vary considerably in diameter in different stages of growth,
often increasing very considerably in diameter as they pass over into
reproductive condition. The chromatophores of the different species
seem distinctly different in the earlier vegetative conditions but lose
their character as the cells pass on towards reproductive stages. Wille
(1901 and 1906) has made the more recent and more considerable
studies of the marine species and has brought out many new points
of view, showing how necessary it is to have plants for study in practically all stages of development. Dried plants are often very unsatisfactory in that chromatophore structure is generally difficult of exact
determination. Specimens preserved in liquid are much more favorable for investigation.

KEY TO THE SPECIES

1. Filaments always free, no branches.. 2
1. Filaments often grown together, occasional branches........4. **U. laetevirens** (p. 286)
  2. Cells (especially fertile) much shorter than broad...................................... 3
  2. Cells (including fertile) usually as long as, or longer than broad...................
    ....1. **U. implexa** (p. 283)
3. Chromatophore a complete ring, fertile filaments broad (up to 60μ or even 80μ
    diam.)..2. **U. flacca** (p. 284)
3. Chromatophore an incomplete ring, fertile filaments much narrower (not over
    38–40μ diam.)..3. **U. pseudoflacca** (p. 285)

1. Ulothrix implexa Kuetz.

Plants light green, forming soft masses, cells 6–15μ diam., sometimes slightly swollen at the middle, nearly quadrate, chromatophore,
when young, occupying only the middle part of the cell, often an
incomplete ring; fertile cells nearly quadrate, not swollen or increased
in width.

Growing on rocks near the mouths of streams, and on wood in
quiet water, in the littoral belt. Alaska (St. Michael) to California
(San Francisco).

Kuetzing, Sp. Alg., 1849, p. 349; Collins, Green Alg. N. A., 1909,
p. 185; Setchell and Gardner, Alg. N.W. Amer., 1903, p. 217. *Ulothrix
subflaccida* Wille, Stud. ueb. Chloroph., I–VII, 1901, p. 27, pl. 3,
f. 90–100 (?) ; Collins, Green Alg. N. A., 1909, p. 186; Collins, Holden
and Setchell, Phyc. Bor.-Amer. (Exsicc.), no. 1275.

Wille (1901, p. 22) has raised the question as to the exact nature of the type of this species and as to the status of other plants referred to it. Hazen (1902, p. 155) thinks, however, that Wille's doubts are not founded on sufficient basis for rejecting the name and that the type of *U. implexa* may be reckoned among the marine species.

We have followed the usual fashion of referring here our most slender species whose cells are usually as long as, or often longer than, broad. The chromatophore in young cells in active vegetative condition forms a more or less complete band about the middle of the outer cell wall. Many of our specimens, however, seem to have a complete chromatophore clothing the entire outer wall. We are inclined to consider the cells of such specimens as probably passing over into the fertile condition, but not, as yet, having undergone division to form zoospores or gametes. The fertile cells in these species are neither enlarged (i.e., broadened) nor swollen. A careful study of living material in various stages will be very helpful in clearing up these matters.

We follow Hazen in placing the *Ulothrix subflaccida* Wille (*loc. cit.*) as a synonym under *U. implexa*.

2. **Ulothrix flacca** (Dillw.) Thur.

Plants forming bright or dark green, often much entangled, masses or skeins; cells 10–25μ diam., 0.25–0.75 as long as broad, when producing spores up to 50μ diam., and swollen in the middle; chromatophore occupying the whole of the cell wall with 1 to 3, occasionally more, pyrenoids.

Growing on other algae, on rocks and on wood, in the littoral belt. From Alaska to California.

Thuret, *in* Le Jolis, Liste Alg. Mar., 1863, p. 56; Collins, Green Alg. N. A., 1909, p. 185; Saunders, Alg. Harriman Exp., 1901, p. 412; Setchell and Gardner, Alg. N.W. Amer., 1903, p. 217; Hazen, Uloth. and Chaetoph. U. S., 1902, pl. 20, f. 7–9. *Conferva flacca* Dillw., Brit. Conf., 1809, pl. 49.

We can not feel certain whether the *Ulothrix flacca*, as it has been finally limited by Wille (1901, p. 18 *et seq.*) is the *Conferva flacca* of Dillwyn (*loc. cit.*) or not. There is a certain strong suggestion to our minds of *Ulothrix pseudoflacca* Wille in both the figures and the description of Dillwyn. It seems best, however, to follow the present conception of the species and assign under *U. flacca* marine *Ulotriches* which, while comparatively slender (10–25μ diam.) in the vegetative condition, thicken very considerably (up to 60μ or even 80μ) when producing zoospores or gametes. The cells when younger may be almost or quite quadrate, but become in the fertile condition very much shorter than the diameter.

3. Ulothrix pseudoflacca Wille

Filaments 8–32μ diam., free from one another, attached by an elongated, downwardly gradually tapering cell; cells from 0.25 as long as broad to nearly quadrate; cell walls thin; chromatophore parietal, completely covering the outer cell wall, thickened in the region of the single pyrenoid, fertile cells not exceeding the vegetative in diameter, from flattened to nearly globular.

On rocks and algae, upper littoral belt. Alaska (Sitka) to California (San Francisco).

Wille, Stud. ueb. Chloroph., 1901, p. 22, pl. 2, f. 64–81.

Ulothrix pseudoflacca varies as follows: f. *minor* Wille (*loc. cit.*, p. 23, pl. 2, f. 67–69), 8–16μ diam.; f. *major* Wille (*loc. cit.*, p. 23), 10–22μ diam.; f. *maxima* Setchell and Gardner (*in* Gardner, 1919, p. 488, pl. 42, f. 6), 28–32μ, up to 40μ diam. Plate 9, fig. 6 A, B.

Of these f. *minor* has been collected at Sitka, Alaska, f. *major* and f. *maxima* at Lands End, San Francisco, California (all by Gardner).

Ulothrix pseudoflacca bears a considerable resemblance to *U. flacca* except that in its fertile condition it does not reach so considerable a diameter and the fertile cells are usually more or less rounded. The chromatophore in *U. pseudoflacca* is a broken ring while that of *U. flacca* is a complete ring.

Ulothrix pseudoflacca varies much in diameter and, as shown above, may be separated more or less readily into three overlapping forms, the f. *minor* approximating forms of *U. implexa* in slenderness, while f. *maxima* approaches forms of *U. flacca*. As filaments of *U. pseudoflacca* approach the fertile condition, it is difficult to detect the gap in the chromatophore.

Jónsson (1904, pp. 55–57) has reviewed this species as well as others of the genus and has made valuable suggestions. He has also described a related new species (*loc. cit.*, pp. 57–60, f. 8, 9), *Ulothrix scutata*, which has not been thus far detected among the specimens available from our territory.

4. **Ulothrix laetevirens** (Kuetz.) Collins

Filaments 10–25μ diam., two or three often firmly grown together laterally, more or less entangled and creeping; with not infrequent branches, issuing at a wide angle, and usually much more slender than the main filament, of many cells, which are generally 1–3 diam. long; filaments tapering towards the base, the lower cells of the densely packed filaments often subparenchymatously united; cells 0.25–0.75 diam. long, rarely more; chromatophore covering nearly or quite all of the cell wall, but thicker at one side, where the pyrenoid is situated; zoospores usually 8 in a cell; akinetes formed singly from the cells.

On woodwork between tides. Alaska (Unalaska and Sitka) to California (fide Collins).

Collins, Green Alg. N. A., 1909, p. 186. *Schizogonium laetevirens* Kuetzing, Phyc. Germ., 1845, p. 194. *Ulothrix consociata* Wille, Stud. ueb. Chloroph., I–VII, 1901, p. 25, pl. 2, f. 82–89, Algol. Unters., I–VII, 1906, p. 12, pl. 1, f. 30, 31 (*fide* Collins).

Collins (1909, p. 186) considers that *Schizogonium laetevirens* Kuetz. and *Ulothrix consociata* Wille are identical. He formed his opinion after examining a topotype of Kuetzing's species. We have adopted this opinion, but have formed our idea of the species upon the description and figures of Wille. Our specimens show fairly frequent branches and filaments laterally coalescent. In dimensions, both of vegetative and fertile segments and cells, our specimens agree closely with the plant described and figures by Wille. The plant has a considerable likeness to *U. pseudoflacca* Wille, but it is to be distinguished by its tendency to branch and to have coalescent filaments.

FAMILY 12. CHAETOPHORACEAE DE-TONI AND LEVI

Frond of more or less branched filaments, erect or prostrate, at times enclosed in a more or less gelatinous envelope; cells uninucleate, green, without haematochrome, the terminal often acute or forming a colorless hair; chromatophore parietal, band-shaped, at times annular, containing one or more pyrenoids; propagation vegetative, by akinetes or by aplanospores; zoospores of two sorts, macrozoospores and microzoospores, 2- 4-ciliated, usually produced from modified cells; 2-ciliated isogametes and heterogametes known in some genera; setae of various kinds present in some genera.

De-Toni and Levi-Morenos, Fl. Alg. Venez., III, 1888, p. 171
(Repr.); Blackman and Tansley, Class. Green Alg., 1902, p. 138.
Chaetophoroideae Harvey, Man. Brit. Alg., 1841, p. 10 (in part).

The Chaetophoraceae form a family of both fresh-water and marine
species. The latter are nearly all prostrate epiphytes or endophytes,
living upon the larger Chlorophyceae, Melanophyceae and Rhodophyceae. The species of one genus on our coast bores into shells.
There are doubtless a number of genera and species of this family on
our coast still awaiting discovery.

The Chaetophoraceae are generally divided into five tribes, four
of which are represented on the Pacific Coast. *Bulbocoleon* represents
the Chaetophoreae, *Entocladia* belongs to the Leptosireae, *Ulvella*
Pseudulvella and *Pseudopringsheimia* belong to the Ulvelleae, *Gomontia* belongs to the Gomontieae, while *Internoretia, Endophyton* and
Pseudodictyon are not definitely placed as yet owing to a lack of
knowledge of the details of the processes of reproduction.

25. **Bulbocoleon** Pringsheim

Thallus minute, epi- or endophytic in various lubricous or
gelatinous algae of loose tissues; filaments creeping, branching, of
irregularly shaped cells, rounded or somewhat elongated horizontally;
bearing on the upper sides of the filaments single or clustered rounded
cells prolonged into long, hyaline, unseptate hairs; chromatophore of

the non-piliferous cells; plate-like, perforate, with 5–10 pyrenoids, that of the piliferous cells, irregular and toothed, with 2 pyrenoids; zoospores (?) 2-ciliated, produced from non-piliferous cells somewhat enlarged on the upper side.

Pringsheim, Beitr. z. Morph d. Meeresalg., 1862, p. 1; Collins, Green Alg. N. A., 1909, p. 283.

The species of *Bulbocoleon* are those members of the Chaeto-phoraceae whose cells are in branching filaments which bear groups of specialized cells produced into continuous (i.e., non-septate) hairs. They are generally endophytic in gelatinous brown and red algae and have 2-ciliated zoospores. Little is known of this genus on our coast and it seems more than likely that related genera provided with hairs may also be found. Careful search and examination should be made for *Bulbocoleon* and other piliferous genera of the family Chaeto-phoraceae in and on the gelatinous or lubricous species of brown and red algae of our coast.

Bulbocoleon piliferem Pringsheim

Non-piliferous cells 12–16μ diam., 2–3 times as long as broad.

On *Cumagloia Andersonii* (Farlow) S. and G. Southern Califor-nia (San Pedro).

Pringsheim, Beitr. z. Morph. d. Meeresalg., 1862, p. 8, pl. 1; Hazen, Ul. and Chaet. U. S., 1902, p. 227; Collins, Green Alg. N. A., 1909, p. 283.

Thus far, *Bulbocoleon piliferum* has been detected only in the fronds of *Cumagloia Andersonii* in southern California (San Pedro, Miss S. P. Monks, Mrs. H. D. Johnston). It is likely to occur along our coast northward to the region of Puget Sound at least, and also on other species of algae such as those of *Mesogloia, Leathesia, Scyto-siphon, Chorda, Ralfsia*, etc. It occurs at times so abundantly as to discolor the host, but often is to be found in small quantity, giving no outward (or macroscopic) indication of its presence.

26. Entocladia Reinke

Plants microscopic, composed of creeping, irregular, much branched filaments, without hairs, growing on or within aquatic plants; growth mostly by division of terminal cells; chromatophore a parietal band with one or more pyrenoids; reproduction by 2-4-ciliated zoospores.

Reinke, Zwei par. Algen, 1879, p. 476. *Endoderma* Lagerheim, Bidr. Sver. Alg., 1883, p. 74. *Entoderma* Wille, Chlorophyceae, *in* Engler and Prantl, Natürl. Pflanzenfam., 1890, p. 94.

It seems desirable to restore the earlier name of *Entocladia*, since it differs in one letter from the earlier *Endocladia* J. Ag. (1841) and belongs to an entirely different class of algae. There seems, therefore, to be little likelihood of any serious confusion. The genus is now credited with several species, some of which occasionally possess hairs. It seems best to us to restrict the generic name to species without hairs. One characteristic of the genus is supposed to be the habitat, viz., growing within the cell membranes (or cuticula?) of various green, brown and red algae. It does not seem to us that this manner of growth ought to receive too great emphasis in determining the generic limits. Removal of this, as a criterion of generic distinction, would probably result in the combination of the genus *Epicladia* Reinke with the genus *Entocladia* Reinke. Since we have not as yet detected a species of *Epicladia* on the Pacific Coast, we may leave discussion of this point to others, but we may instance the genus *Coleochaete* as possessing a species endophytic in the membranes of the segments of *Nitella* while most of the rest of the species are epiphytic.

KEY TO THE SPECIES

1. Filaments scarcely if at all coalescing..1. **E. viridis** (p. 289)
1. Filaments coalescing, at least at the center... 2
    2. Free filaments numerous and long..............................2. **E. codicola** (p. 290)
    2. Free filaments few, short...3. **E. cingens** (p. 291)

1. **Entocladia viridis** Reinke

Filaments branching freely, 3–8μ diam., cells 1–6 diam. long, cylindrical, or more often irregularly swollen and contorted; chromatophore nearly covering the cell wall and containing a single pyrenoid.

Growing on *Callithamnion Pikeanum*. Central California (Moss Beach, San Mateo Co.).

Reinke, Zwei par. Algen, 1879, p. 476, pl. 6, f. 6–9. *Endoderma viride* Lagerheim, Bidr. till Sver. Alg., 1883, p. 74; Collins, Green Alg. N. A., 1909, p. 279; Collins, Holden and Stechell, Phyc. Bor.-Amer. (Exsicc.), no. 2236.

We have not been able to make a satisfactorily extensive study of the *Entocladia* species apparently not uncommon in the cell and other membranes of our various marine algae. Consequently we have neither determined, to our satisfaction, whether true *E. viridis* occurs on our coast nor the number of species of its general type ultimately to be found in our territory. We have, therefore, merely recorded the specimens already distributed in the Phycotheca Boreali-Americana (no. 2236).

2. **Entocladia codicola** S. and G.

Plate 19, fig. 7

Filaments light green, branching profusely, at maturity forming a continuous layer in the center of the mass with tapering free ends around the margin; young cells 3–4µ diam., 1–2.5 times as long, terminal cells slender and conical; cells in the center of the thallus 5–8µ diam.; pyrenoids single; reproduction unknown.

Growing in the membrane, at the tips of the utricles of *Codium fragile.* Central and southern California.

Setchell and Gardner, Phyc. Cont. I, 1920, p. 293, pl. 24, fig. 7 a, b.

Entocladia codicola seems closely related to *Entocladia viridis* Reinke (1879, p. 476, pl. 6, f. 6–9), found growing in the membrane of *Derbesia;* but it is a larger plant with the filaments much more compact in the center, forming, in fact, a pseudo-parenchymatous disk with free filaments around the margin. The cells are shorter than those of *E. viridis,* some being even shorter than the diameter. In the pseudo-parenchymatous character of the center of the disklike frond it resembles *Epicladia Flustrae* Reinke (1888, p. 241, nomen nudum, 1889, p. 31, pl. 24, 1889a, p. 86), but the dimensions given for that species are greater in general than those in ours. Reproductive bodies have been observed in the cells of the central portion of the disk in *E. codicola,* but the nature of these, their method of escape, and their subsequent behavior have not been determined. Until more is known concerning these later phases of the plant, its proper placing must remain somewhat in doubt. It is here placed provisionally with *Entocladia* on account of its endophytic habit of growth, rather than with *Epicladia,* which has the habit of growing on the outside of the host. This habit of growth seems to be the only one by which the two genera are distinguished, so far as the diagnoses reveal. Little, however, is known concerning the reproduction in *Epicladia,* and until that matter can be cleared up it can have but little claim to generic distinction. Reinke expressed doubt as to the validity of the genus when he diagnosed it (1889). Collins (1909) has retained both genera, and under *Endoderma* (*Entocladia*) has included two species, viz. *E. Pithophorae* West and *E. polymorphum* West, which are epiphytic, and thus, as he remarks (*loc. cit.*, p. 280), "connects *Endoderma* with *Epicladia,* but the filaments do not unite to form a definite disk."

Entocladia codicola seems to be confined to the coast of California and to the above mentioned host plant, at least examination of considerable material of different species of *Codium* in different localities, ranging from Sitka, Alaska, to southern California, has not revealed its presence elsewhere.

3. Entocladia cingens S. and G.

Plate 18, fig. 7

Thallus early forming a pseudo-parenchymatous tissue surrounding the filaments of the host within the membrane, having a few marginal filaments extending parallel with the long diameter of the host; cells in the center of the thallus nearly isodiametric, 5–8μ diam., enlarging later to form sporangia; cells of the free marginal filaments 3–4μ diam., 2–3 times as long as the diameter, terminal cells long, conical.

Growing within the membrane of *Chaetomorpha californica* Wille. Southern California (Ocean Beach, near San Diego).

Setchell and Gardner, Phyc. Cont. I, 1920, p. 292, pl. 23, fig. 7.

The plants of this species seem to be nearing maturity in December, since a few empty cells in the center of the thallus were found from which reproductive bodies probably had escaped. Aside from this condition, nothing further is known of its method of reproduction.

Entocladia cingens is placed in this genus on account of the resemblance of the vegetative development to that of the type species, *E. viridis* Reinke, and because of the same endophytic habit as that species. It differs from *E. viridis* in having the branching filaments more closely coalescent, the enlarging cells in the main part of the thallus soon forming a pseudo-parenchymatous tissue, leaving only a few free marginal filaments.

E. viridis, E. codicola, and *E. cingens* form a well connected series, using the vegetative characters as a basis. The first named species has a rather wide-spreading thallus, composed of relatively sparsely branching filaments, scarcely, if at all, coalescing in the center. In the second, the filaments coalesce freely in the center so that at least

half of the thallus is formed into a pseudo-parenchymatous tissue at
the time of reproduction, but leaving an abundance of free branching
marginal filaments.   The thallus of the third is almost wholly transformed into a pseudo-parenchymatous tissue at maturity, leaving only
a few free marginal filaments.

27. **Endophyton** Gardner

Filaments sparingly and irregularly branched within the medulla
of the host, but branching more freely near the surface, sending off
in addition to horizontal branches many short, erect branches perpendicular to the frond, the end cells of the perpendicular branches
becoming the sporangia and growing out to the surface of the host;
chromatophore band-shaped; each cell containing a single pyrenoid;
zoospores pyriform, 2-cyliated.

Endophytic within the fronds of various species of red algae.

Gardner, New Chlorophyceae, 1909, p. 371; West, Algae I, 1916,
p. 304.

This is a form genus, whose cytology is at present too little known.
In its habit, it is more thoroughly endophytic than any other genus
of the Chaetophoraceae, since its filaments traverse the medullary as
well as the cortical tissues of its hosts.   The absence of hairs and
specialized zoosporangia seem to mark it as distinct.

Endophyton ramosum Gardner
Plate 11, figs. 3, 4

Filaments 4–6μ diam., tortuous, often very irregular in shape;
cells 6–8 times longer than broad, cross-walls distinct; sporangia club-
shaped, tapering to a point at the outer end, when young 10–12μ
diam.; zoospores numerous, 3μ diam., escaping through openings in
the upper ends of the sporangia which grow to the surface of the host;
plants occupy small areas a few millimeters in diameter, usually near
the base of the frond of the host, but may spread promiscuously over
the entire area of the host.

Endophytic within the fronds of various species of red algae, e.g.,
Iridaea laminarioides, Gigartina exasperata, etc.   Central California
(San Francisco) and probably farther north.

Gardner, New Chlorophyceae, 1909, p. 372, pl. 14, f. 3, 4; Collins,
Green Alg. N. A., 1909, p. 282; Collins, Holden and Setchell, Phyc.
Bor.-Amer. (Exsicc.), no. 1627.

The presence of this endophyte in the fronds of species of *Iridaea*, *Gigartina* and possibly also of *Nitophyllum* is indicated externally by spots of a much redder color than that of the uninfested portions of the host. Its detection at other points than about San Francisco will probably follow careful search for it on similar hosts.

28. **Pseudodictyon** Gardner

Thallus much branched, the main filaments comparatively long and tortuous, creeping among the cells of the cortical layer of the host plant (*Laminaria* sp.), branching freely; the branches, usually arising at right angles to the main filaments and bending backward among the cortical cells, unite into a sort of net; a very short erect branch composed of two or three cells arises from practically every cell of the horizontal filaments in the central portion of the thallus; the uppermost cell of each erect branch becomes a sporangium and grows to the surface of the host; each cell contains a single peripheral chromatophore with one pyrenoid; reproduction unknown.

Gardner, New Chlorophyceae, 1909, p. 374; West, Algae I, 1916, p. 304.

Since neither the cytology nor the reproduction of the type and only species thus far referred to this genus, has been studied, it exists merely as a form genus of probably close relationship to *Entocladia*. It is to be hoped that some one having favorable opportunity may add much that is desirable to our knowledge of this seemingly very distinct endophytic genus.

Pseudodictyon geniculatum Gardner

Plate 11, figs. 5, 6

The young, creeping filaments 3–4μ diam., becoming larger with age; tips of the geniculate filaments seem to coalesce with neighboring cells, giving the young plant the appearance of a fine net, the meshes enclosing 4–8 cells of the host plant; sporangia 8–12μ diam.; cell walls thin and cross walls distinct.

Growing in abundance in the terminal parts of the blade of *Laminaria Sinclairii*. San Francisco Bay, California.

Gardner, New Chlorophyceae, 1909, p. 374, pl. 14, f. 5, 6; Collins, Green Alg. N. A., 1909, p. 283; Collins, Holden and Setchell, Phyc. Bor.-Amer. (Exsicc.), no. 1628.

The host plant of ''a'' of the above distribution is *Laminaria Sin-clairii;* and the host of ''b'' of the same distribution is *Dictyoneuron californicum.* The plants under ''b'' are not quite typical, but with our present knowledge seem too closely related to the typical form to warrant separation from it. Its life history has not been studied. The ''net'' character of the thallus is much less regular than in the typical form. This may be due to the character of the cells of the host plant.

29. **Internoretia** S. and G.

Thallus endophytic, consisting of profusely branched filaments, at first of a single series of cells increasing by apical divisions perpendicular to the long diameter, but later, by oblique and longitudinal divisions, building up cylindrical threads composed of numerous cells in cross diameter; branching at right angles, anastomosing, forming a network; chromatophore parietal, with one pyrenoid; reproduction unknown.

Setchell and Gardner, Phyc. Cont. I, 1920, p. 294.

The genus *Internoretia* was proposed for a peculiar endophyte found by Professor T. C. Frye, growing within the membranes of *Porphyra.* Its reproduction not having been determined, it is added to the numerous form genera of uncertain position and placed provisionally among the Chaetophoraceae. It resembles *Pseudodictyon* Gardner and *Zygomitus* B. and F. From the former it differs in forming solid filaments several cells in thickness. From *Zygomitus, Internoretia* differs in the greater regularity of its solid portions, and in the more uniform network brought about by the regular giving off of branches at right angles.

Internoretia Fryeana S. and G.
Plate 18, figs. 3–6

Cells of the terminal filaments 3.5–5μ diam., 3–5 times as long, apical cell conical; cells of the older part of the thallus isodiametric, angular; otherwise as the genus.

Growing within the membrane of *Porphyra Naiadum.* Friday Harbor, Washington.

Setchel land Gardner, Phyc. Cont. I, 1920, p. 295, pl. 23, figs. 3–6.

This most interesting little plant is as yet known only from the collections and observations of Professor T. C. Frye. It generally occurs in such abundance as to discolor the host plant. In some years it is very common, discoloring most of the plants of *Porphyra* in the neighborhood of the Puget Sound Marine Station, while in other years it is difficult to find any plants at all.

30. **Ulvella** Crouan

Thalli forming small disks on larger plants or other objects, firmly attached by the under surface, unistratose at first, later, at times, pluristratose, of radiating, laterally united, dichotomous filaments; segments multinucleate, with parietal chromatophore and no pyrenoid; 2-ciliated zoospores formed in the central cells, 4–8–16 in a cell, escaping by an opening at the top.

Crouan, Notice sur quelq. nouv. algues mar., 1859, p. 288, pl. 22, f. E; Wille, Nachträge, 1909, p. 89.

The genus *Ulvella* is based upon *Ulvella Lens,* found growing upon bits of porcelain and glass at Brest, France. The original specimens have been examined by Huber (1893, p. 295), who has figured the type (*loc. cit.,* pl. 11, f. 4–6) and added to our knowledge of its morphology and cell structure. It is a prostrate, unistratose or paucistratose, discoid plant composed of radiating filaments with no trace of hairs and whose terminal cells fork before dividing. The chromatophore is single and devoid of a pyrenoid. Nothing is known of the reproduction of the type of the species. It has come to be generally accepted that *Dermatophyton radians* Peter, a species inhabiting the carapaces of fresh water tortoises in southwestern Europe, is also to be referred to *Ulvella* (cf. Schmidle, 1899, Wille, 1909, p. 89) and the characters of multinucleate "cells" and 2-ciliated zoospores considered to be characteristic for the genus. Unfortunately, we do not know whether these characters hold for the type species, *Ulvella Lens.* We also unfortunately do not know whether they hold for the single species of our coast which we are inclined to refer under this genus and also to *Ulvella Lens.* It is to be hoped that further information may be obtained at some favorable opportunity.

Ulvella Lens Crouan

Plate 33

Thallus orbicular, bright green, 150–250μ, up to 1.5 mm. diam.; marginal segments usually cuneate, 3.5–4.5μ diam., 15–25μ long with terminal growth, central segments 8–15μ diam.

Growing firmly attached to *Laurencia* sp. in the upper sublittoral belt. Central California (Pacific Grove).

Crouan, *loc. cit.;* Huber, Contributions des Chaetophorées, 1892, p. 294 *et seq.*, pl. 11, f. 4–6; Collins, Green Alg. N. A., 1909, p. 286; Wille, Nachträge, 1909, p. 89.

We feel compelled to refer here a plant found in abundance on a species of *Laurencia* at Pacific Grove, California, in July. In general appearance, in structure, in dimensions of the cells and manner of peripheral growth, the specimens correspond exactly with the figures and description of Huber (*loc. cit.*) for the type material of *Ulvella Lens.* Our plants, however, have only been detected, thus far, as epiphytes. We have seen, in our preserved material, the central cells both empty and divided into 8 rounded bodies, probably zoospores. Unfortunately we have not preserved specimens in a condition suitable for the study of the number of nuclei in the cells.

31. **Pseudulvella** Wille

Thalli forming small disks, composed of radiating filaments closely placed, coalescent and finally more or less loosely pluristratose towards the center, unistratose and more or less free and dichotomous towards the periphery; cells uninucleate, with a single parietal chromatophore containing a single pyrenoid; zoospores 4-ciliated.

Wille, Nachträge, 1909, p. 90. *Ulvella* Snow, *Ulvella americana* 1899, p. 309 (in part).

The genus *Pseudulvella* is said to differ from *Ulvella* in that the cells of its species have only a single nucleus, the chromatophore has a single pyrenoid and the zoospores are 4-ciliated. It is said to differ from *Pseudopringsheimia* in the lack of rhizoid-like filaments which penetrate the host. It is placed by Wille (*loc. cit.*, p. 79) among the genera whose species lack hairs, although Snow (1899, p. 310) says that "in quite a number of cases, long gelatinous hairs extended from the surface" in the type species, *Ulvella americana* Snow. This statement as to hairs undoubtedly led Collins (1909, p. 289) to refer the *Ulvella americana* Snow to *Chaetopeltis.* Since Snow does not figure the hairs and the statement seems little convincing, we are inclined to follow Wille in presuming that no hairs are present.

Key to the Species

1. **Pseudulvella prostrata** (Gardner) S. and G.

Plate 11, figs. 1, 2

Thallus 2–3 mm. diam. consisting of 2–3 layers of cells in the center, but of a single layer at the margin, formed by branching filaments growing out radially on the host, coalescing in the central part of the thallus but free from each other around the margin, and adhering very firmly to the host; filaments branched, 6–7μ diam., nearly uniform in diameter throughout their entire length; cells quadrate in the center of the thallus, those at the outer ends of the filaments 1.5–2.5 times as long as broad; color very dark green; end cells blunt, each containing a single peripheral chromatophore and one pyrenoid.

Growing on the basal portion of *Iridaea laminarioides*. Central California (Lands End, San Francisco).

Setchell and Gardner, Phyc. Cont. I, 1920, p. 295. *Ulvella prostrata* Gardner, New Chlorophyceae, 1909, p. 373, pl. 14, f. 1, 2; Collins, Green Alg. N. A., 1909, p. 287; Collins, Holden and Setchell, Phyc. Bor.-Amer. (Exsicc.), no. 1629.

Although we know too little of the *Ulvella prostrata* Gardner both from the point of view of its cytology and of the nature of its zoospores, it cannot remain in *Ulvella* because of the presence of a pyrenoid in the chromatophore. It does not possess the penetrating rhizoidal filaments supposed to be characteristic of *Pseudopringsheimia*. It does, however, resemble *Pseudulvella americana* (Snow) Wille sufficiently closely in its general structure to be referred to the same genus for the present at least.

2. **Pseudulvella consociata** S. and G.

Plate 19, figs. 4–6

Thallus irregular in size and in outline, 100–140μ thick, increasing by irregular and obscurely radiating filaments early coalescing and becoming parenchymatous; color dark green; erect filaments firmly coalescent, 7–10μ diam., cells nearly cylindrical, slightly irregular in form, 1–2 times as long as the diameter; zoosporangia (?) terminal, pyriform to spherical, producing 8 zoospores.

Growing on the shells of *Ilyanassa obsoleta* Say. Central California (Bay Farm Island, Alameda).

Setchell and Gardner, Phyc. Cont. I, 1920, p. 296, pl. 24, fig. 4–6.

The shells of *Ilyanassa obsoleta* were introduced some years ago along with oysters from the Atlantic coast of North America, and possibly the plant here described was introduced with the host.

The comparison of this species with others will be found included in the discussion under *P. applanata.*

3. **Pseudulvella applanata** S. and G.

Thallus thin, parenchymatous, spreading by marginal growth, expanded to several mm. in diam., smooth and glossy, 45–55μ thick; color grass green; cells in fairly regular vertical rows nearly iso-diametric, sharply angled, 6–7.5μ diam.; chromatophore covering the cell wall, provided with one pyrenoid; zoosporangia (?) slightly modi-. fied surface cells; reproduction unknown.

Growing on the shells of *Littorina planaxis* Nutt.   Central California.

Setchell and Gardner, Phyc. Cont. I, 1920, p. 295.

Littorina planaxis is very abundant in small tide pools and in moist places along high-tide level from Sitka to San Diego. *Pseudulvella applanata* has been studied only on material found along the coast of central California, but it is presumed to have a much wider distribution.

Its presence on the host is readily recognizable by its expanded, grass green, glossy appearance.   Microscopically it may readily be distinguished from all other described species of the genus by its very small, closely compact, parenchymatous cells, and its seeming absence of radiating filaments composing the basal layer.   It spreads over the host by tangential and by radial divisions of the peripheral cells, at least it can thus be stated when the plant is of considerable size. It probably starts on the very young host, and doubtless many plants early coalesce to form a confluent thallus.   We have never been able to ascertain the nature of the early developmental stages, although even very small shells have been examined.

Reproductive bodies have been seen to escape from the surface cells.   Whether they are zoospores or gametes, the number of cilia they posses, and their behavior after escaping are subjects for further investigation.

The three species of *Pseudulvella* treated of in this paper may be arranged, so far as the basal layer is concerned, in the following sequence: *P. prostrata,* with distinctly radiating basal filaments which branch rather frequently and which are comparatively loosely coalescent; *P. consociata,* with indistinctly radiating basal filaments closely coalescent; and *P. applanata,* with a parenchymatous basal layer.

32. **Pseudopringsheimia** Wille

Thallus cushion-shaped, epiphytic, often penetrating the hosts at intervals, or growing upon the shells of mollusks; increase in diameter produced by terminal growth of radiating filaments branching frequently and coalescing to form a parenchymatous basal layer; increase in thickness produced by horizontal divisions of the cells of the basal layer forming a mass of erect filaments sometimes branching in turn; cells without hairs containing one chromatophore in the outer end, and one pyrenoid; zoosporangia mostly terminal on the erect filaments or occasionally subterminal as well.

Wille, Nachträge, 1909, pp. 88, 89.

The genus Pseudopringsheimia was founded on the two species, *Ulvella confluens* Rosenv. and *U. fucicola* Rosenv. The particular characteristic separating the genus, especially from *Pseudulvella* Wille is the presence of rhizoidal outgrowths from the base penetrating the host plant. In other respects the species of *Pseudopringsheimia* closely resemble those of *Pseudulvella.* It is a question as to how important such structures as penetrating rhizoids should be estimated as being in separating genera, but since Wille has separated the two genera and since we have altogether too little knowledge of their reproductive processes, it seems best to recognize both for the present.

Pseudopringsheimia apiculata S. and G.

Plate 17, figs. 1, 2

Thallus minute, 145–160μ thick, hemispherical when alone, but often with many crowded closely together forming a continuous stratum 2–3 mm. diam.; color bright green; erect filaments 8–12μ diam., composed of 9–12 cylindrical or slightly swollen cells; rhizoids aggregated into short conical fascicles; zoosporangia (?) producing 8 zoospores, terminal, slightly swollen, varying from convex to decidedly apiculate; zoospores (?) 4-ciliated.

Growing on the rhachis and the cysts of *Egregia Menziesii.* Central California.

Setchell and Garnder, Phyc. Cont. I, 1920, p. 297, pl. 22, figs. 1, 2.

Pseudopringsheimia apiculata is closely related to *P. confluens* (Rosenv.) Wille. The most conspicuous difference is to be found in the shape and size of the zoosporangia, if the terminal reproductive cells are to be designated as such. Those of *P. confluens* are long and comparatively narrow, and produce 30–40 zoospores, while in *P. apiculata* they are shorter, somewhat swollen, mostly with a pronounced terminal projection, and produce about 8 zoospores. These reproductive bodies are very small, and it is exceedingly difficult to determine the number of cilia. On one occasion four cilia were observed but the reproductive bodies seemed a little larger and somewhat more irregular in form than the average. These may have been the zygotes formed by the fusion of 2-ciliated gametes, but had not yet come to rest.

33. Gomontia Born. and Flah.

Thallus consisting of creeping, freely branched, septate filaments, from the under side of which many erect, more or less branched filaments arise; cells irregular in shape and size, uninucleate, occasionally multinucleate, with parietal, band-form or shield-shaped chromatophore covering the whole or part of the cell, or a network extending through the cell, with one or two pyrenoids; reproduction by zoosporangia, producing a few egg-shaped zoospores with 4 cilia, by large, irregularly shaped, thick walled, gametangia (?) with rhizoidal outgrowths, producing 2-ciliated gametes (?) whose conjugation is unknown, and by similar large cells producing aplanospores.

Bornet and Flahault, Note sur deux algues, 1888, p. 163 (Repr., p. 5); Sur quelq. pl. viv. dans le test calc., 1889, pp. clii–clx, pl. 6–8, 10, fig. 3.

The genus *Gomontia* is composed of species having the peculiar and presumably distinctive habit of boring into calcareous material. They are usually to be found boring into shells of mollusks, chiefly lamellibranchs, of both fresh and salt waters. One species (*G. codiolifera* (Chodat) Wille), however, bores (?) into limestone rocks. We have considered, as chief characteristic of this genus, the formation of large ''sporangia'' (gametangia (?) and aplanosporangia) which are usually provided with one or more rhizoid-like appendages. In some species, however, there seem to be no such appendages (e.g., *G. arrhiza* Hariot). The ''sporangia'' arise as segments of the branching filaments. These segments enlarge and give off processes, finally the walls thicken more or less and the "rhizoids" often thicken so as to become entirely solid or, at least, throughout the greater portion. The "sporangia" vary considerably in shape and size, even in specimens inhabiting the same shell, and at times exhibit extraordinary differences, but at other times closely follow a particular type. It has been puzzling to us to interpret this variation. As a result of considerable experience, however, we are inclined to lay more stress on distinctions of form and size in the "sporangia" than has been done by others.

In connection with any attempt to examine carefully and critically the various specimens of *Gomontia* as to their identity with, or distinctness from, the eight or more species already described, it has become necessary to examine carefully the type of the species, viz., *Gomontia polyrhiza* (Lager.) B. and F. We have two sources of information as to the type species, viz., the original description and figures of Lagerheim (1885) and the careful and detailed description and figures of Bornet and Flahault (1888). For reasons which we shall give below, we are inclined to suspect that the plant of Bornet and Flahault is a different species from that of Lagerheim, and our suspicion is so strong that we have felt it necessary (cf. Setchell and Gardner, 1920, p. 298) to bestow upon it a new specific name, viz., *G. Bornetii* S. and G. In carrying out the idea that the "sporangia" furnish diagnostic characters, we have separated several species which present difficulties, to be sure, in narrow diagnosis, but which seem rather more satisfying than the attempt to lump all under the one name. The explanation of the variability may possibly be that the different texture of the various species of shells, or of different portions of the same shell, affect the size and shape, but there does not seem to be evidence forthcoming to support this idea. It seems possible that two, or even more, species may be inhabitants of the same shells and thus cause an intermingling of different types of "sporangia." In our attempt to clear up the situation, we have assumed the possibilities of distinct species, at times, intermingled.

KEY TO THE SPECIES

1. "Sporangia" with 2 to several rhizoids.. 2
1. "Sporangia" usually with a single rhizoid.............................4. **G. caudata** (p. 304)
    2. "Sporangia" longer than broad (Codiolum-type)....1. **G. polyrhiza** (p. 302)
    2. "Sporangia" broader than long (Acarid-type) 3
3. Rhizoids nearly simple, stout, blunt..................................2. **G. Bornetii** (p. 302)
3. Rhizoids branched, slender, acute............................3. **G. habrorhiza** (p. 304)

1. **Gomontia polyrhiza** (Lagerh.) B. and F.

Plate 19, fig. 1

''Sporangia'' irregularly and broadly clavate to nearly cylindrical, up to 150μ diam., and 240μ long, producing usually several blunt, at times slightly branched rhizoids at the smaller (proximal) end.

Growing in clam shells. Neah Bay, Washington.

Bornet and Flahault, Note sur deux nouveaux genres d'algues perforantes, 1888, pp. 161–163, Sur quelq. pl. viv. dans le test. calc., 1889, pp. clii-clx, pl. 6–8 (as to combination only); Setchell and Gardner, Phyc. Cont. I, 1920, p. 298, pl. 24, fig. 1. *Codiolum polyrhizum* Lagerheim, Cod. poly. n. sp. etc., 1885 (at least in greater part).

The above description is taken in part from the original of Lagerheim and in part from the material collected at Neah Bay. The material from which Lagerheim drew his description was apparently in the sporangial stage exclusively, as least he did not recognize a sterile, or vegetative stage. The Neah Bay material, collected in May, is likewise in a reproductive stage, or if the vegetative stage is present, it and the sporangial stage could not be identified as belonging to the same species, hence the incompleteness of the description.

The ''sporangia'' approximate so closely to the figures (especially figs. 10, 11) and the description of Lagerheim as to make it sufficiently safe to ally our plant with his and to keep it distinct from the *G. polyrhiza* of Bornet and Flahault (*G. Bornetii* S. and G.).

The filaments of this species have been examined by us in a specimen distributed by Reinbold from Keil. Reinbold's specimens have ''sporangia'' largely of the *Codiolum*-type, both old and young, but it also has an occasional ''sporangium'' of the Acrid-type (apparently good *G. Bornetii*). Since Reinbold's locality is not far distant from Lagerheim's type locality, it seems extremely probable that his plant is true *G. polyrhiza*. The filaments in Reinbold's specimens are so close to those of *G. Bornetii* as figured by Bornet and Flahault (*loc. cit.*) as to be indistinguishable.

2. **Gomontia Bornetii** S. and G.

Horizontal filaments irregular, much branched, erect filaments with clavate ends, less branched; cells 4–12μ, most frequently 6μ diam., 15–55μ long, cylindrical to more or less swollen and crooked; ''sporangia'' variable and irregular in form, 80–125μ wide, 150–200μ long, having numerous blunt, mostly simple rhizoids arising principally on one side, but occasionally promiscuously scattered all over the sporangia; zoospores of two sorts, one 3.5μ wide and 5μ long, the other 5–6μ wide and 10–12μ long, development unknown; aplanospores 4μ diam.

Growing in clam shells. Neah Bay, Washington.

Setchell and Gardner, Phyc. Cont. I, 1920, p. 298.

Gomontia polyrhiza Bornet and Flahault, Notes sur deux nouveaux genres d'algues perforantes, 1888, pp. 161–163 (pp. 3–5, repr.), Sur quelq. pl. viv. dans le test. calc., 1889, pp. clii–clx, pl. 6–8 (not *Codiolum polyrhiza* Lagerheim).

Bornet and Flahault (1889) distinctly state (p. clv) that the greatest dimensions of the ''sporangia'' in their specimens are 120μ for the height and 75μ for the width and mention that Lagerheim found ''sporangia'' in his specimens up to 240μ in height and 60μ in breadth. We judge, therefore, that the *Codiolum*-type of ''sporangium'' which Lagerheim figures (1885, pl. 28, figs. 10, 11 in particular) and describes (''*plerumque plus minus elongatis*,'' *loc. cit.* p. 22) was not to be found in the French material and certainly is not illustrated by Bornet and Flahault, unless figure 9 on plate 7 may represent it. The type of ''sporangium'' illustrated by Bornet and Flahault (1889, pl. 7, 8) belongs to the shorter and broader type, the Acarid-type as it may be called, and has blunt, simple or slightly branched rhizoids. Lagerheim (*loc. cit.*, pl. 28, figs. 7, 8, 12, 13) has also figured ''sporangia'' of the true Acarid-type and probably found a mixture of species in the shells he examined. Since, however, he emphasizes the elongated, or *Codiolum*-type of ''sporangia,'' it seems best to reserve his specific name for the species with the *Codiolum*-type of ''sporangium'' and assign the new specific name (*Bornetii* S. and G.) to the species having the Acarid-type of ''sporangium'' and with blunt, rather stout, simple or, at most, slightly branched rhizoids.

The filaments of *G. Bornetii* are well represented by Bornet and Flahault (1889, pl. 6, figs. 1–8) and by their usually large number of short blunt or almost bulbously enlarged branchlets and their compact massing, make a characteristic appearance after decalcification. They are very similar to those of *G. polyrhiza* so far as we may determine, but somewhat different from those of *G. habrorhiza*, although this difference is not readily described.

While we find what seems referable to *G. Bornetii* in the Puget Sound region and that of central California, we desire more abundant and more decisive material to make us certain.

3. **Gomontia habrorhiza** S. and G.

Plate 19, figs. 2, 3

Filaments repeatedly and irregularly branched; cells very variable in form and size, typically cylindrical, 4–7μ diam., 2–8 times as long; chromatophore without pyrenoids, filling the cell; "sporangia" narrow to wide, bluntly conical, 50–70μ high, 25–60μ wide, developing many very slender, attenuate, dendritically branched rhizoids from the lower side; reproduction unknown.

Growing in dead clam shells. Neah Bay, Washington.

Setchell and Gardner, Phyc. Cont. I, 1920, p. 299, pl. 24, figs. 2, 3.

In certain shells from Neah Bay we have found all the "sporangia" of the Acarid-type and with the processes or rhizoids slender, branched, and attenuated to a point. The "sporangia" seem so distinct from those of *G. Bornetii* that we described the plant possessing them as new. In some shells, we have found the "sporangia" of this species intermingled with others. The vegetative filaments of *G. habrorhiza* seem less entangled and slightly larger than those of either *G. polyrhiza* or *G. Bornetii*.

4. **Gomontia caudata** S. and G.

Plate 18, figs. 1, 2

Filaments short, sparsely branched; cells 5.5–6.5μ diam., 2–12 times as long; chromatophore covering the terminal cells and young "sporangia," broken in the older cells; pyrenoids inconspicuous; "sporangia" clavate, 50–70μ diam., 160–200μ long, tapering to a single rhizoid below with thick, hyaline, homogeneous wall at maturity; rhizoid often becoming much thickened and striated.

Growing in shells of *Mytilus californicus*. Neah Bay, Washington.

Setchell and Gardner, Phyc. Cont., 1920, p. 300, pl. 23, figs. 1, 2.

We have found in shells of the larger edible mussel of our coast a *Gomontia* with filaments seemingly less abundantly branched and "sporangia" (aplanosporangia?) with very thick walls and with a single long rhizoid (cf. pl. 18, figs. 1, 2). These "sporangia" bear a certain resemblance to the "cells" figured by Lagerheim (1885, pl. 28, fig. 4, 6) but are, at least, thicker walled. The fact which seemed to indicate distinctness was that only this type of "sporangium" was found in the shells examined.

Thalli filamentous, branched, forming diffuse or feltlike tufts or expansions, in some genera compact and discoidal; cells with one to several nuclei; chromatophore single and band-shaped or several and lenticular, the chlorophyll masked by haematochrome; zoosporangia borne on geniculate or hooked cells, usually deciduous and producing 2-ciliated zoospores; gametangia terminal or intercalary producing 2-ciliated gametes.

De-Toni, Consp. gen. Chloroph., 1888, p. 449; Hansgirg, Ueber Gatt. *Herposteiron*, etc., 1888, p. 222; Collins, Green Alg. N. A., 1909, p. 315; West, Algae I, 1916, p. 305. *Chroolepidaceae* Borzi, Stud. Alg., fasc. 1, 1883, p. 25 (in part).

More recent authors agree in separating Trentepohliaceae from Chaetophoraceae, principally on account of habitat and the presence of haematochrome in the cells, masking the green of the chlorophyll and giving an orange-red or yellow color to the members of this family. The plants belonging to Trentepohliaceae are usually epiphytic, or partially endophytic, but some species of *Trentepohlia* grow upon rocks. Only one member of the family affects, at times, a habitat subject to the direct action of the sea water. Since that member has been found in a marine situation on our coast, we feel compelled to include an account of it.

34. **Trentepohlia** Mart.

Frond composed of dichotomously or irregularly branched, erect filaments of a single series of cells, arising from irregularly branched creeping filaments; branches arising either from the middle or from the upper ends of the cells; color greenish at times in active vegetative condition, changing to yellowish or red, fading to white when dried and dead; cells cylindrical to spherical with thick hyaline walls; chromatophore without pyrenoid, band-shaped or broken; reproduction by 2-ciliated zoospores in sporangia borne on special hooked or curved cells, and by 2-ciliated gametes in lateral, terminal or intercalary transformed vegetative cells.

Martius, Flora Cryptog. Erlang., 1817, p. 351.

The members of this genus are usually found upon wood or the trunks of trees, although certain species are found upon rocks. The rock-inhabiting species as well as those found upon trees, are of frequent occurrence in the maritime region, especially where fogs are frequent. Only one species, however, is said to be found where actually immersed at high tide and that grows upon wood. The aerial species on rocks and wood are often conspicuous, forming broad patches of yellow or deep orange-red upon the rocks or tree trunks. Those visiting the shore in search of marine algae are very likely to encounter some of these species.

Trentepohlia odorata var. umbrina (Kuetz.) Hariot

Filaments forming a more or less dense, at times pulverulent or tomentose stratum, without marked distinction between prostrate and erect positions, flexuous, somewhat torulose, with short branches; color green, varying to brownish or orange-red; cells varying from cylindrical to ovoid, 10–30μ diam., 1–1.5μ times as long; cell wall thin when young, becoming thick and lamellate with age; gametangia spherical to ellipsoid, lateral, terminal or intercalary, 20–30μ diam.; sporangia similar to gametangia.

Growing on piles of Douglas Fir (*Pseudotsuga taxifolia*) along high-tide level. Breakwater, San Pedro Harbor, Los Angeles County, California.

Hariot, Notes sur le genre *Trentepohlia*, 1889, pp. 400–403; Collins, Green Alg. N. A., 1909, p. 319. Collins, Holden and Setchell, Phyc. Bor.-Amer. (Exsicc.), no. 2288. *Chroolepus umbrinum* Kuetzing, Phyc. Gen., 1843, p. 283, pl. 7, f. 2. *Trentepohlia umbrina* var. *quercina* Collins, Holden and Setchell, Phyc. Bor.-Amer. (Exsicc.), no. 662.

One of us (Gardner) has found what certainly seems to be this species growing on piles of Douglas Fir (*Pseudotsuga taxifolia*) along high-tide level and above, but where frequently immersed in or splashed by the sea water. The zone affected by the tide was of green color, but above, where contact with salt water was less frequent, the color was deep orange. The locality was observed on two occasions several years apart, and the *Trentepohlia* found persisting. The filaments are to be found mostly within the empty wood cells of the piles.

BIBLIOGRAPHY

AGARDH, C. A.

1817. Synopsis algarum Scandinaviae, adjecta dispositione universali algarum. Lund.

1817–1825. Aphorismi botanici. Decades I–XVI. Lund.

1820–1828. Species algarum rite cognitae cum synonymis, differentiis specificis et descriptionibus succinctis.
    1820. Vol. I, part I, pp. 1–268.
    1822. Vol. I, part II, pp. 269–531.
    1828. Vol. II, part I, pp. i-lxxvi and 1–189.

1824. Systema algarum. Lund.

1828–1835. Icones algarum europaearum. Representation d' algues euro-péennes suivi de celle des espèces exotiques les plus remarquables récemment découvertes. Leipzig.
    1828. Vol. 1, pls. 1–10.
    1828 (or 1829?), vol. 2, pls. 11–20.
    1829. Vol. 3, pls. 21–30.
    1835. Vol. 4, pls. 31–40.
    ["'Literatur Bericht zur Linnaea, für das Jahr 1830,''' p. 86, has a notice of the work: "Livraisons 1–3, nos. 1–30, Leipsic 1828 u. 29.'' This seems to us to imply that part 1 was issued in 1828, covering plates 1–10, part 3 in 1829 covering plates 21–30, while part 2, plates 11–20 may be either 1828 or 1829. Collins in a letter of June 16, 1919, where he writes us the above information, says also, "safe to assume that part 4 was issued in 1835,'' i.e., plates 31–40.]

AGARDH, J. G.

1842. Algae maris Mediterranei et Adriatici, observationes in diagnosin specierum et dispositionem generum. Paris.

1846. *Anadema*, ett nytt slägte bland Algerne. Kongl. Sv. Vet.-Akad. Handl., pp. 1–16. Stockholm.

1847. Nya Alger från Mexico. Oefvers af Kongl. Vet.-Akad., Förhandl. Arg. 4, no. 1. Stockholm.
    [Often quoted as "Alg. Lieb.'' Volume 4 of the "Oefversigt'' is dated on title page and final page, 1848, but the individual parts are all dated "1847'' on the title page; the exact date of imprint, however, is given at the bottom of the final page of each number. Numbers 1–9 were issued in 1847, but number 10 and the title page and index were delayed until 1848.]

1873–1890. Till Algernes systematik, Nya bidrag.
    1873. Första afdelningen, vol. 9. Lunds Univ. Arssk.
    1882. Andra afdelningen, vol. 17. Lunds Univ. Arssk.
    1883. Tredje afdelningen, vol. 19. Lunds Univ. Arssk.
    1885. Fjerde afdelningen, vol. 21. Lunds Univ. Arssk.
    1887. Femte afdelningen, vol. 23. Lunds Univ. Arssk.
    1890. Sjette afdelningen, vol. 26. Lunds Univ. Arssk.

AHLNER, K.

1877. Bidrag till Kännedomen om de Svenska formerna af Algslägtet *Entero-morpha*. Akademisk aufhandling. Stockholm.

ANDERSON, C. L.
1891. List of California marine algae with notes. Zoe, vol. 2, pp. 217–225.

ARDISSONE, F.
1883–1886. Phycologia mediterranea. Part I, Floridee.
1883. Mem. Soc. Critt. Ital., vol. 1.
Ibid. Part IIa, Ooosporee, Zoosporee, Schizosporee.
1886. *Ibid.*, vol. 2.

ARESCHOUG, J. E.
Algae scandinavicae exsiccatae.
1840. Edition 1.
1861–1879. Series nova, Fasc. I–IX. Upsala.
1843. Algarum (Phycearum) minus rite cognitarum pugillus secundus. Linnaea, vol. 17, pp. 256–269, pl. 9.
1850. Phyceae Scandinaviae marinae, sive Fucacearum nec non Ulvacearum, quae in maribus paeninsulam scandinavicam affluentibus crescunt, descriptiones. (Fucaceae, ex Act, Upsal., vol. 13. Ulvaceae, ex Act, Upsal., vol. 14.) Upsala.
1866. Observationes phycologicae. Die confervaceis nonnullis, part 1. Upsala.

BATTERS, E. A. L.
1891. Hand-list of the algae *in* the algae of the Clyde Sea area. Journal of Botany, vol. 29, pp. 212–214; 229–236; 273–283.
Ibid. Reprinted, with additions, from the Journal of Botany for 1891, pp. 1–25.
1894. New or critical British algae, Grevillea, vol. 22, p. 114.

BENTHAM, G.
1843. *Tetranema mexicanum, in* Lindley, Edwards's Botanical register or ornamental flower-garden and shrubbery, consisting of coloured figures of plants and shrubs cultivated in British gardens, accompanied by their history, best method of treatment in cultivation, propagation, etc. Vol. 6, new series. London.

BLACKMAN, F. F., and TANSLEY, A. G.
1902. A revision of the classification of the green algae. The New Phytologist, vol. 1, pp. 17, 47, 67, 89, 114, 133, 163, 189, 213, 238.

BONNEMAISON, T.
1822. Essai d'un classification des Hydrophytes loculées ou plantes marines qui croissent en France. Journ. de Phys., vol. 94, pp. 174–203, 6 pl.

BÖRGESEN, F.
1902. The marine algae of the Faeröes, *in* Warming, Botany of the Faeröes, part II, p. 339. Copenhagen.

BORNET, E.
1892. Les algues de P.-K.-A. Schousboe, récoltées au Maroc & dans la Mediterranée de 1815 a 1829. Mém. Soc. Nat. Sci. Nat et Math. de Cherbourg. Vol. 28, p. 165. Paris.

BORNET, E., and FLAHAULT, C.
1888. Note sur deux nouveaux genres d'algues perforantes. Journal de Botanique, vol. 2, p. 161.
1889. Sur quelques plantes vivant dans le teste calcaire des mollusques. Bull. Soc. Bot. de France, vol. 36, p. cxlvii, pls. 6–12.

Bory de Saint Vincent, J. B. (M. A. G.)

1804. Voyage dans les quatre principales îles des mers d'Afrique, fait par ordre du gouvernment pendant les années IX et X de la république (1801 et 1802), avec l'histoire de la traversée du Capt. Baudin jusqu'au Port Louis de l'isle Maurice. 3 vols. Paris.

1827–1829. Cryptogamie, in Voyage autour du monde—sur la Corvette de Sa Majesté, La Coquille, par M. L. I. Duperrey, pp. 1–96, 1827; 97–200, 1828; 201–300, 1829; according to Sherborn and Woodward, *in* Journal of Botany, vol. 39, p. 206. The title page is dated 1828. (Atlas, 1826.)

1828a. *In* Dictionaire classique d'histoire naturelle, vol. 13. Paris.

Borzi, A.

1883. Studi algologici saggio di ricerche sulla biologia delle alghe. Fasc. 1. Messina.

Brand, F.

1904. Ueber die Ahnheftung der Cladophoraceen und ueber verschiedene polynesischen Formen dieser Familie. Beihefte Bot. Centralb., vol. 18, pp. 165–193, pls. V, VI.

1908. Zur Morphologie und Biologie des Grenzgebietes zwischen den Algengattungen *Rhizoclonium* und *Cladophora*. Hedwigia, vol. 48, pp. 45–73.

Braun, A.

1855. Algarum unicellularum genera nova et minùs cognita. Leipzig.

Briquet, J.

1906. Règles internationales de la nomenclature botanique adoptées par le congrès international de botanique de Vienne, 1905.

1912. *Ibid.* Deuxième édition mise au point d'après les décisions du congrès international de botanique de Bruxelles, 1910. Jena.

Chodat, R.

1902. Algues vertes de la Suisse. Beiträge zur Kryptogamen-flora der Schweitz. Vol. 1, part 3.

Cohn, F.

1872. Ueber parasitische Algen. Beitr. Biol. Pflanzen., vol. 1, part 2, p. 87.

Collins, F. S.

1902. The marine *Cladophoras* of New England. Rhodora, vol. 4, no. 42.

1903. The Ulvaceae of North America. Rhodora, vol. 5, p. 1, pls. 41–43.

1906. New species, etc., in the Phycotheca Boreali-Americana. Rhodora, vol. 8, no. 90, pp. 104–113.

1907. Some new green algae. Rhodora, vol. 9, p. 155.

1909. The green algae of North America. Tufts College Studies, vol. 2, no. 3, Scientic Series. Mass.

1909a. New species of *Cladophora*. Rhodora, vol. 11, pp. 17–20, pl. 78.

1912. The green algae of North America, Supplement I. Tufts College Studies, vol. 3, no. 2, Scientific Series. Mass.

1913. The marine algae of Vancouver Island. Canada Geological Survey. Victoria Memorial Museum, Bulletin no. 1, pp. 99–137. Victoria, B. C.

1918. The green algae of North America, Supplement II, Tufts College Studies, vol. 4, no. 7, Scientific Series. Mass.

Collins, F. S., Holden, I., and Setchell, W. A.

1895–1917. Phycotheca Boreali-Americana. Fasc. 1–45 and A–E. (Exsicc.). Malden, Mass.

COTTON, A. D.
1912. Clare Island Survey, part 15, Marine Algae. Proc. Roy. Irish Acad., vol. 31, pp. 1–178, pls. 1–11. Dublin.

COVILLE, F. V., and ROSE, J. N.
1898. List of plants collected by Dr. and Mrs. Leonard Stejneger on the Commander Islands during 1895–1897, *in* Jordan, Fur seals and fur seal islands, vol. 4, p. 352.

CROUAN, P. L., and H. M.
1859. Notes sur quelques espèces et genres nouveaux d'algues marino de la rade de Brest. Ann. Sci. Nat., 4 sér., Bot., vol. 12, p. 288.

DAVIS, B. M.
1908. Spore formation in *Derbesia*. Annals of Botany, vol. 22, p. 1.

DECAISNE, J.
1841. Plantes de l'Arabie heureuse recueillés par M. P. E. Botta. Arch. du. Mus., vol. 2, p. 89. Paris.
1842. Essai sur une classification des algues et des polypiers calcifères de Lamouroux. Ann. Sci. Nat., 2 sér. Bot., vol. 17, p. 297.
1842a. Mémoire sur les corallines ou polypiers calcifères. Ann. Sci. Nat., 2 sér. Bot., vol. 18, p. 96.

DE CANDOLLE, A. P.
1801. Extrait d'un rapport sur les Conferves, fait à la Société Philomathique par C. C. Decandolle. Bull. Sci. Soc. Phil., no. 51, p. 17. Paris.

DELILE, A. R.
1813. Flore d'Egypte. Paris.

DE NOTARIS, G.
1846. Novita algologiche (*Prospetto della Flora ligustica* nella *Descrizione di Genova e del Genovesato*, publicata per cura del Municipio di Genova in occasione del Congresso degli scienziati italiani tenuto in questa città nell'anno 1846), con 1 tav. Genoa.

DESMAZIÈRES, J. B. H. J.
1825. Plantes cryptogames du nord de la France. Fasc. 1 (Exsicc.). Lille.

DESVAUX, N. A.
1813–1814. Journal de botanique appliquée à l'agriculture, à la pharmacie, à la médecine et aux arts. 4 vols. Paris.

DE-TONI, J. B. (or G. B.)
1888. Conspectus generum Chlorophycearum omnium hucusque cognitorum. Notarisia, ann. III, no. 10, pp. 447–453. Venice.
1889. Sylloge Algarum. Vol. 1, sec. I, II, Chlorophyceae.
1895. Phyceae Jajonicae novae, addita enumeratione algarum in ditione maritima Japoniae hucusque collectarum. Mem. R. Ist. Veneto, sci. lett. ed arti. Vol. 25, no. 5. Venice.

DE-TONI, J. B. (or G. B.), and LEVI-MORENOS, D.
1888. Flora algologica della Venezia, part III, Le Cloroficee. Atti R. Ist. Veneto, ser. 4, vol. 5, pp. 1511–1593; vol. 6, pp. 95–155 and 289–350. Venice.

DILLENIUS, J. J.
1741. Historia muscorum, in qua circiter sex centae species veteres et novae ad sua genera relatae describunter, et iconibus genuinis illustrantur; com appendice et indice synonymorum. Oxonii.

DILLWYN, L. W.

    1802–1809. British Confervae or coloured figures and descriptions of the British plants referred by Botanists to the genus *Conferva*. London.

    [The title page as above is dated 1809, then follow two pages of the "Preface to the First Fasciculus," dated June 1, 1802, which is also a sort of advertisement. There are 87 consecutively numbered pages of "Introduction," discussions of the systems of Roth, Vaucher, De Candolle and Hudson, and a "Synopsis of the British Confervae." The five pages of Index of this portion are unnumbered. These 87 pages, index and title page were issued probably late in 1809. There are 109 plates consecutively numbered, and mostly dated as to issue, together with one or more pages of text for each species figured. The plates and leaflets of text were undoubtedly issued in fascicles, some of which were numbered, and at various dates. Finally there are five "supplementary plates," done in a very different style from the others and labelled from A–G. They are not dated neither are they accompanied by leaflets of text as are the other 109 plates. .From the dates on the plates and the fascicle numbers at the bottom of some of the descriptive leaflets, the work was probably issued as follows:

        1802, June 1, "Preface to the First Fasciculus."
        1802, July 1, Fasciculus 1, text and plates 1–12.
        1802, Nov. 1, Fasciculus 2, text and plates 13–20.
        1803, June 1, Fasciculus 3, text and plates 21–32.
        1803, Nov. 1, Fasciculus 4, text and plates 33–38.
        1804, Dec. 1, Fasciculus 5, text and plates 39–44.
        1805, Sept. 1, Fasciculus 6, text and plates 45–50.
        1805, Dec. 1, Fasciculus 7, text and plates 51–56.
        1806, Mar. 1, Fasciculus 8, text and plates 57–62.
        1806, June 1, Fasciculus 9, text and plates 63–68.
        1806, Sept. 1, Fasciculus 10, text and plates 70–75.
        1806, Dec. 1, Fasciculus 11, text and plates 76–81.
        1807, Mar. 1, Fasciculus 12, text and plates 82–87.
        1807, June 1, Fasciculus 13, text and plates 88–93.
        1808, July 1, Fasciculus 14, text and plates 94–99.
        1909, (Feb. 20?), Fasciculus 15, text and plates 100–105.
        1809, ?, Fasciculus 16, text and plates 69 and 106–109.
        1909, ?, Fasciculus 16?, supplementary plates A–G.
        1809, Title page, 87 pages of text, and 5 pages of index.]

DON, G.

    1831–1838. A general history of the diclamydeous plants comprising complete descriptions of the different orders, etc.

        1831, vol. 1.
        1832, vol. 2.
        1834, vol. 3.
        1838, vol. 4.

    [The first three volumes are entitled "A general system of gardening and botany, etc.," the fourth volume only having the above title. The title of the first three volumes is frequently quoted as "Gen. Syst."]

DUMORTIER, B. C.

    1822. Commentations botanicae (Observationes botaniques) (1823). Tournay.

    1829. Analyse des familles des plantes avec l'indication des principaux genres, qui s'y rattachent. Tournay.

ENGLISH BOTANY, or coloured figures of British plants with their essential char-
    1790–1814. acters, synonymes, and places of growth, to which will be added
        occasional remarks by James Edward Smith. The figures by
        James Sowerby. Vols. 1–36. London.

FARLOW, W. G.
    1881. Marine algae of New England and adjacent coast. Report of the
        U. S. Fish Comm. for 1879. Washington.

FARLOW, W. G., ANDERSON, C. L., and EATON, D. C.
    1877–1889. Algae Exsiccatae Americae-Borealis. Fasc. 1–5. Boston.

FORSTER, J. R., and G.
    1776. Characteres generum plantarum quas in itinere ad insulas maris Aus-
        tralis collegerunt, descripserunt delinearunt annis MDCCLXXII–
        MDCCLXXV. London.

FOSLIE, M.
    1890. Contributions to the knowledge of the marine algae of Norway I, East-
        Finmarken. Tromsö Museums Aarshefter, vol. 13.
    1891. *Ibid.*, II. Vol. 14.

FREEMAN, E. M.
    1899. Observtaions on *Constantinea*. Minnesota botanical studies, ser. 2,
        part 2, pp. 175–190, pl. 2. Minneapolis.
    1899a. Observations on *Chlorochytrium*. Minnesota botanical studies, ser. 2,
        part 3, pp. 195–204, pl. 3. Minneapolis.

FRIES, E.
    1825. Systema orbis vegetabilis. Primas lineas novae constructionis peri-
        clitatur Elias Fries. Part I, Plantae homonemeae. Lund.
    1835. Corpus florarum provincialium Sueciae. I. Floram Scanicam scripsit
        Elias Fries. Upsala.
    1846. Summa vegetabilum Scandinaviae, Sect. 1.

FRYE, T. C., and ZELLER, S. M.
    1915. *Hormiscia tetraciliata* sp. nov. Puget Sound Marine Station Publica-
        tions, vol. 1, no. 2, pp. 9–13, pl. 2.

GARDNER, N. L.
    1909. New Chlorophyceae from California. Univ. Calif. Publ. Bot., vol. 3,
        pp. 371–375, pl. 14.
    1917. New Pacific Coast marine algae I. Univ. Calif. Publ. Bot., vol. 6,
        no. 14, pp. 377–416., pls. 31–35.
    1919. New Pacific Coast marine algae IV. Univ. Calif. Publ. Bot., vol. 6,
        no. 18, pp. 487–496, pl. 42.

GOBI, C.
    1879. Bericht ueber die algologischen Forschungen im finnischen Meerbusen
        im Sommer 1877 ausgeführt. St. Petersb. Gesellsch. d. Naturf., vol.
        10, p. 83.

GRAY, J. E. (see S. F. Gray).
    1821. A natural arrangement of British plants. Vol. 1. London.

GRAY, S. F.
    1821. A natural arrangement of British plants. Vol. 1. London.
        [J. E. Gray, son of S. F. Gray, did the systematic work in the
        two volumes which came out under the above title, with S. F. Gray
        as the author. J. E. Gray is sometimes cited as the author.]

GREVILLE, R. K.

    1823–1829. Scottish Cryptogamic Flora, or colored figures and descriptions of cryptogamic plants belonging chiefly to the order Fungi, and intended to serve as a continuation of English Botany. Vols. 1–6. Edinburgh.

        [We have consulted the six volume edition, each volume containing sixty plates and dated as follows: vol. 1, 1823; vol. 2, 1824; vol. 3, 1825; vol. 4, 1826; vol. 5, 1827; vol. 6, 1828. (Cf. Pritzel, 1872, p. 128, and Jackson, 1881, p. 246 for dates, 1823–1829.)]

    1830. Algae Britannicae, or descriptions of the marine and other inarticulated plants. Edinburgh.

HAGEM, O.

    1908. Beobachtungen ueber die Gattung *Urospora* in Kristianiafjord. Nyt Mag. fur Natur., vol. 46, p. 289. Christiania.

HANSGIRG, A.

    1886–1888. Prodromus der Algenflora von Böhmen. Erster Theil enthaltend die Rhodophyceen, Phaeophyceen und einer Theil der Chlorophyceen. Arch. naturw. Landesdurchf. v. Böhmen.

        1886, vol. 5, part 1, no. 6.

        1888, vol. 6, part 2, no. 6.

    1888. Ueber die Gattungen *Herposteiron* Näg. und *Aphanochaete* Berth. non A. Br. nebst einer systematischen Uebersicht aller bisher bekannten oogamen und anoogamen Confervoiden-Gattungen. Flora, pp. 211–223.

HARIOT, P.

    1889. Algues, *in* Mission scientifique du Cap Horn, 1882, 1883. Vol. 5, Botanique, pp. 1–109, pls. 1–9. Paris.

    1889–1890. Notes sur le genre *Trentepohlia* Martius. Journ. de Bot., vol. 3, and vol. 4.

HARVEY, W. H.

    1834. Notice of a collection of algae, communicated to Dr. Hooker by the late Mrs. Charles Telfair from ''Cap Malheureux'' in Mauritius, with descriptions of some new and little known species. The Journal of Botany (Hooker), vol. 1, p. 147. London.

    1838. The genera of South African plants, arranged according to the natural system. Cape Town.

    1846–1851. Phycologia Britannica. London.

        [According to a memorandum from E. M. Holmes to F. S. Collins, this work was issued in parts of six plates each, part 1 having been issued January, 1846, then monthly parts to part 42, issued June 1, 1849. After that the issues were irregular, the last part (60) having been issued August, 1851. The issues may be summarized, so far as we have the information, as follows:

        1846, plates 1–72.

        1847, plates 73–144.

        1848, plates 145–216.

        1849, plates 217–258.

        1849–1851, plates 259–354.

        1851, plates 355–360.

    The title pages of the volumes are dated as follows:

        1846, vol. 1 (plates 1–120).

        1849, vol. 2 (plates 121–240).

        1851, vol. 3 (plates 241–360)].

1849. A manual of British marine algae, containing generic and specific descriptions of all the known British species of sea-weeds, with plates to illustrate all the genera. London.

1852–1858. Nereis Boreali-Americana.
    1852, Part I, Melanospermeae.
    1853, Part II, Rhodospermeae.
    1858, Part III, Chlorospermeae.

1859. Characters of new algae, chiefly from Japan and adjacent regions collected by Charles Wright in the North Pacific Exploring Expedition under Captain John Rodgers. Proc. Amer. Acad., vol. 4, pp. 327–334.

1862. Notice of a collection of algae made on the northwest coast of North America, chiefly at Vancouver's Island by Dr. David Lyall, 1859–1861. Jour. Proc. Linn. Soc., Bot., vol. 6, pp. 157–177. London.

HASSALL, A. H.
1843. Observations on the growth, reproduction, and species of the branched freshwater Confervae. Ann. Magaz. Nat. Hist., vol. 11, p. 359.

1845, Ed. 1.   A history of the British fresh water algae including descriptions
1852, Ed. 2.   of the Desmideae and Diatomaceae with upwards of one
1857, Ed. 3.   hundred plates illustrating the various species. Vol. 1, 2. London.

HAUCK, F.
1885. Die Meeresalgen Deutschlands und Oesterreichs, *in* Rabenhorst's Kryptogamen-Flora von Deutschland, Oesterreich und der Schweiz. Vol. 2, part 10, Schizophyceae. Leipzig. (Issued in 10 ''Lieferungen'' between 1883 and 1885.)

HAUCK, F., and RICHTER, P.
1885–1896. Phykotheka Universalis. Fasc. 1–15 (Exsicc.). Leipzig.

HAZEN, T. E.
1902. The Ulrothricaceae and Chaetophoraceae of the United States. Mem. Torr. Bot. Club, vol. 11, no. 2, pp. 135–245, pls. 21–42.

HOOKER, W. J.
1833. The English Flora of Sir James Edward Smith. Class XXIV, Cryptogamia. Vol. 5 (vol. 2 of Dr. Hooker's British Flora) part 1, comprising the Mosses, Hepaticae, Lichens, Characeae and Algae. London.

HOOKER, J. D.
1844–47. The Botany of the Antarctic Voyage of H. M. Discovery Ships Erebus and Terror, etc. Vol. 1, Flora Antarctica.
    [The reference in the text is to part 23, issued in 1847, probably on Feb. 2 of that year. For dates of issuance of the various parts see B. Daydon Jackson, *in* the Journal of Botany, 1912, pp. 284, 285.]

HORNEMANN, J. W.
1818. Icones plantarum sponte nascentium in regno Daniae, et in ducatibus Slesvici, Holsatiae et Lauenburgiae ad illustrandum opus de iisdem plantis, regio jussu exarandum, Florae Danicae nomine inscriptum. (Usually quoted Flora Danica.) Vol. 9, fasc. 25–27, pl. 1441–1620. Date on title page of volume 9 is 1818.
    1813, fasc. 25.
    1816, fasc. 26.
    1818, fasc. 27.

HOWE, M. A.

1893. A month on the shores of Monterey Bay. Erythea, vol. 1, pp. 63–68.

1907. Phycological studies III. Further notes on *Halimeda* and *Avrainvillea.* Bull. Torr. Bot. Club, vol. 34, pp. 491–516, pls. 25–30.

1911. Phycological studies V. Some marine algae of Lower California, Mexico. Bull. Torr. Bot. Club, vol. 38, pp. 489–514, pls. 27–34.

1914. The marine algae of Peru. Mem. Torr. Bot. Club, vol. 15, pp. 1–185, pls. 1–66.

HUBER, J.

1892 (1893, repr.). Contributions à la connaissance des Chaetophorées épiphytes et endophytes et de leurs affinités. Ann. Sci. Nat., 7 sér. Bot., vol. 16, pp. 265–359, pls. 8–18.

HUDSON, G.

1778. Flora Anglica. Vols. 1, 2. 2nd edition. London.

HURD, ANNIE MAY.

1916. *Codium dimorphum.* Puget Sound Mar. Stat. Publ., vol. 1, no. 19, p. 211. Seattle.

1916a. *Codium mucronatum.* Puget Sound Mar. Stat. Publ., vol. 1, no. 12, pp. 109–123. Seattle.

HUS, H. T. A.

1902. An account of the species of *Porphyra* found on the Pacific coast of North America. Proc. Calif. Acad. Sci., 3rd ser. Bot., vol. 2, no. 6, pp. 173–236, pls. 20–22.

IMHÄUSER, L.

1889. Entwicklungsgeschichte und Formenkreis von *Prasiola.* Flora, p. 233, pls. 10–13.

JACKSON, B. D.

1881. Guide to the literature of botany; being a classified selection of botanical works including nearly 600 titles not given in Pritzel's ''Thesaurus.'' London.

1912. Index to the Linnean Herbarium, with indications of the types of species marked by Carl von Linné. Forming a supplement to the proceeding of the society for the 124th session. 1911–1912. London.

JESSEN, C. F. G.

1848. Prasiolae generis algarum monographia. ''Dissertatio inauguralis botanica.'' Kilia (Kiliae).

JÓNSSON, H.

1903. The marine algae of Iceland. (III Chlorophyceae, IV Cyanophyceae.) Botanisk Tidsskrift, vol. 25, part 3, p. 337.

1904. The marine algae of East Greenland. Meddelelser om Grönland. Vol. 30. Copenhagen.

JUERGENS, G. H. B.

1816–22. Algae aquaticae quas in littore maris Dynastiam Jeveranam et Frisiam orientalem alluentis rejectas et in harum terrarum aquis habitantes collegit, etc. Decades 1–20. Jever.

KJELLMAN, F. R.

Om Spetsbergens marina Klorofyllförande Thallophyter.

1875, I, vol. 3, no. 7.

1877, II, vol. 4, no. 6.

Bihang till K. Sv. Vet.-Akad. Handl. Stockholm.

1877a. Ueber die Algenvegetation des Murmanschen Meeres an der westküste von Nowaja Semlja und Wajgatsch. Nova Acta Reg. Soc. Sci., ser. 3, vol. extra ord. (no. 12). Upsala.

1883. The algae of the Arctic Sea. Kongl. Sv. Vet.-Akad. Handl., vol. 20, no. 5. Stockholm.

1883a. Norra ishafvets Algflora. Stockholm.

1889. Om Beringhafvets Algflora. Kongl. Sv. Vet.-Akad. Handl., vol. 23, no. 8. Stockholm.

1893. Studier öfver chlorophycéslägtet *Acrosiphonia* J. G. Ag., och dess skandinaviska arter. Bihang till Kongl. Sv. Vet.-Akad. Handl., vol. 18, afd. 3, no. 5. Stockholm.

1897. *Blastophysa polymorpha* och *Urospora incrassata.* Tva nya Chlorophyceer fran sveriges vestra kust. Bihang till Kongl. Sv. Vet.-Akad. Handl., vol. 23, afd. 3, no. 9. Stockholm.

1897a. Marina chlorophyceer från Japan. Bihang till Kongl. Sv. Vet.-Akad. Handl., vol. 23, afd. 3, no. 11. Stockholm.

KLEBS, G.
1883. Ueber die Organization einiger Flagellaten-Gruppen und ihre Beziehungen zu Algen und Infusorien. Untersuch. aus. dem Botan.-Inst. Tübingen. p. 233.

KLEEN, E. (A. G.)
1874. Om Nordlandenes högra hafsalger. Akademisk afhandling. Oefversigt af Kongl. Vet.-Akad. Förhandl., no. 9. Stockholm.

KUCKUCK, P.
1894. Bemerkungen zur marinen Algenvegetation von Helgoland. Wissenschaftliche Meeresuntersuchungen herausgegeben von der Kommission zur wissenschaftlichen Untersuchung der deutschen Meere in Kiel und der Biologischen Anstalt auf Helgoland. Neue Folge, vol. 1, part 1, pp. 223–263. Kiel und Leipzig.

1907. Abhandlungen ueber Meeresalgen I. Ueber den Bau und die Fortpflanzung von *Halicystis* Areschoug und *Valonia* Ginnani. Botan. Zeitung, vol. 65, p. 139.

KUETZING, F. T.
1833. Algologische Mittheilungen. Flora, vol. 16, pp. 513–521.

1843. Phycologia generalis, oder Anatomie, Physiologie und Systemkunde der Tange. Leipzig.

1843a. Ueber die systematische Eintheilung der Algen. Linnaea, vol. 17, p. 75.

1845. Phycologia germanica d. i. Deutschlands Algen in bündigen Beschreibungen, nebst einer Anleitung zum Untersuchen und Bestimmen dieser Gewächse für Anfänger. Nordhausen.

1845–1871. Tabulae Phycologicae.

1845–49, vol. 1.	1856, vol. 6.	1861, vol. 11.	1866, vol. 16.
1850–52, vol. 2.	1857, vol. 7.	1862, vol. 12.	1867, vol. 17.
1853, vol. 3.	1858, vol. 8.	1863, vol. 13.	1868, vol. 18.
1854, vol. 4.	1859, vol. 9.	1864, vol. 14.	1869, vol. 19.
1855, vol. 5.	1860, vol. 10.	1865, vol. 15.	1871, index.

[Judging from Pritzel (Thesaurus, 1851, p. 145) vol. 1 was published in several parts, and probably at different dates. In quoting in the text we have used the date 1845, the first date given in the volume, without discriminating as to the exact date of the publication of the species referred to.]

1849. Species Algarum. Leipzig.

KUNTZE, O.
    1891. Revisio generum plantarum vascularium omnium atque cellularium mul-
        tarum secundum leges nomenclaturae internationales cum enumer-
        atione plantarum exoticarum in itinere mundi collectarum. Parts I
        and II. Würzburg. (Algae, part II, pp. 877–930.)

KYLIN, H.
    1907. Studien ueber Algenflora der schwedischen Westküste. Akademische
        Abhandlung. Upsala.

LAGERHEIM, G.
    1883. Bidrag till Sveriges algflora. Oefv. Kongl. Sv. Vet.-Akad. Förhandl.,
        no. 2. Stockholm.
    1885. *Codiolum polyrhizum* n. sp. Ett bidrag till kännedomen om slägtet
        Codiolum A. Br. Oefversigt af Kongl. Vet.-Akad. Förhand., no. 8,
        p. 21, pl. 28. Stockholm.

LAGERSTEDT, N. G. W.
    1869. Om algslägtet *Prasiola*. Försök till en monographi. Akademisk afhand-
        ling. Upsala.

LAMARK, J. B. DE et DECANDOLLE, A. P.
    1805. Flore française ou déscriptions succinctes de toutes les plantes qui
        croissent naturellement en France, disposées selon une nouvelle
        méthodè d'analyse et précédées par un exposé des principes élémen-
        taires de la botanique. Ed. III, vol. 2. Paris.

LAMOUROUX, J. V.
    1809 (May). Observations sur la physiologie des algues marines. Nouveau
        Bull. des sci. par la Soc. Phil. de Paris, vol. 1, p. 331.
    1809a. Mémoire sur trois nouveaux genres de la famille des algues marines.
        Journ. de Bot., vol. 2, p. 129.
    1812. Sur la classification des Polypes corallines. Bull. Soc. Philom., vol. 3,
        p. 186.
    1813. Essai sur les genres de la famille des thalassiophytes non articulées.
        Ann. du Mus. d'Hist. Naturelle par les professeurs de cet établisse-
        ment, vol. 20, pp. 21–47, 115–139, 267–293, pls. 7–13. Paris.
        [Usually quoted from the reprint which is paged successively
        from 1 to 84. Plates are numbered the same in both.]

LEAVITT, CLARA K.
    1904. Observations on *Callymenia phyllophora* J. Ag. Minnesota botanical
        studies, ser. 3, part 3, p. 291. Minneapolis.

LE JOLIS, A.
    1863. Liste des algues marines de Cherbourg. Mém. Soc. Imp. Sci. Nat. de
        Cherbourg, vol. 10, p. 1. Paris.
            Reprint of above, 1880. The paper was published separately in
        1863 before it appeared in the Mémoires 1864.

LIEBMANN, F. M.
    1869. Chênes de l'Amerique tropicale. Iconographie des espèces nouvelles
        ou peu connues. Ouvrage posthume de F. M. Liebmann, achevé et
        augmenté d'un aperçu sur la classification des chênes en general par
        A. S. Oersted. Leipzig.

LINK, H. F.
  1820. Epistola de algis aquaticis in genera disponendis. Nees, Horae Physi-
        cae, p. 1.

LINNAEUS, C.
  1737. Genera plantarum, eorumque characteres naturales secundum numerum,
        figuram, situm et proportionem omnium fructificatonis partium.
  1753. Species plantarum, exhibentes plantas rite cognitas, ad genera relatas,
        cum differentiis specificis, nominibus trivialibus, synonymis selectis,
        locis natalibus secundum systema sexuale digestas. Ed. I. Stock-
        holm.
  1755. Flora suecica, exhibens plantas per regnum Sueciae crescentes, sys-
        tematice cum differentiis specierum, synonymis autorum, nominibus
        incolarum, solo locorum, usu pharmacopoeorum. Ed. II. Stockholm.

LUDWIG, C. G.
  1737. Definitiones generum plantarum in usum auditorum. Leipzig.

LYNGBYE, H. C.
  1819. Tentamen hydrophytologiae Danicae, continens omnia hydrophyta
        cryptogamia Daniae, Holsatiae, Faeroae, Islandiae, Groenlandiae
        hucusque cognita, systematice disposita, descripta et iconibus illus-
        trata, adjectis simul speciebus Norvegicis. Copenhagen.

MCCLATCHIE, A. J.
  1897. Seedless plants of Southern California. Proc. So. Calif. Acad. Sci.,
        vol. 1, pp. 337–395.

MACKAY, J. T.
  1836. Flora hibernica, comprising the Flowering plants, Ferns, Characeae,
        Musci, Hepaticae, Lichens and Algae of Ireland, arranged according
        to the natural system, with a synopsis of the genera according to
        the Linnaean system. Dublin.

MARTIUS, C. F. P. VON
  1817. Flora cryptogamica Erlangensis, sistens vegetabilia e classe ultima
        Linn. in agro Erlangensi hucusque detecta. Nuremberg.

MEDICUS, F. C.
  1789. Philosophische Botanik mit kritischen Bemerkungen. Part I. Mann-
        heim.

MENEGHINI, G.
  1838. Cenni sulla organografia e fisiologia delle alghe. Nuovi saggi dell'
        I. R. Academia di scienze, lettere ed arti, vol. 4, p. 324. Padova.

MÖNCH, K.
  1794. Methodus plantas horti botanici et agri Marburgensis a staminum situ
        describendi.

MONTAGNE, J. F. C.
  1845. Plantes cellulaires, *in* Botanique, *in* D'Urville, Voyage au pôle sud et
        dans l'océanie.
  1849. Sexième centurie de plantes cellulaires nouvelles, tant indigènes qu'
        éxotiques. Décades III A VI (1). Ann. Sci. Nat., 3 sér., Bot., vol.
        11, pp. 33–66.
  1850. Cryptogamia Guyanensis, seu plantarum cellularium in Guyana gallica
        annis 1835–1849 a Cl. Leprieur collectarum enumeratio universalis.
        Ann. Sci. Nat., 3 sér., Bot., vol. 14, p. 283.

Müller, O. F.

1782. Icones plantarum sponte nascentium in regnis Daniae et Norvegiae et in ducatibus Slesvici, Holsatiae et Oldenburg, ad illustrandum opus de iisdem plantis, regio jussu exarandum, Florae Danicae nomine inscriptum. Vol. 5, fasc. 13–15, pl. 721–900. Usually quoted as Flora Danica.
    1778, fasc. 13.
    1780, fasc. 14.
    1782, fasc. 15.

Naegeli, C.

1847. Die neuern Algensysteme und Versuch zur Begründung eines eigenen System der Algen und Florideen.

Okamura, K.

1899–1903. Algae Japonicae Exsiccatae. Fasc. 1, 2.

1912. Icones of Japanese algae, vol. 2, no. 9. Tokyo.

1915. Icones of Japanese algae, vol. 3, no. 7 and no. 8. Tokyo.

Olivi, G.

1792. *Lamarckia* novum plantarum cryptogamarum genus, *in* Olivi, Zoologia Adriatica ossia catalogo ragionato degli animali del golfo e delle lagune di Venezia; preceduto da una dissertazione sulla storia fisica e naturale del golfo; e accompagnato da memoirie ed osservazione di fisica storia naturale ed economia.

1794. *Ibid., in* Usteri, Ann. der Bot., vol. 3, part 7, p. 76.

Oltmanns, F.

1904–1905. Morphologie und Biologie der Algen. Jena.
    1904, vol. 1.
    1905, vol. 2.

Pascher, A.

1907. Studien ueber die Schwärmer einiger Süsswasseralgen. Bibliotheca Botanica, vol. 15, part 67. Stuttgart.

Pfeiffer, L.

1873–1875. Nomenclator botanicus. Nominum ad finem anni 1858 publici juris factorum, classes, ordines, tribus, familias, divisiones, genera, subgenera vel sectiones designantium enumeratio alphabetica. Adjectis auctoribus, temporibus, locis systematicis apud varios, notis literariis atque etymologicis et synonymis.
    1873, vol. 1, part 1.
    1875, vol. 1, part 2.
    1874, vol. 2, part 1.
    1874, vol. 2, part 2.

Postels, A., and Ruprecht, F.

1840. Illutrationes algarum in oceano Pacifici inprimio septemtrionali collectarum. St. Petersburg.

Pringsheim, N.

1862. Beiträge zur Morphologie der Meeresalgen. Physikal. Abhandl. der Kön. Akad. d. Wissensch. Berlin.

PRITZEL, G. A.
  Thesaurus literaturae botanicae omnium gentium inde a rerum botanicarum initiis ad nostra usque tempora, quindecim millia operum
  recensens.
      1851, Ed. I.
      1872, Ed. II.

RABENHORST, L.
  1864–1868. Flora Europaea Algarum aquae dulcis et submarinae. Leipzig.
      1864, vol. 1.
      1865, vol. 2.
      1868, vol. 3.

REED, MINNIE.
  1902. Two ascomycetous fungi parasitic on marine algae. Univ. Calif. Publ.
      Bot., vol. 1, pp. 141–161, pls. 15, 16.

REINBOLD, T.
  1899. Die Chlorophyceen (Grüntange) der Kieler Föhrde. Schriften Natur-
      wis. Ver. Schleswig Holstein, vol. 8, p. 109.
  1893. Revision von Juergens' algae aquaticae. La Nouva Notarisia, ser. 4.
      pp. 192–206.

REINKE, J.
  1879. Zwei parasitische Algen. Botan. Zeitung, vol. 37, p. 473, pl. 6.
  1888. Einige neue braune und grüne Algen der Kieler Bucht. Berichte der
      deutsch. Botan. Gesellsch., vol. 6, p. 240.
  1889. Atlas deutscher Meeresalgen. Part I. Berlin.
  1892. *Ibid.* Part II. Berlin.
  1889a. Algenflora der westlichen Ostsee deutschen Antheils. Sexter Bericht
      der Kommission zur wissenschaftlichen Untersuchung der deutschen
      Meere in Kiel. I Heft. Berlin.

REINSCH, P. F.
  1890. Die Süsswasseralgenflora von Süd-Georgien. Deutsch Polar Exped.,
      vol. 2, no. 14, 15.

ROSENVINGE, L. K.
  1893. Grönlands Havalger. Meddelelser om Grönland, vol. 3, pp. 765–981.
      Copenhagen.
  1894. Les algues marine du Grönland. Ann. des. Sci. Nat., 7 sér., Bot., vol.
      19, pp. 53–164.

ROTH, A. G.
  1797–1806. Catalecta botanica quibus plantae novae et minus cognitae de-
      scribuntur atque illustrantur. Leipzig.
      1797, Fasc. 1.
      1800, Fasc. 2.
      1806, Fasc. 3.

ROUSSEL, H. F. A. DE.
  1806. Flora du Calvados et des terreins adjacens, composée suivant la méthode
      de M. Jussieu. Ed. 2.

RUMPHIUS, G. E.
  1743. Herbarium Amboinense, etc. Amsterdam.
      1741–1755. 6 vols.

Ruprecht, F. J.
    1851. Tange des Ochotskischen Meeres. Middendorff's Sibirische Reise, vol. 1, part 2, "lieferung" 2, p. 193.
        [We have been puzzled by the variety of quotation, both as to title and date, given by different writers for this very important work of Ruprecht. Our own copy seems to be the regular issue of "Dr. A. Th. Middendorff's Sibirische Reise" designated as "Band I, Theil 2, Botanik, Zweite Lieferung," etc., whose title page states place of publication as "Buchdruckerei der Kaiserlichen Akademie der Wissenschaften. St. Petersburg, 1851." It is stated that it is to be obtained from "Eggers et Co." of St. Petersburg and from "Leopold Voss" of Leipzig. On the reverse of the front cover is printed the following: "Auf Verfügung der Kaiserlichen Akademie der Wissenschaften. Fuss, beständiger Secretär. December, 1850." The title is the German title given above, viz., "Tange des Ochotskischen Meeres," etc. Pritzel (2nd ed.) gives the date as 1850 and the title as follows: "Algae Ochotenses. Die ersten sichern Nachrichten über die Tange des Octoskischen Meeres." In the account of Ruprecht's life and works in the Bulletin of the St. Petersburg Academy of Sciences for 1871 (Suppl.) the date is given as 1850. Heinsius gives title and date as recorded by Pritzel. The Botanische Zeitung (vol. 9, 1851, p. 443) gives the date as 1850, but later (vol. 14, 1856, p. 553) states that it was issued in 1851. The copy of Middendorff's Reise in the Gray Herbarium has been examined by F. S. Collins, who writes that the Ruprecht portion of "Band I" is distinctly stated to have been "Gedruckt 1856" and he has given this date in his "Green Algae of North America." Kjellman in "The Algae of the Arctic Sea" (1883, p. 331) has given the date as 1848. Under the circumstances, we have retained the date 1851 as best substantiated.]

Saunders, De A.
    1899. Four siphonaceous algae of the Pacific Coast. Bull. Torr. Bot. Club, vol. 26, pp. 1–4, pl. 350.
    1901. Papers from the Harriman Alaska Expedition, 25, The Algae. Proc. Wash. Acad. Sci., vol. 3, pp. 391–486. Washington.
    1904. Harriman Alaska Expedition with coöperation of Washington Academy of Sciences. Alaska, vol. 5, Cryptogramic Botany. The Algae of the expedition, pp. 153–212, pls. 10–29. New York, Doubleday, Page Co.
        [This is a reprint of the above paper, 1901, with change in title, volume, pages and plates. Otherwise it is the same.]

Schmidle, W.
    1899. Algologische Notizen VIII–XIII. Allg. Botan. Zeitschrift, nos. 1, 2.
        [Reviewed in Just's Botanischer Jahresbericht.]

Schmitz, F.
    1878. Ueber grüne Algen aus dem Golf von Athens. Sitzungs. d. Naturf. Gesell. zu Halle. Halle.
    1879. Beobachtungen ueber die vielkernigen Zellen der Siphonocladiaceen. Festschrift d. Naturf. Gesell. zu Halle. Halle.
    1896. Kleinere Beiträge zur Kenntniss der Florideen VI. La Nuova Notarisia, vol. 7, pp. 1–32.

SETCHELL, W. A.
  1899. Algae of the Pribilof Islands. *In* Jordan, Fur seals and fur-seal islands of the North Pacific Ocean, vol. 3, pp. 589–596, pl. 95. Washington.

SETCHELL, W. A., and GARDNER, N. L.
  1903. Algae of Northwestern America. Univ. Calif. Publ. Bot., vol. 1, no. 3, pp. 165–418, pls. 17–27.
  1920. Phycological Contributions I. Univ. Calif. Publ. Bot., vol. 7, no. 9, pp. 279–324, pls. 21–31.

SMITH, J. E.
  1790–1814. See ''English Botany.''

SNOW, JULIA W.
  1899. *Ulvella americana* n. sp. Bot. Gazette, 1899, vol. 27, pp. 309–314, pl. 7.

SOLIER, A.
  1846. Sur deux Algues zoosporées, formant le nouveau genre *Derbesia*. Rev. Bot., vol. 1, p. 452.
  1847. Mémoire sur deux algues zoosporées, devant former un genre distinct, le genre *Derbesia*. Ann. Sci. Nat., 3 sér., Bot., vol. 7, pp. 157–166, pl. 9.

STACKHOUSE, J.
  1795–1801. Nereis Britannica; continens species omnes Fucorum in insulis Britannicis crescentium. Bath.
      1795, Fasc. 1, pp. i–viii.
      1797, Fasc. 2, pp. ix–xviii.
      1801, Fasc. 3, pp. xix–xl and 1–112, with appendix, pls. 1–17 and A–G.

STOCKMAYER, S.
  1890. Ueber die Algengattung *Rhizoclonium*. Verhandl. der k.-k. zool.-bot. Ges. Wien, vol. 40, p. 571.

STROEMFELT, H. F. G.
  1886. Om Algenvegetationen vid Islands Kuster. Akademisk Afhandling Köngl. Vet. och Vitt. Samh. Handl. Göteborg.

SURINGAR, W. F. R.
  1867. Algarum Japonicarum Musei Botanici Lugduno-Batavi, Index praecursorius. Annales Bot. Musei Bot. L. B. (Sept.).
  1868. *Ibid.*, Hedwigia, vol. 7, p. 53.
  1870. Algae Japonicae Musei Lugduno-Batavi (edidit societas scientiarum Hollandica quae Harlemi est) (cf. also Hedwigia, vol. 9, p. 129).

SVEDELIUS, N.
  1900. Algen aus den Ländern der Magellansstrasse und Westpatagonien. I Chlorophyceae. Svenska Exped. till Magellansländerna, vol. 3, no. 8, pp. 283–316, pls. 16–18. Stockholm.

THURET, G.
  1850. Recherches sur les zoospores des algues et les anthéridies des cryptogames. Ann. Sci. Nat., 3 sér., Bot., vol. 14, p. 214.
  1854. Note sur la synonymie des *Ulva Lactuca* et *latissima* L. suivie de quelques remarques sur la tribu des Ulvacées. Mém. Soc. Sci. Nat. de Cherbourg, vol. 2, pp. 17–32.

THURET, G., and BORNET, E.
  1878. Études phycologiques. Paris.

TILDEN, JOSEPHINE E.
  1894–1909. American Algae (Exsicc.). Centuries 1–7, fasc. I.

TRELEASE, W. (with P. A. SACCARDO and C. H. PECK).
  1904. The Fungi of Alaska (issued by the Harriman Alaska Expedition with
     coöperation of Washington Academy of Sciences), vol. 5, pp. 11–64,
     pls. 2–7.

TREVISAN, V.
  1842. Prospetto della Flora Euganea. (Cf. Flora, oder allgemeine botanische
     Zeitung, vol. 26, 1843 (July) p. 464, for further information.)

TURNER, D.
  1802. A synopsis of the British Fuci. 2 vols. London.
  1808–1819. Fuci sive plantarum fucorum generi a botanicis ascriptarum
     icones descriptiones et historia. London.
       1808, vol. 1.
       1809, vol. 2.
       1811, vol. 3.
       1819, vol. 4.

TYSON, W.
  1909–1910. South African marine algae. Fasc. 1, 2. (Exsicc.)

VICKERS, ANNA.
  1908. Phycologia Barbadensis. Iconographie des algues marines récoltées à
     l'ile Barbade (Antilles). (Chlorophycées et Phéophycées). Paris.

WEBER, F., and MOHR, E. M. H.
  1804. Naturhistorische Reise durch einen Theil Schweden. Göttingen.

WEST, G. S.
  1904. A treatise on the British freshwater algae. Cambridge.
  1916. Algae I. Myxophyceae, Peridinieae, Bacillarieae, Chlorophyceae. Cam-
     bridge Botanical Handbooks.

WIGGERS, F. H.
  1780. Primitiae florae Holsatiae. Kiliae.

WILLE, N.
  1897. Conjugatae and Chlorophyceae, *in* Engler and Prantl, Natürl. Pflanzen-
     fam., 1 Th., 2 Abt. Leipzig.
       1890, Lief. 40, pp. 1–48.
       1890, Lief. 41, pp. 49–96.
       1890, Lief. 46, pp. 97–144.
       1891, Lief. 60, pp. 145–175.
     [The date of the title page of the volume on Algae is 1897.]
  1899. Meddelelser om sein Undersögelser angaaende Cellekjaenernes Forhold
     hos Slaegten *Acrosiphonia* (J. Ag.) Kjellm. Botaniska Notiser, p.
     281. Lund.
  1900. Die Zellkerne bei *Acrosiphonia* (J. Ag.) Kjellm. Botan. Centralb.,
     vol. 81, p. 238.
  1901. Studien ueber Chlorophyceen I–VII. Medd. f. d. Biol. Station Drobak
     No. 2 Vidensk. Skrifter. I Math. naturv. Klasse, 1900, no. 6. Chris-
     tiania.
  1906. Algologische Untersuchungen an der biologischen Station in Drontheim
     I–VII. Det Kgl. Norske Vidensk. Selk. Skrifter, no. 3.
  1909. Conjugatae und Chlorophyceae, *in* Engler and Prantl, Natürl. Pflanzen-
     fam., Nachträge zum 1 Th., 2 Abt.

WITTROCK, V. B.
  1866. Försök till en monographi öfver algslägtet *Monostroma.* Akademisk afhandling.
  1880. Points-fortekning öfver Skandinaviens växter. Part 4. Lund.

WITTROCK, V. B., and NORDSTEDT, O.
  1877–1903. Algae aquae dulcis exsiccatae praecipue Scandinavicae quas adjectis algis marinis chlorophyllaceis et phycochromaceis. Upsala.
    1877–1889, Fasc. 1–21.
    1893–1903, Fasc. 22–35.
    In 1896 G. Lagerheim was added as one of the distributors.

WOODWARD, T. J.
  1797. Observations upon the generic characters of *Ulva,* with the descriptions of some new species. Trans. Linn. Soc., vol. 3, pp. 46–58. London.

WYATT, MARY.
    Algae Danmonienses, or dried specimens of marine plants principally collected in Devonshire. 4 vols. and supplement. Tor Quay.

YENDO, K.
  1903. Three species of marine *Ecballocystis.* The Bot. Mag., vol. 17, p. 199, pl. 8, figs. 1–15. Tokyo.
  1914. Notes on algae new to Japan II. The Bot. Mag., vol. 27, p. 263. Tokyo.
  1916. Notes on algae new to Japan V. The Bot. Mag., vol. 30, p. 243. Tokyo.

ZANARDINI, G.
  1843. Saggio di classificazione naturale delle ficee. Venice. (Also Botan. Zeit., 1844, pp. 401–408.)
  1860–1871. Iconographia Phycologica Adriatica, vols. 1–3. Venice.
    [From the printed evidence, it is difficult to determine the exact dates of the publication of this large work of Zanardini. The title pages of the volumes bear dates as follows: vol. 1, 1860; vol. 2, 1865; and vol. 3, 1871. These dates are not sufficiently exact for reference as to priority of publication. The three volumes of the Iconographia are made up of reprints from the various parts published in the "Memorie dell' I. R. Istituto Veneto," where they appeared under the title "Scelta di Ficee nuove o piu rare del Mare Adriatico." The general distribution of pages, plates and approximate dates have been collated by Dr. W. G. Farlow and confirmed, so far as possible, by us. They are as follows:
    Volume 9 of "Memorie," dated 1860, has part 1, plates 1–8, pages 41–78, was presented April 16, 1860, and is the same as volume 1 of the "Iconographia," plates 1–8, Preface and pages i-iv and 1–34.
    Vol. 10, 1861, parts 1?, 2?, pls. 9–16, pp. 93–124, presented June 17, 1861, is vol. 1, pls. 9–16, pp. 35–66.
    Vol. 10, 1861, part 3, pls. 26–33, pp. 449–484, presented Apr. 23, 1862, is vol. 1, pls. 17–24, pp. 67–102.
    Vol. 11, 1862, part 2, pls. 11–18, pp. 271–306, presented Apr. 15, 1863, is vol. 1, pls. 25–32, pp. 103–138.
    Vol. 12, 1864, part 1, pls. 1–8, pp. 9–43, presented May 30, 1864, is vol. 1, pls. 33–40, pp. 139–175.
    Vol. 12, 1864, part 2, pls. 14–21, pp. 377–410, presented May 22, 1865, is vol. 2, pls. 41–48, preface, index and pp. iii-viii, 1–32.

Vol. 13, 1866, part 1, pls. 2–9, pp. 143–176, no date of presentation, is vol. **2, pls. 49–56, pp. 33–66.**

Vol. 13, 1866, parts 2?, 3?, pls. 10–17, pp. 403–434, presented May 27, 1867, is vol. 2, pls. 57–64, pp. 67–98.

Vol. 14, 1869, part 2, pls. 4–11, pp. 181–216, presented June 22, 1868, is vol. 2, pls. 65–72, pp. 99–134.

Vol. 14, 1870, part 3, pls. 26–33, pp. 437–472, presented June 21, 1869, is vol. 2, pls. 73–80, pp. 135–168.

Vol. 15, 1871, part 2, pls. 10–17, pp. 425–460, presented June 20, 1870, is vol. 3, pls. 81–88, pp. 5–36.

Vol. 17, 1873, part 3, pls. 14–21, pp. 429–460, presented Dec. 22, 1872, is vol. 3, pls. 89–96, pp. 37–68.

Vol. 18, 1874, part 1, pls. 2–9, pp. 255–286, presented Dec. 22, 1873, is vol. 3, pls. 97–104, pp. 69–100.

Vol. 19, 1876, part 3, pls. 23–30, pp. 513–544, presented June 18, 1876, is vol. 3, pls. 105–112, pp. 101–132.]

EXPLANATION OF PLATES

PLATE 9

Hormiscia sphaerulifera S. and G.

Fig. 2. A, basal portion of filament showing intramatrical rhizoids. B, vegetative cells in the median portion of the filament. C, an empty sporangium. × 25.

Hormiscia grandis (Kylin) S. and G.

Fig. 3. A, basal portion of filament showing numerous intramatrical rhizoids. B, two zoosporangia of moderate size. C, a long typical zoosporangium, empty. D, moderate sized vegetative segments. × 80.

Hormiscia penicilliformis (Roth) Fries

Fig. 4. A, sporeling with rhizoid penetrating a filament of *Ulothrix flacca*. B, basal portion of a young filament, showing a few intramatrical rhizoids. C, characteristic zoosporangia. D, young vegetative segments. × 80.

Rhizoclonium lubricum S. and G.

Fig. 5. A, terminal portion of a young filament with short segments. B, terminal portion of a filament with long, "resting segments." × 160.

Ulothrix pseudoflacca f. *maxima* S. and G.

Fig. 6. A, B, vegetative filaments. C, sporangia. × 250.

Codium dimorphum Svedelius

Fig. 7. A young utricle with an empty sporangium. × 80.

Fig. 8. A terminal portion of a utricle, showing the extremely thick, lamellate and tuberculate end wall. × 100.

Codium intertextum var. *cribosum* M. A. Howe

Fig. 9. Portion of the end wall of a utricle, showing internal modifications. × 100.

Codium Setchellii Gardner

Fig. 10. Typical utricle and sporangium, showing the scars of three previous sporangia. × 80.

Fig. 11. Showing different forms of utricles and sporangia.

This plate is from Gardner, New Pac. Coast Mar. Alg. IV, 1919, pp. 487–496, pl. 42. Fig. 9 was labeled *Codium adhaerens*.

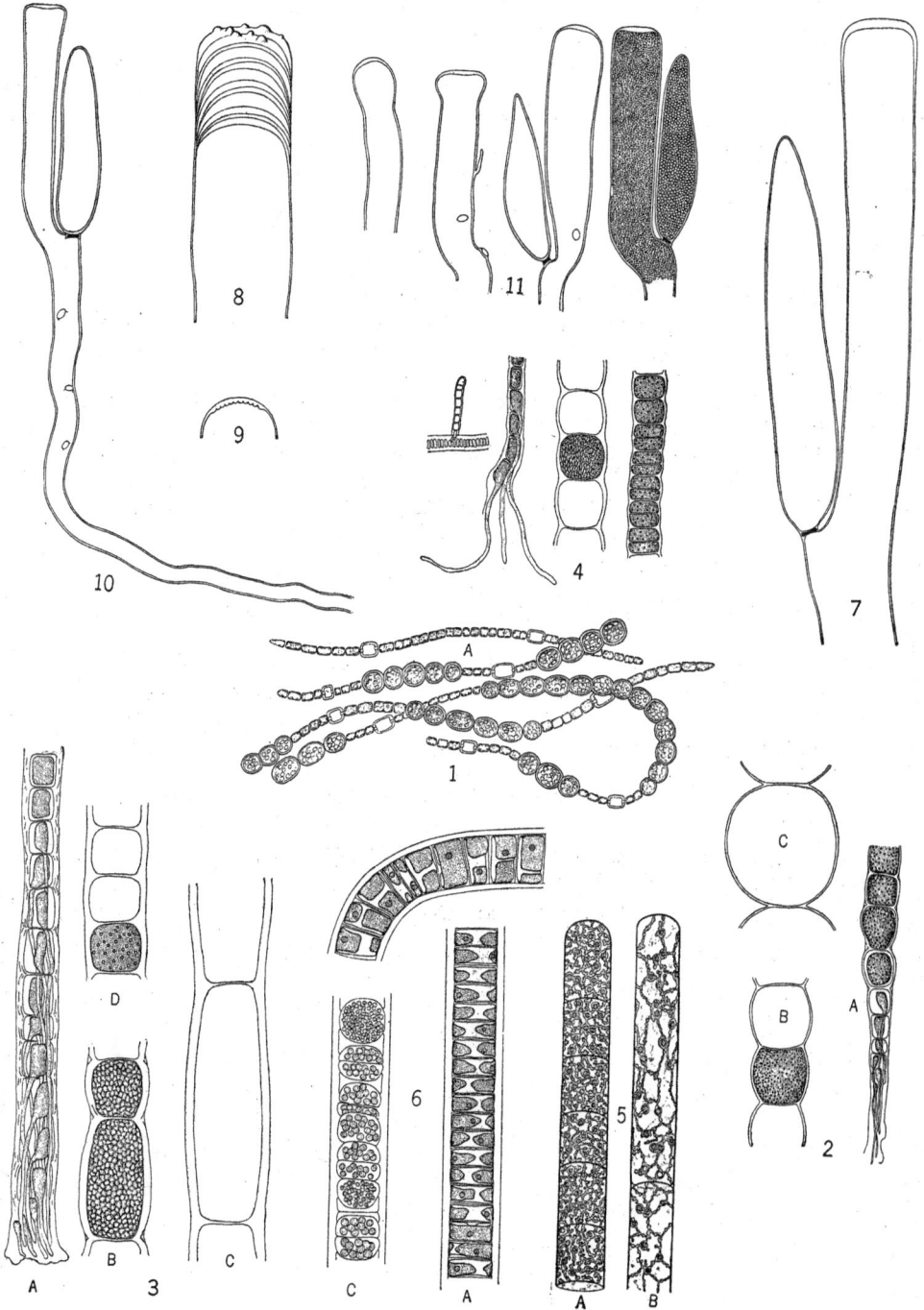

PLATE 10

Prasiola borealis Reed

Fig. 1. A group of whole plants infested by the parasitic fungus, *Guignardia alaskana* Reed. × 3.5.

Fig. 2. A cross section of the frond. × 375.

Fig. 3. A surface view at the margin of the frond showing the grouping of the cells. × 285.

Figures 1–3 are from Reed, Two new Asco. Fungi, 1902, pp. 141–164, pls. 15, 16.

Collinsiella tuberculata S. and G.

Fig. 4. Habit sketch of a group of plants. × 4.

Fig. 5. A vertical section through one of the fronds.

Fig. 6. Dissection of a part of the vertical section which has been treated with Chloriodide of Zinc to show the branching. The cell contents are much shrunken.

Fig. 7. A tangential section at the surface to show the division planes.

Fig. 8. A terminal cell showing a diminished chromatophore and two pyrenoids.

Fig. 9. A terminal cell showing the vacuolated appearance of the chromatophore.

Fig. 10. A young terminal cell showing a complete parietal chromatophore and one pyrenoid.

Figures 4–10 are from Setchell and Garder, Alg. N.W. Amer., 1903, pp. 165–418, pl. 17.

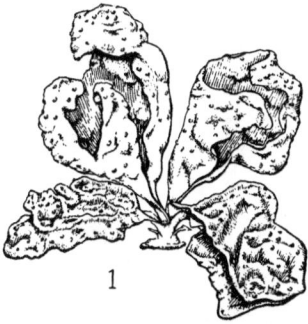

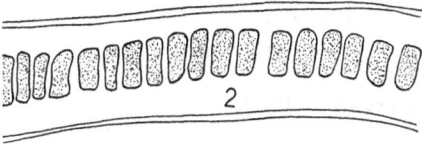

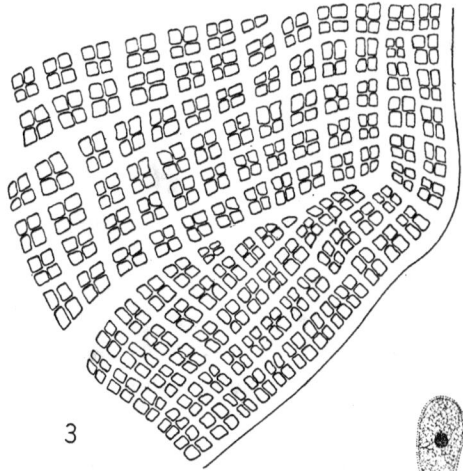

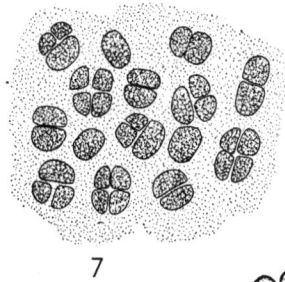

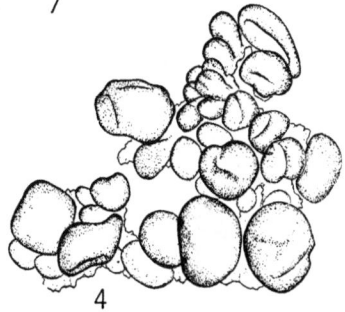

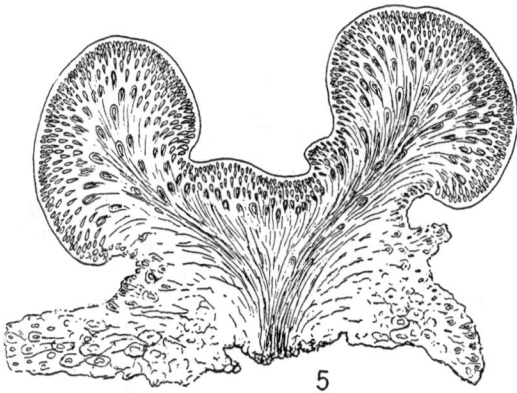

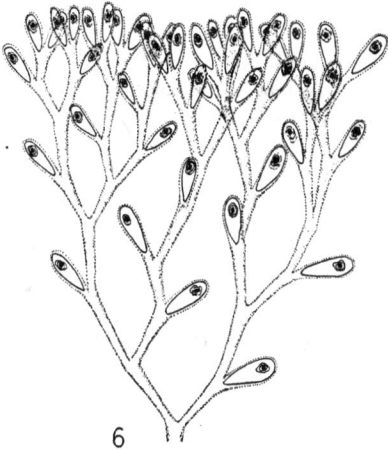

PLATE 11

Pseudulvella prostrata (Gardner) S. and G.

Fig. 1. Showing a small section of the margin of the thallus on the surface of *Iridaea laminarioides*.

Fig. 2. A vertical section of fig. 1, perpendicular to the long diameter of the filaments of *Ulvella*: the drawing is imperfect, it should show a common enclosing cuticle.

Endophyton ramosum Gardner

Fig. 3. Showing the narrow filaments within the central part of the host, *Iridaea laminarioides*, terminating in sporangia at the surface.

Fig. 4. Showing pieces of irregularly shaped filaments.

Pseudodictyon geniculatum Gardner

Fig. 5. A part of a thallus showing a few terminal filaments penetrating among the cortical cells of *Laminaria Sinclairii*.

Fig. 6. A cross section of fig. 5, showing terminal sporangia on short, erect filaments.

All highly magnified.

This plate is from Gardner, New Chlorophyceae from California, 1909, pp. 371–376, pl. 14.

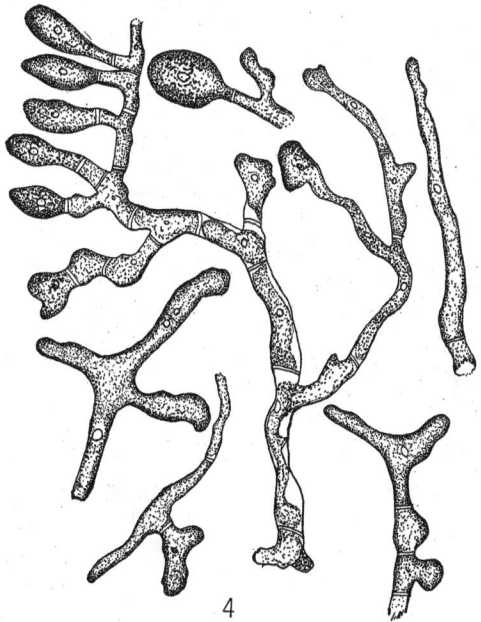

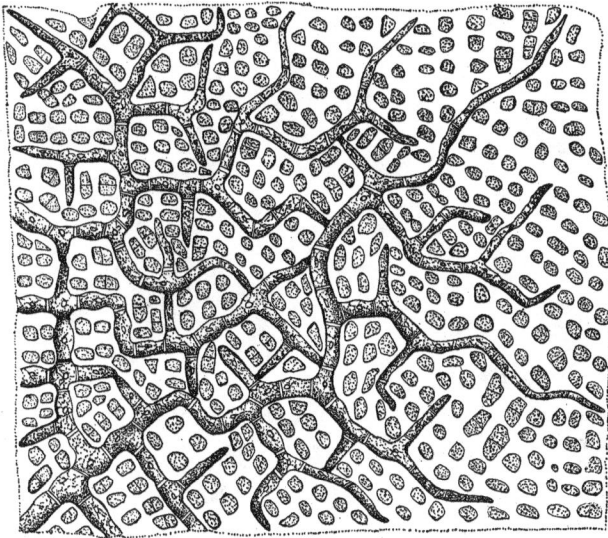

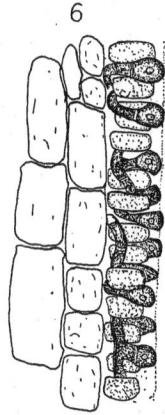

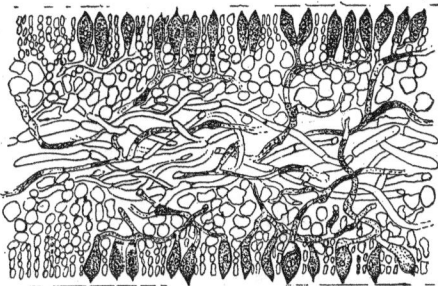

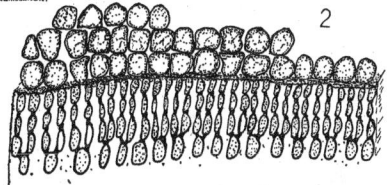

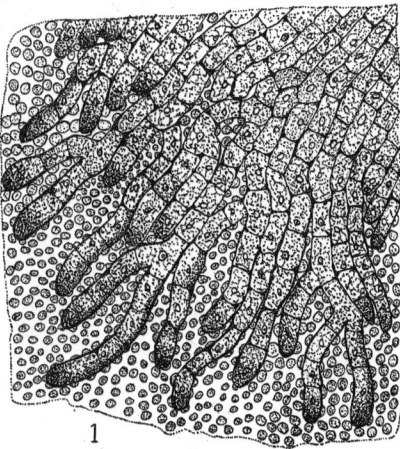

PLATE 12

Ulva Linza L.

Fig. 1. A vertical section through the margin of a frond showing the separation of the two layers. Plant from Friday Harbor, Washington. × 250.

Fig. 2. A surface view of fig. 1. × 250.

Fig. 3. The same view as fig. 1. Plant from Admiralty Island, Alaska. × 250.

Fig. 4. A surface view of fig. 3.

Gayella constricta S. and G.

Fig. 5. Habit sketch of a whole plant. × 30.

Fig. 6-8. Showing the development of rhizoids.

Fig. 9. Two enlarged segments showing shapes and arrangement of the cells in surface view. × 100.

Fig. 10. A cross section of a large segment showing the radial arrangement of the cells.

Figures 5-10 are from Gardner, New Pac. Coast Mar. Alg. I, 1917, pp. 377-416, pl. 33.

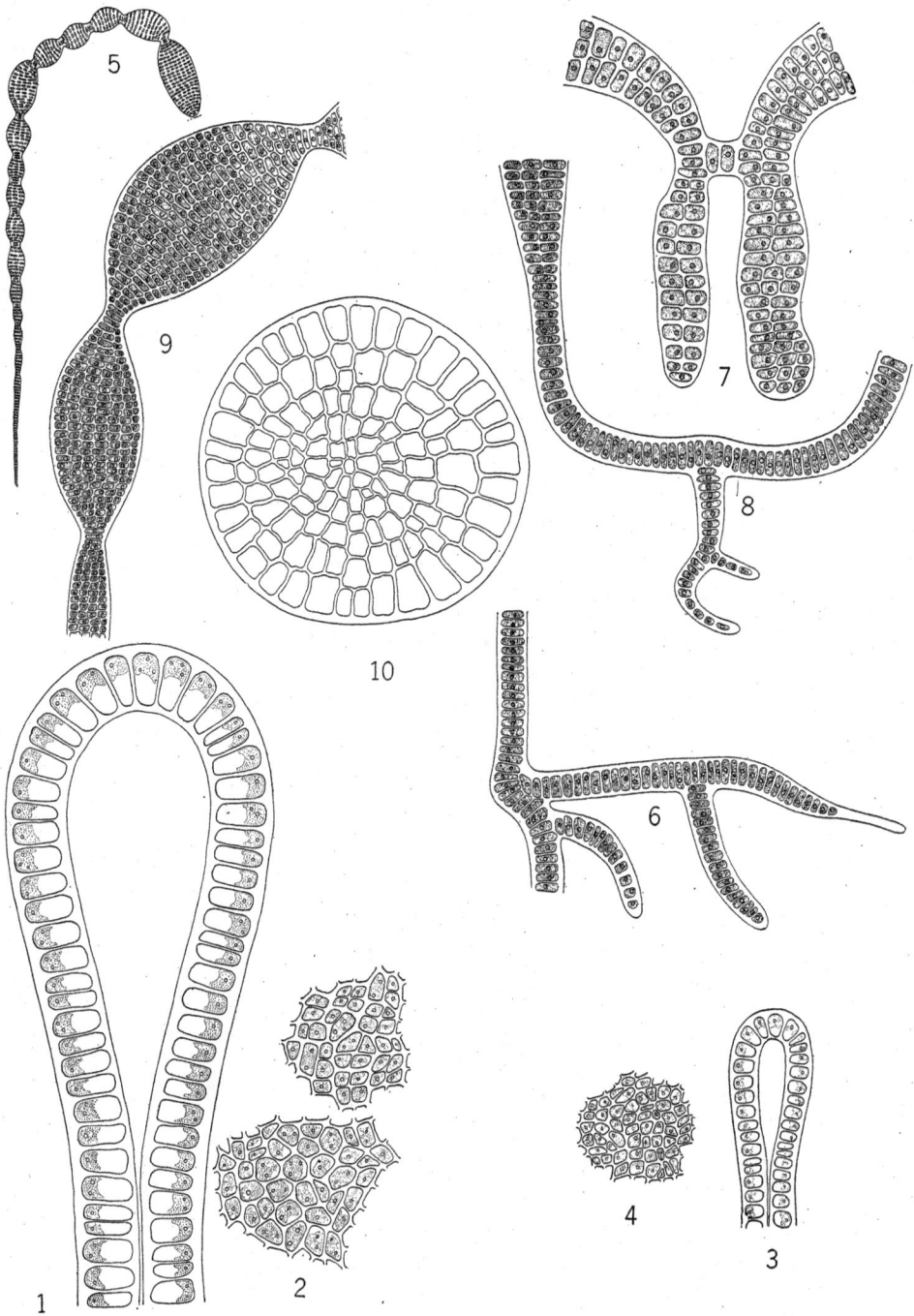

PLATE 13

Chlorochytrium inclusum Kjellm.

Fig. 1. A vertical section through the host, showing the penetration of the endophyte to the medulla. × 250.

Cladophora microcladioides Collins

Fig. 2. Showing the method of branching of the terminal ramuli. × 8.

Halimeda discoidea Dec'ne

Fig. 3. A habit sketch of a small plant. × 1.

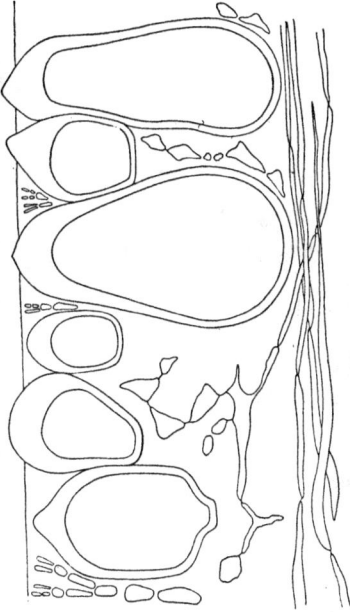

1

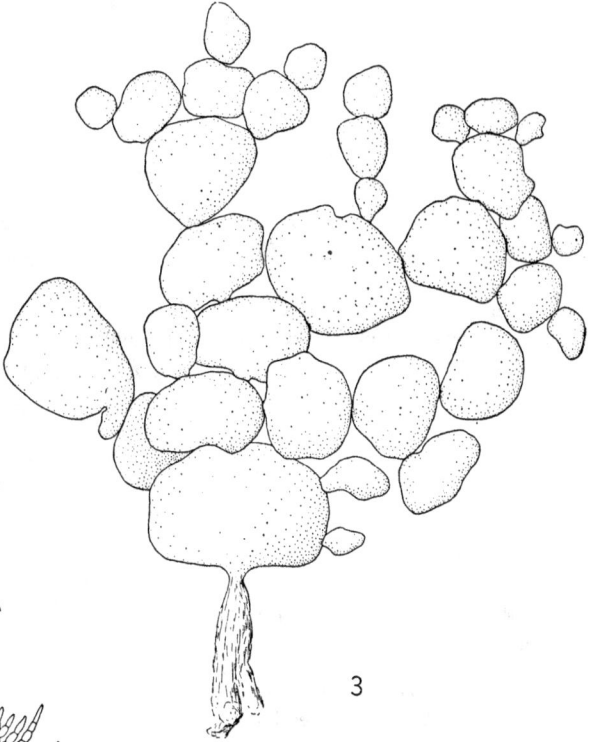

3

2

PLATE 14

Bryopsis plumosa (Huds.) Ag.

Fig. 1. Diagrammatic illustration of a whole plant showing the method of branching. × 1.

Fig. 2. A sketch of a branch showing the arrangement, the relative length and the constriction of the bases of the pinnules. × 40.

Halicystis ovalis (Lyngb.) Aresch.

Fig. 3. A habit sketch of a whole plant except the rhizoidal base. × 2.

Enteromorpha tubulosa Kuetz.

Fig. 4. A portion of a small frond showing the linear arrangement of cells. × 250.

Fig. 5. A section of a frond. × 250.

Percursaria percursa (Ag.) Rosenv.

Fig. 6. A portion of a frond showing one row of cells at one end and two rows at the other. × 250.

Enteromorpha compressa (L.) Grev.

Fig. 7. A vertical section through the frond. × 250.

Fig. 8. A surface view. × 250.

Chaetomorpha aerea (Dillw.) Kuetz.

Fig. 9. A group of young plants showing long basal cells with rhizoids. × 35.

Figs. 10, 11. Sketches showing different ages of the cells in vegetative stages. × 35.

Monostroma zostericola Tilden

Fig. 12. A surface view. × 500.

Fig. 13. A vertical section. × 500.

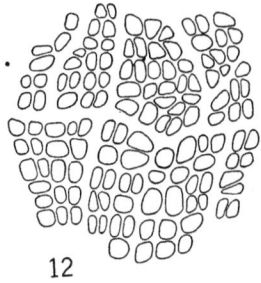

12

13

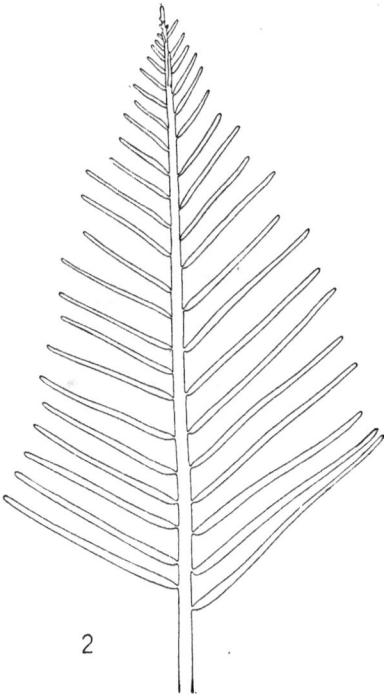

2

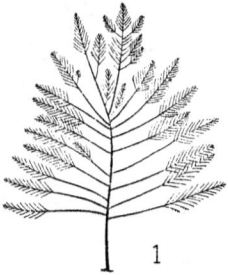

1

3

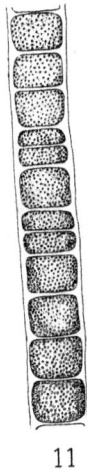

11

10

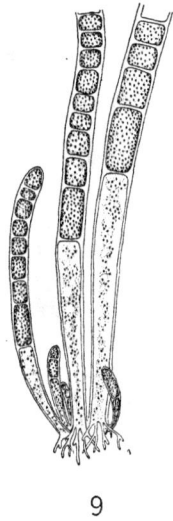

9

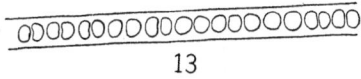

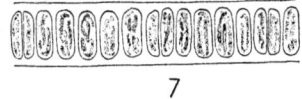

7

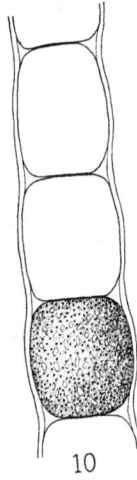

4

5

8

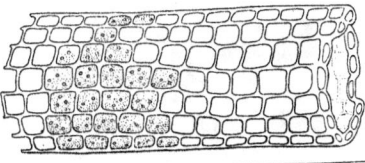

6

PLATE 15

Chlorochytrium Porphyrae S. and G.

Fig. 1. A vertical section through the host, showing plants of the endophyte in various stages of development and of embedding. × 140.

Codiolum gregarium A. Br.

Fig. 2. A group of three plants. × 120.

Derbesia marina (Lyng.) Kjellm.

Fig. 3. A part of a filament showing the method of branching, and one sporangium. × 65.

Bryopsis corticulans Setchell

Figs. 4, 5. Sketches showing origin and method of development of the corticating filaments. × 25.

Codium latum Suring.

Fig. 6. Sketch of a single utricle showing the position of the sporangia and of the hairs. × 100.

Figure 1 is from Gardner, New Pac. Coast Mar. Alg. I, 1917, pp. 377–416, pl. 32, fig. 6.

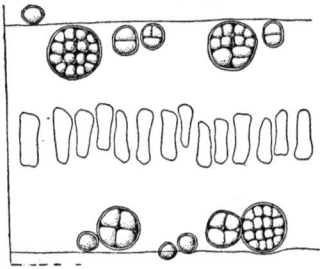

1

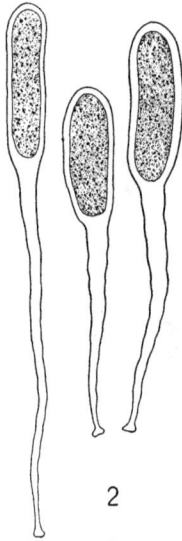

2

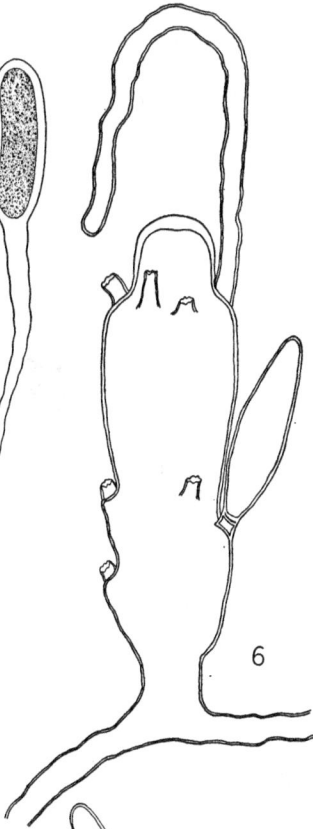

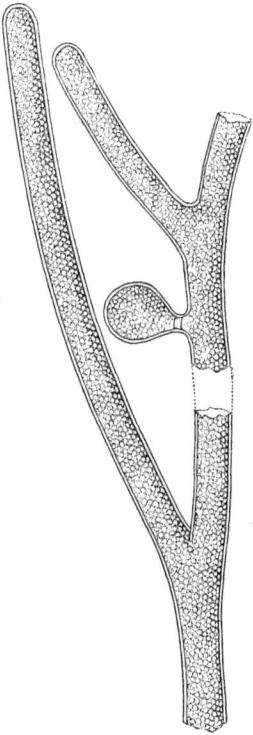

3

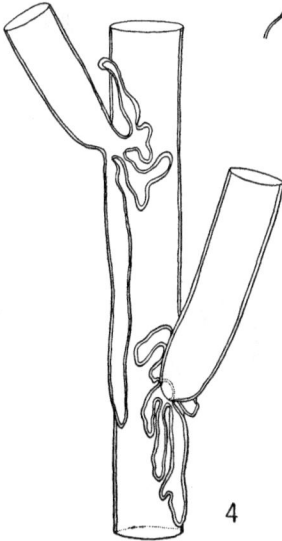

4

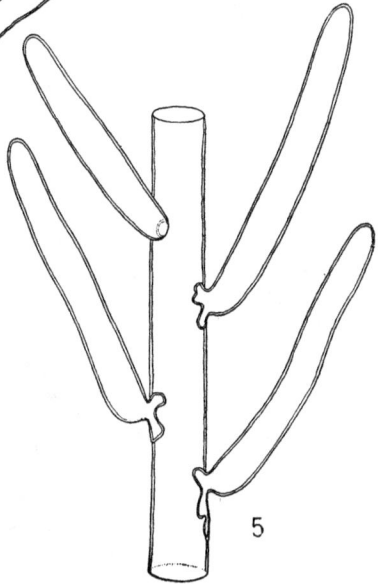

5

6

PLATE 16

Enteromorpha micrococca var. *subsalsa* Kjellm.

Fig. 1. A habit sketch of a piece of a frond showing the method of branching. × 20.

Cladophora trichotoma (Ag.) Kuetz.

Fig. 2. A habit sketch of a few terminal ramuli.

Enteromorpha compressa (L.) Grev.

Fig. 3. A habit sketch of a frond. × 1.

Spongomorpha coalita (Rupr.) Collins

Fig. 4. Sketch showing the characters of the hooked branches. × 10.

Codium Ritteri S. and G.

Fig. 5. A sketch of a group of utricles. × 25.

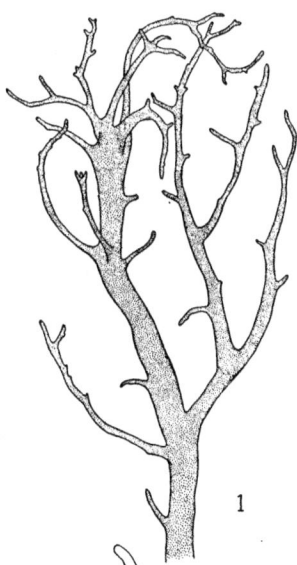

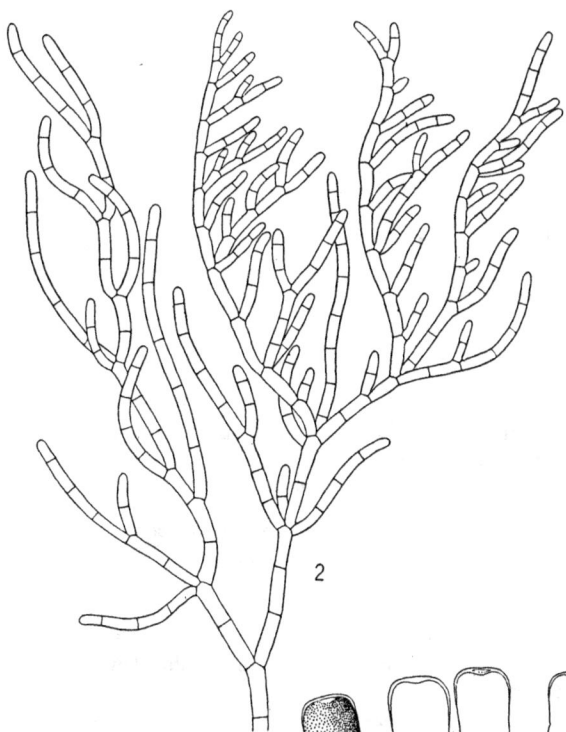

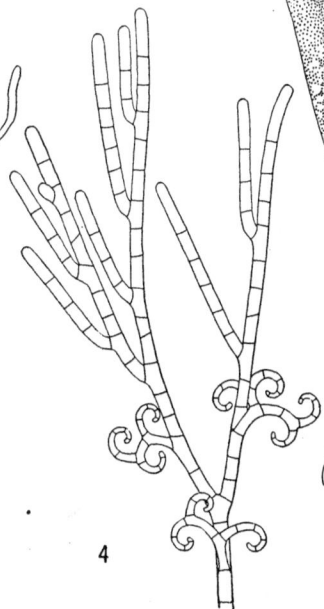

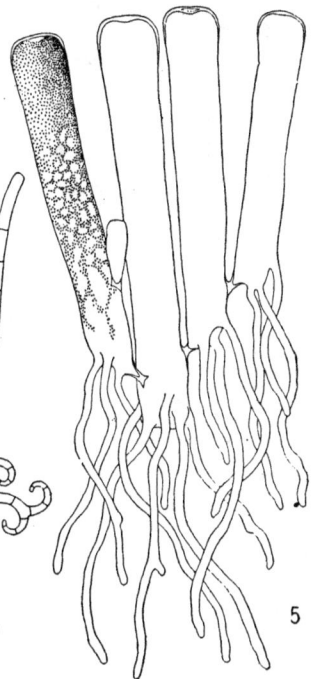

PLATE 17

Pseudopringsheimia apiculata S. and G.

Fig. 1.   A section through the thallus of a mature plant perpendicular to the host.   × 250.

Fig. 2.   A section through the thallus of a young plant.   × 250.

Prasiola delicata S. and G.

Fig. 3.   a-f, Series of different forms of plants.   × 10.

Ulva vexata S. and G.

Fig. 4.   A group of plants showing different shapes and sizes.   × 1.

Fig. 5.   A group of mature plants showing the presence of the parasitic fungus, *Guignardia Ulvae* Reed.   × 3.

Fig. 6.   A cross section showing the presence of fungal hyphae in the medulla. × 250.

Fig. 7.   A surface view.   × 250.

This plate is from Setchell and Gardner, Phyc. Cont. I, 1920, pp. 279–324, pl. 22.

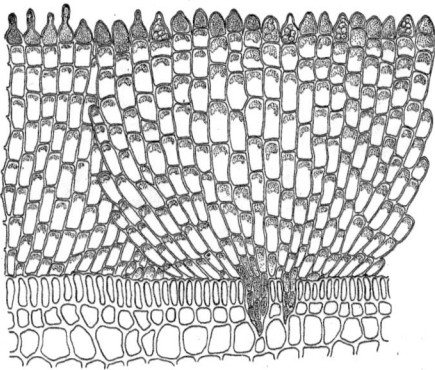

1

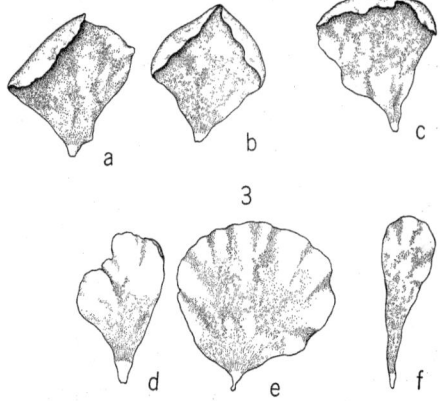

a   b   c

3

d   e   f

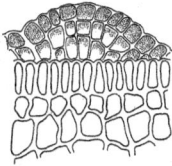

2

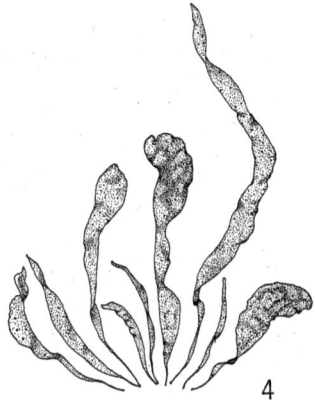

4

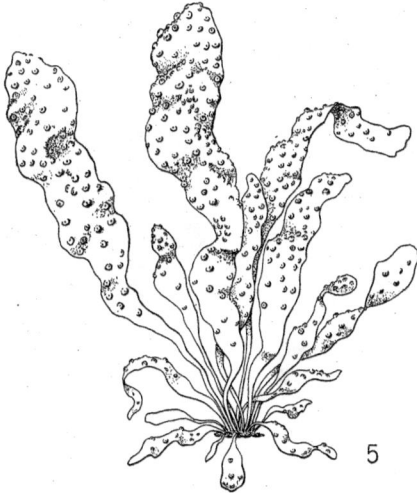

5

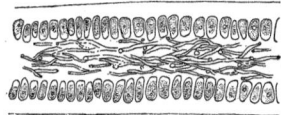

6

7

PLATE 18

Gomontia caudata S. and G.

Fig. 1.   Two pieces of filaments.   × 400.

Fig. 2.   a-d, Different stages and forms of "sporangia."   × 400.

Internoretia Fryeana S. and G.

Fig. 3.   A surface view of the host plant, showing the method of permeation and branching of a few terminal filaments.   × 375.

Fig. 4.   A stage slightly in advance of fig. 3, showing cell divisions in planes parallel to the long diameter of the cells.   × 375.

Fig. 5.   A stage in development nearing maturity.   × 375.

Fig. 6.   A cross section of the host cutting the filaments of *Internoretia* at right angles to their long diameter.   × 375.

Entocladia cingens S. and G.

Fig. 7.   A plant growing in the membrane of *Chaetomorpha californica* and nearing maturity.   × 250.

This plate is from Setchell and Gardner, Phyc. Cont. I, 1920, pp. 279–324, pl. 23.

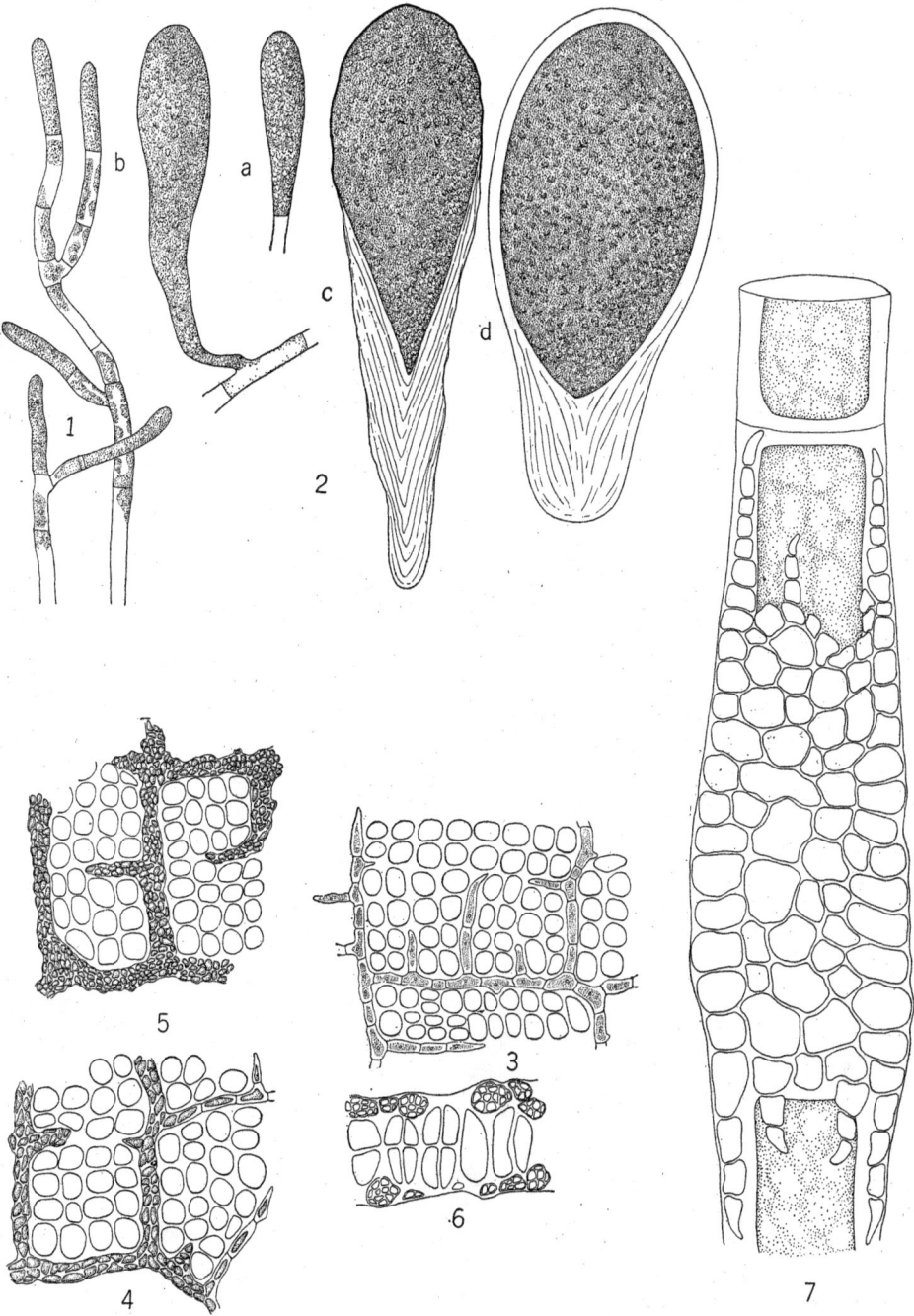

PLATE 19

Gomontia polyrhiza (Lagerh.) B. and F.

Fig. 1. A group of three ''sporangia,'' the two larger nearing maturity.
× 175.

Gomontia habrorhiza S. and G.

Fig. 2. A young thallus. × 375.

Fig. 3. a-e, Illustrating three forms of the ''sporangia.''

Pseudulvella consociata S. and G.

Fig. 4. A surface view of a young thallus. × 375.

Fig. 5. A section of a mature thallus. × 375.

Fig. 6. A vertical filament near the surface of a young thallus showing
branching. × 225.

Entocladia codicola S. and G.

Fig. 7. a, A young thallus, showing the method of branching of the filaments and of their radiation from a center. × 125. b, A mature thallus with
sporangia in the center. × 125.

Prasiola delicata S. and G.

Fig. 8. A surface view showing typical arrangement of cells. × 500.

This plate is from Setchell and Gardner, Phyc. Cont. I, 1920, pp. 279-324,
pl. 24.

[346]

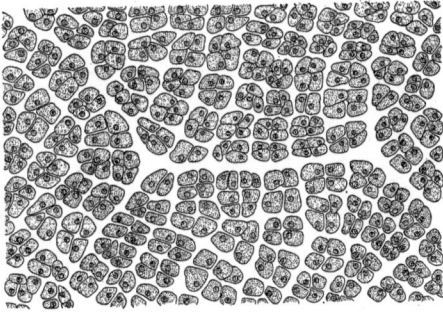

8

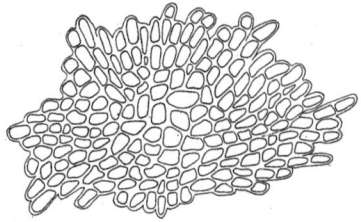

4

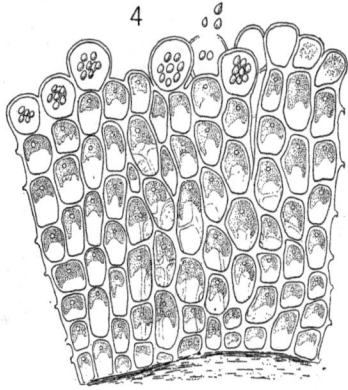

5

6

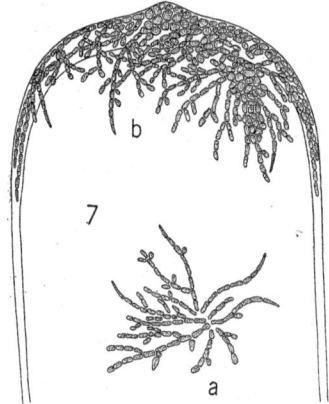

7

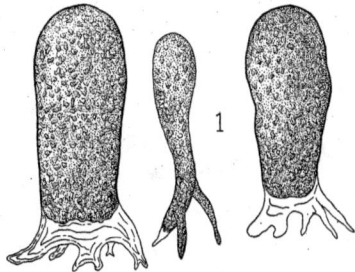

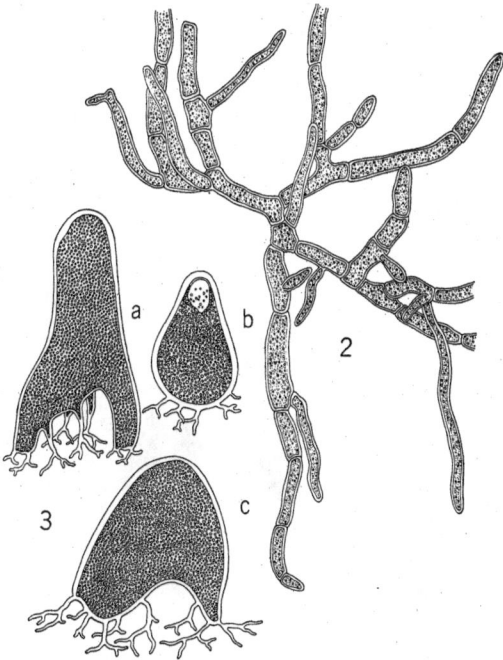

PLATE 20

Prasiola delicata S. and G.

Fig. 1. A micro-photograph of a marginal segment, surface view showing the arrangement of the vegetative cells. × 442.

Prasiola meridionalis S. and G.

Fig. 2. A micro-photograph of a portion of the surface, showing vegetative cells and interspersed aplanospores (?). × 442.

This plate is from Setchell and Gardner, Phyc. Cont. I, 1920, pp. 279–324, pl. 25.

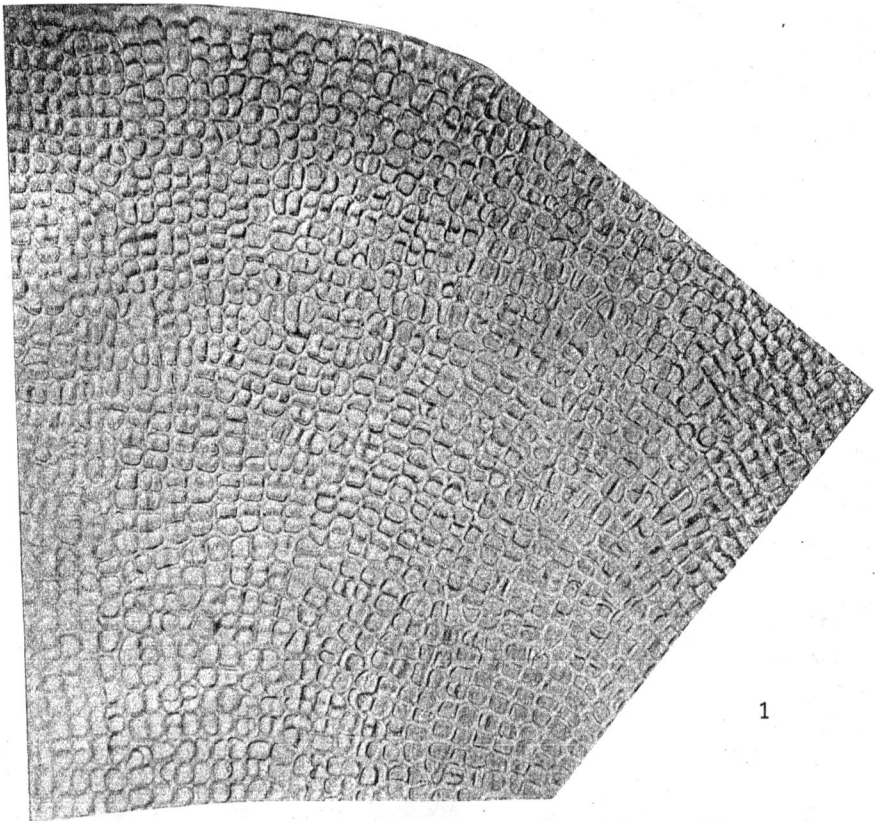

1

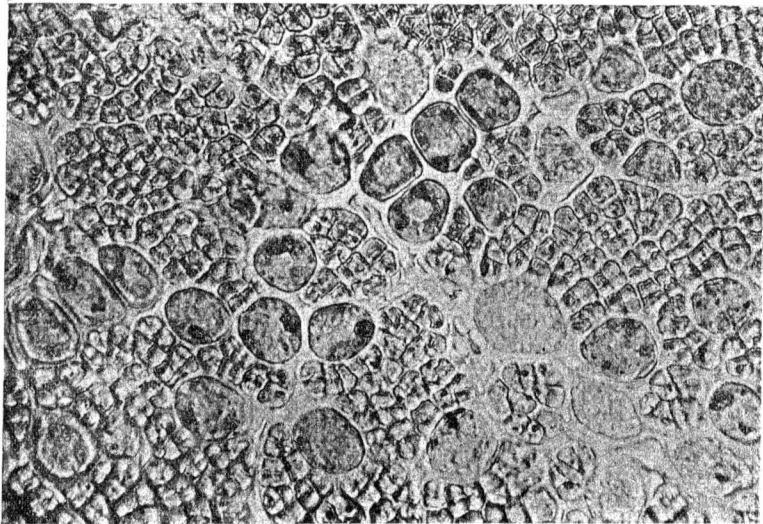

2

PLATE 21

Ulva dactylifera S. and G.

Fig. 1. Photograph of a whole plant, with the exception of a portion of the base, the type. ✕ 0.75.

Ulva stenophylla S. and G.

Fig. 2. A micro-photograph of a portion of the surface, showing the rounded angles and relatively thick walls of the cells. ✕ 442.

1

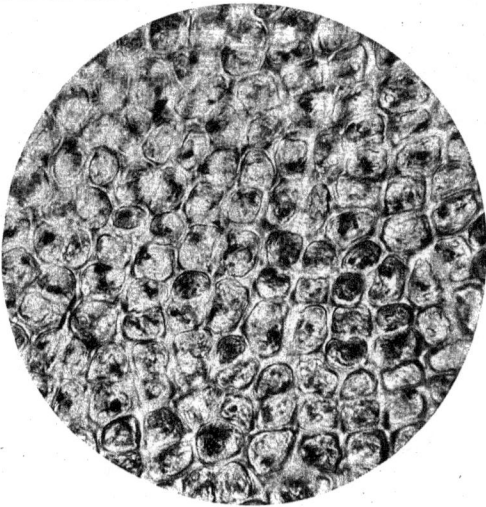

2

PLATE 22

Ulva angusta S. and G.

A photograph of a group of plants, the type.

PLATE 23

Ulva taeniata (Setchell) S. and G.

A photograph of a whole dried plant, showing the extreme crisped nature of the dentate margins at the base. × 0.3.

Ulva stenophylla S. and G.

A photograph of a whole plant, the type.   × 0.3.

PLATE 25

Monostroma areolatum S. and G.

A photograph of a whole dried plant, the type. $\times$ 0.5.

PLATE 26

Ulva angusta S. and G.

Fig. 1. A micro-photograph of a part of the surface. $\times$ 442.

Monostroma areolatum S. and G.

Fig. 2. A micro-photograph of a part of the surface.

Plates 21–26 are from Setchell and Gardner, Phyc. Cont. I, 1920, pp. 279–324, pls. 26–31.

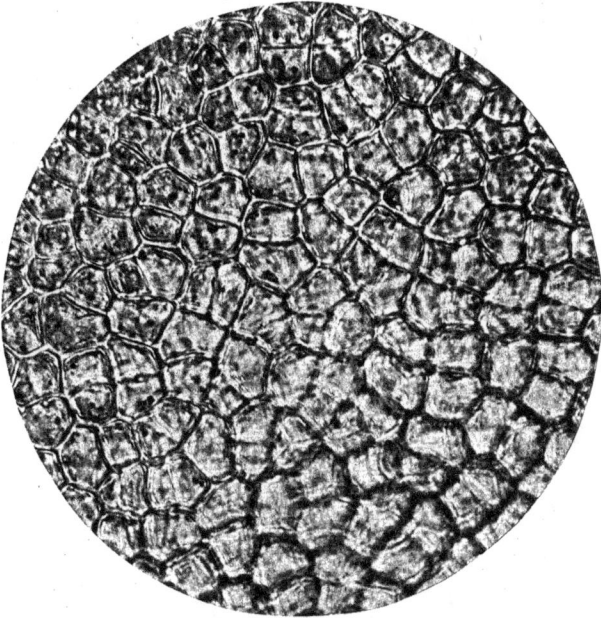

1

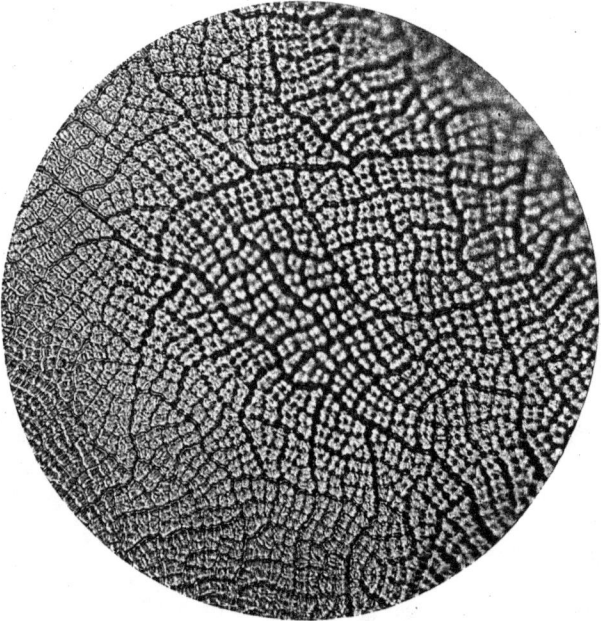

2

Bryopsis corticulans Setchell

A photograph of a whole plant. × 1.

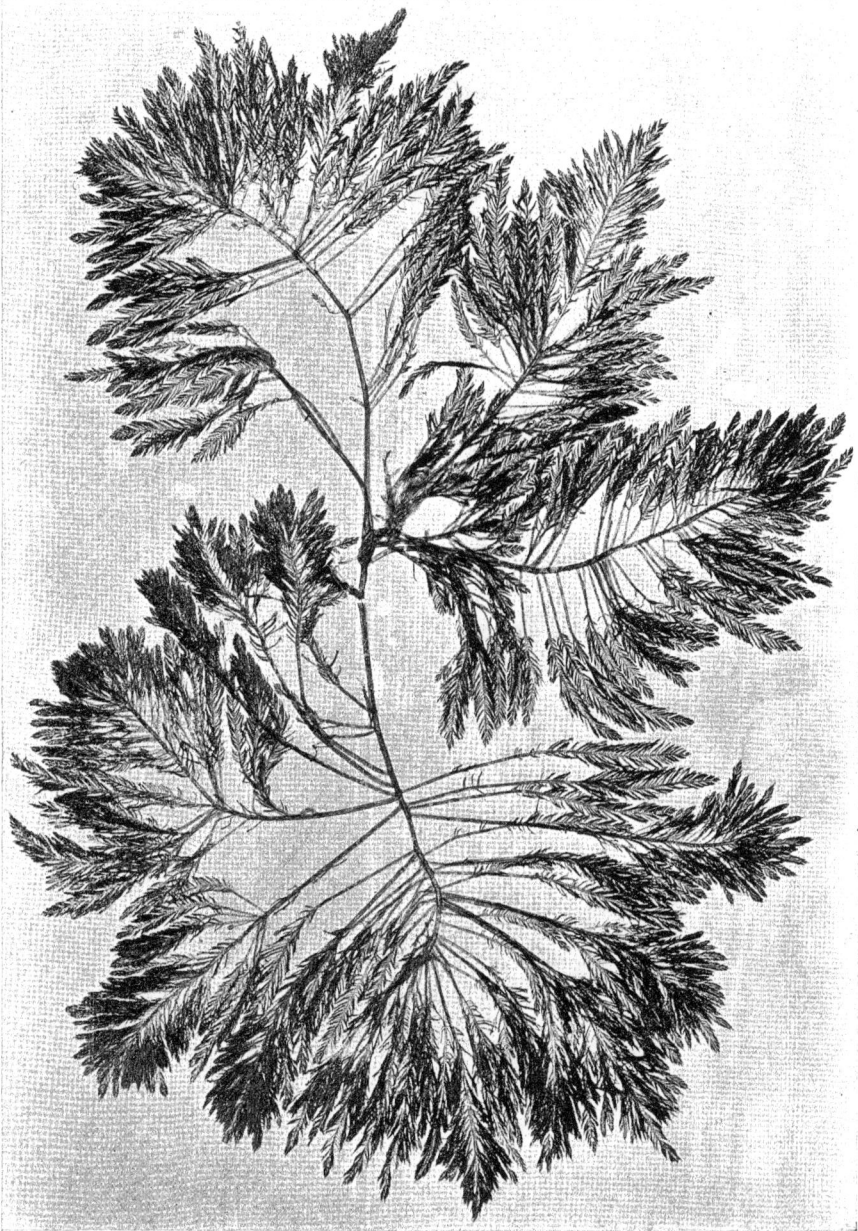

PLATE 28

Codium fragile (Suring.) Hariot

A photograph of an entire dried plant, showing several fronds arising from the same expanded holdfast.   × 0.5.

PLATE 29

Codium fragile (Suring.) Hariot

A photograph of a dried frond, showing long, slender branches and dichotomous branching.   × 0.5.

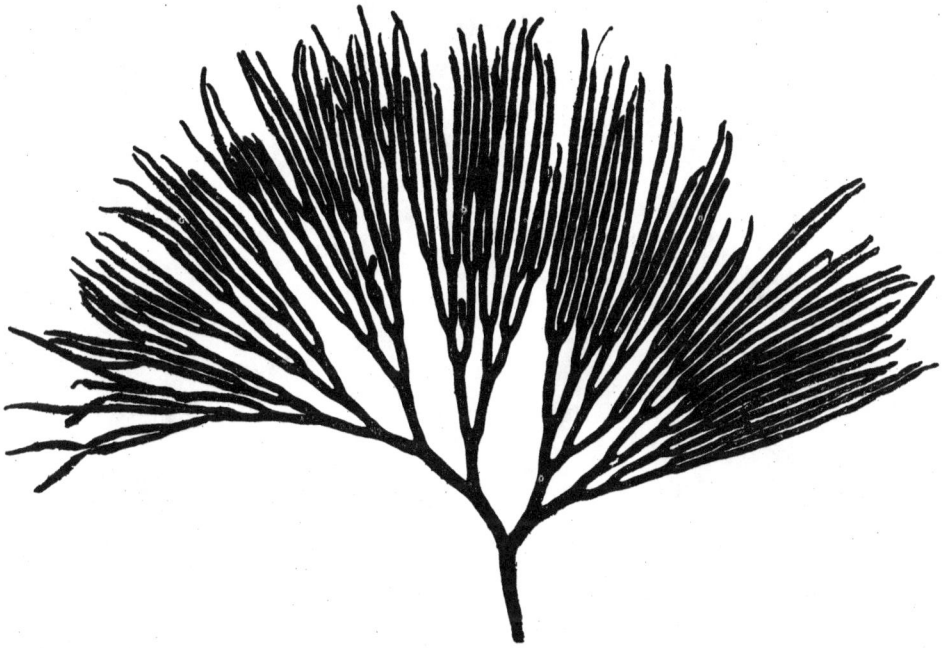

PLATE 30

Codium Setchellii Gardner

A photograph of a part of a thallus. $\times$ 1.

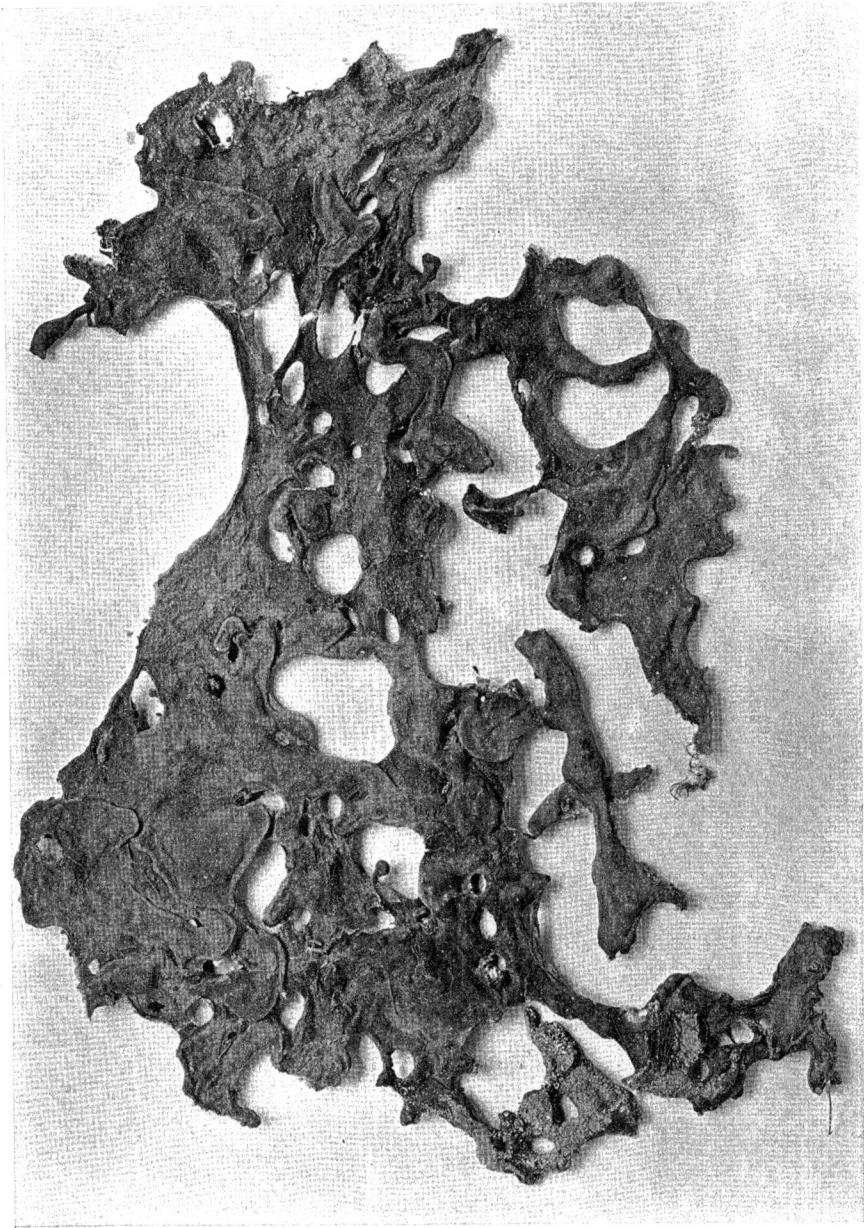

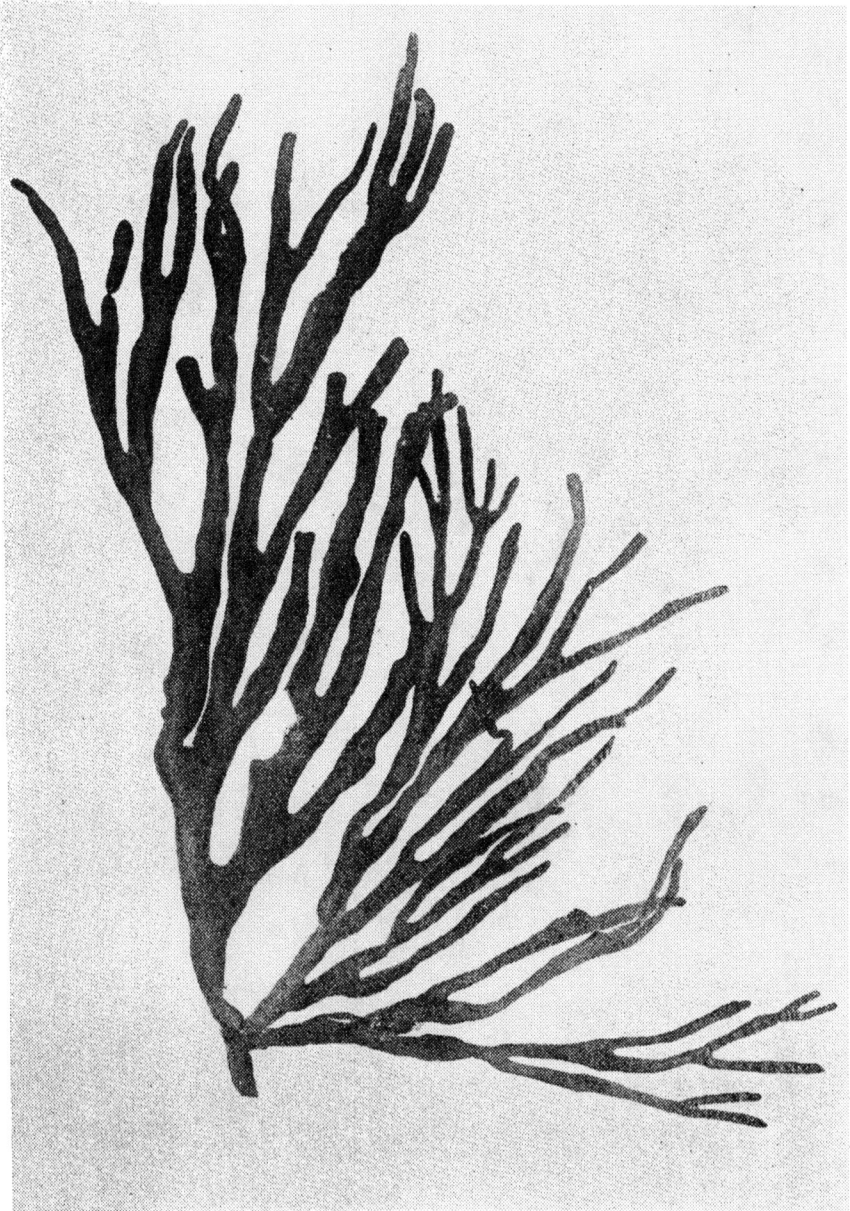

PLATE 32

Spongomorpha coalita (Rupr.) Collins

A photograph of a group of plants showing the method of combining into rope-like masses.   × 0.5.

PLATE 33

Ulvella Lens Crouan

A micro-photograph showing plants in various stages of development. $\times$ 200.

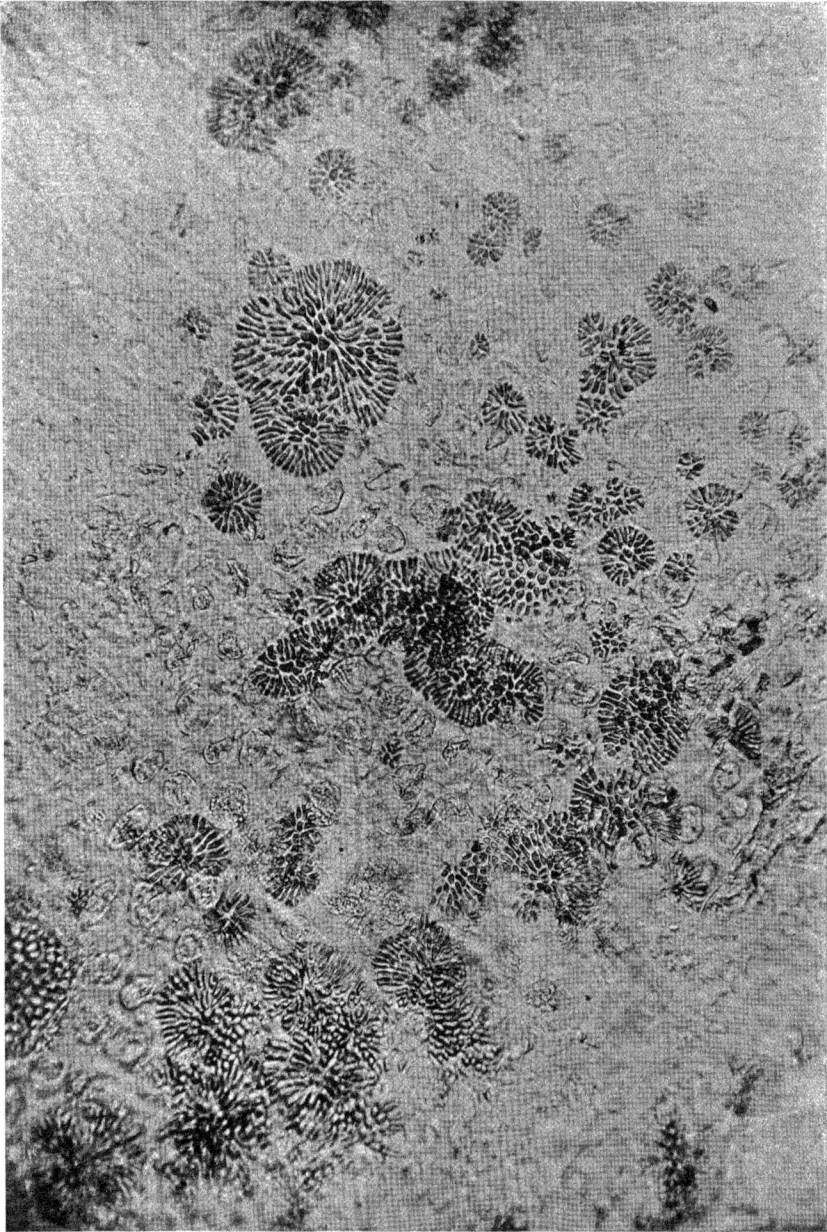

www.ingramcontent.com/pod-product-compliance
Lightning Source LLC
Chambersburg PA
CBHW062153270326
41930CB00009B/1517